Introduction to Logistics Systems Planning and Control

WILEY-INTERSCIENCE SERIES IN SYSTEMS AND OPTIMIZATION

Advisory Editors

Sheldon Ross
Department of Industrial Engineering and Operations Research, University of California, Berkeley, CA 94720, USA

Richard Weber
Statistical Laboratory, Centre for Mathematical Sciences, Cambridge University, Wilberforce Road, Cambridge CB3 0WB

The concept of a system as an entity in its own right has emerged with increasing force in the past few decades in, for example, the areas of electrical and control engineering, economics, ecology, urban structures, automation theory, operational research and industry. The more definite concept of a large-scale system is implicit in these applications, but is particularly evident in such fields as the study of communication networks, computer networks, and neural networks. The *Wiley-Interscience Series in Systems and Optimization* has been established to serve the needs of researchers in these rapidly developing fields. It is intended for works concerned with the developments in quantitative systems theory, applications of such theory in areas of interest, or associated methodology.

Introduction to Logistics Systems Planning and Control

Gianpaolo Ghiani

Department of Innovation Engineering,
University of Lecce, Italy

Gilbert Laporte

Canada Research Chair in Distribution Management,
HEC Montréal, Canada

Roberto Musmanno

Department of Electronics, Informatics and Systems,
University of Calabria, Italy

John Wiley & Sons, Ltd

Copyright © 2004   John Wiley & Sons Ltd, The Atrium, Southern Gate, Chichester,
West Sussex PO19 8SQ, England
Phone   (+44) 1243 779777

Email (for orders and customer service enquiries): cs-books@wiley.co.uk
Visit our Home Page on www.wileyeurope.com or www.wiley.com

Reprinted November, 2004

All Rights Reserved. No part of this publication may be reproduced, stored in a retrieval system or
transmitted in any form or by any means, electronic, mechanical, photocopying, recording, scanning or
otherwise, except under the terms of the Copyright, Designs and Patents Act 1988 or under the terms of
a licence issued by the Copyright Licensing Agency Ltd, 90 Tottenham Court Road, London W1T 4LP,
UK, without the permission in writing of the Publisher. Requests to the Publisher should be addressed
to the Permissions Department, John Wiley & Sons Ltd, The Atrium, Southern Gate, Chichester, West
Sussex PO19 8SQ, England, or emailed to permreq@wiley.co.uk, or faxed to (+44) 1243 770571.

This publication is designed to provide accurate and authoritative information in regard to the subject matter
covered. It is sold on the understanding that the Publisher is not engaged in rendering professional services.
If professional advice or other expert assistance is required, the services of a competent professional should
be sought.

Other Wiley Editorial Offices

John Wiley & Sons Inc., 111 River Street, Hoboken, NJ 07030, USA

Jossey-Bass, 989 Market Street, San Francisco, CA 94103-1741, USA

Wiley-VCH Verlag GmbH, Boschstr. 12, D-69469 Weinheim, Germany

John Wiley & Sons Australia Ltd, 33 Park Road, Milton, Queensland 4064, Australia

John Wiley & Sons (Asia) Pte Ltd, 2 Clementi Loop #02-01, Jin Xing Distripark, Singapore 129809

John Wiley & Sons Canada Ltd, 22 Worcester Road, Etobicoke, Ontario, Canada M9W 1L1

Wiley also publishes its books in a variety of electronic formats. Some content that appears in print may
not be available in electronic books.

Library of Congress Cataloguing-in-Publication Data

Ghiani, Gianpaolo.
    Introduction to logistics systems planning and control / Gianpaolo Ghiani,
    Gilbert Laporte, Roberto Musmanno.
        p. cm. – (Wiley-Interscience series in systems and optimization)
    Includes bibliographical references and index.
    ISBN 0-470-84916-9 (alk. paper) – ISBN 0-470-84917-7 (pbk.: alk. paper)
    1. Materials management. 2. Materials handling. I. Laporte, Gilbert. II. Musmanno, Roberto. III. Title.
IV. Series.

    TS161.G47 2003
    658.7–dc22   2003057594

British Library Cataloguing in Publication Data

A catalogue record for this book is available from the British Library

ISBN 0-470-84916-9 (Cloth)
    0-470-84917-7 (Paper)

To Laura

To Ann and Cathy

To Maria Carmela, Francesco and Andrea

Contents

Foreword

Logistics is concerned with the organization, movement and storage of material and people. The term logistics was first used by the military to describe the activities associated with maintaining a fighting force in the field and, in its narrowest sense, describes the housing of troops. Over the years the meaning of the term has gradually generalized to cover business and service activities. The domain of logistics activities is providing the customers of the system with the right product, in the right place, at the right time. This ranges from providing the necessary subcomponents for manufacturing, having inventory on the shelf of a retailer, to having the right amount and type of blood available for hospital surgeries. A fundamental characteristic of logistics is its holistic, integrated view of all the activities that it encompasses. So, while procurement, inventory management, transportation management, warehouse management and distribution are all important components, logistics is concerned with the integration of these and other activities to provide the time and space value to the system or corporation.

Excess global capacity in most types of industry has generated intense competition. At the same time, the availability of alternative products has created a very demanding type of customer, who insists on the instantaneous availability of a continuous stream of new models. So the providers of logistics activities are asked to do more transactions, in smaller quantities, with less lead time, in less time, for less cost, and with greater accuracy. New trends such as mass customization will only intensify these demands. The accelerated pace and greater scope of logistics operations has made planning-as-usual impossible.

Even with the increased number and speed of activities, the annual expenses associated with logistics activities in the United States have held constant for the last several years around ten per cent of the gross domestic product. Given the significant amounts of money involved and the increased operational requirements, the planning and control of logistics systems has gained widespread attention from practitioners and academic researchers alike. To maximize the value in a logistics system, a large variety of planning decisions has to be made, ranging from the simple warehouse-floor choice of which item to pick next to fulfil a customer order to the corporate-level decision to build a new manufacturing plant. Logistics planning supports the full range of those decisions related to the design and operation of logistics systems.

There exists a vast amount of literature, software packages, decision support tools and design algorithms that focus on isolated components of the logistics system or isolated planning in the logistics systems. In the last two decades, several companies have developed *enterprise resource planning* (ERP) systems in response to the need of global corporations to plan their entire supply chain. In their initial implementations, the ERP systems were primarily used for the recording of transactions rather than for the planning of resources on an enterprise-wide scale. Their main advantage was to provide consistent, up-to-date and accessible data to the enterprise. In recent years, the original ERP systems have been extended with *advanced planning systems* (APSs). The main function of APSs is for the first time the planning of enterprise-wide resources and actions. This implies a coordination of the plans among several organizations and geographically dispersed locations.

So, while logistics planning and control requires an integrated, holistic approach, their treatment in courses and textbooks tends to be either integrated and qualitative or mathematical and very specific. This book bridges the gap between those two approaches. It provides a comprehensive and modelling-based treatment of the complete distribution system and process, including the design of distribution centres, terminal operations and transportation operations. The three major components of logistics systems—inventory, transportation and facilities—are each examined in detail. For each topic the problem is defined, models and solution algorithms are presented that support computer-assisted decision-making, and numerous application examples are provided. The book concludes with an extensive set of case studies that illustrate the application of the models and algorithms in practice. Because of its rigorous mathematical treatment of real-world planning and control problems in logistics, the book will provide a valuable resource to graduate and senior undergraduate students and practitioners who are trying to improve logistics operations and satisfy their customers.

Marc Goetschalckx
Georgia Institute of Technology
Atlanta, May 2003

Preface

Logistics is key to the modern economy. From the steel factories of Pennsylvania to the port of Singapore, from the Nicaraguan banana fields to postal delivery and solid waste collection in any region of the world, almost every organization faces the problem of getting the right materials to the right place at the right time. Increasingly competitive markets are making it imperative to manage logistics systems more and more efficiently.

This textbook grew out of a number of undergraduate and graduate courses on logistics and supply chain management that we have taught to engineering, computer science, and management science students. The goal of these courses is to give students a solid understanding of the analytical tools available to reduce costs and improve service levels in logistics systems. For several years, the lack of a suitable textbook forced us to make use of a number of monographs and scientific papers which tended to be beyond the level of most students. We therefore committed ourselves to developing a quantitative textbook, written at a more accessible level.

The book targets both an educational audience and practitioners. It should be appropriate for advanced undergraduate and graduate courses in logistics, operations management, and supply chain management. It should also serve as a reference for practitioners in consulting as well as in industry. We make the assumption that the reader is familiar with the basics of operations research, probability theory and statistics. We provide a balanced treatment of sales forecasting, logistics system design, inventory management, warehouse design and management, and freight transport planning and control. In the final chapter we present some insightful case studies, taken from the scientific literature, which illustrate the use of quantitative methods for solving complex logistics decision problems.

In our text every topic is illustrated with a numerical example so that the reader can check his or her understanding of each concept before going on to the next one. In addition, a concise annotated bibliography at the end of each chapter acquaints the reader with the state of the art in logistics.

Abbreviations

1-BP	One-Dimensional Bin Packing
2-BP	Two-Dimensional Bin Packing
3-BP	Three-Dimensional Bin Packing
3PL	Third Party Logistics
AP	Assignment Problem
ARP	Arc Routing Problem
AS/RS	Automated Storage and Retrieval System
ATSP	Asymmetric Travelling Salesman Problem
B2B	Business To Business
B2C	Business To Consumers
BF	Best Fit
BFD	Best Fit Decreasing
BL	Bottom Left
CDC	Central Distribution Centre
CPL	Capacitated Plant Location
CPP	Chinese Postman Problem
DC	Distribution Centre
DDAP	Dynamic Driver Assignment Problem
EDI	Electronic Data Interchange
EOQ	Economic Order Quantity
EU	European Union
FBF	Finite Best Fit
FCFS	First Come First Served
FCND	Fixed Charge Network Design
FF	First Fit
FFD	First Fit Decreasing
FFF	Finite First Fit
GIS	Geographic Information System
GDP	Gross Domestic Product
GPS	Global Positioning Systems
IP	Integer Programming

IRP	Inventory-Routing Problem
ITR	Inventory Turnover Ratio
KPI	Key Performance Indicator
LB	Lower Bound
LFND	Linear Fixed Charge Network Design
LMCF	Linear Single-Commodity Minimum-Cost Flow
LMMCF	Linear Multicommodity Minimum-Cost Flow
LP	Linear Programming
LTL	Less-Than-Truckload
MAD	Mean Absolute Deviation
$MAPD$	Mean Absolute Percentage Deviation
MIP	Mixed-Integer Programming
MMCF	Multicommodity Minimum-Cost Flow
MRP	Manufacturing Resource Planning
MSrTP	Minimum-cost Spanning r-Tree Problem
MSE	Mean Squared Error
MTA	Make-To-Assembly
MTO	Make-To-Order
MTS	Make-To-Stock
NAFTA	North America Free Trade Agreement
NF	Network Flow
NLP	Nonlinear Programming
NMFC	National Motor Freight Classification
NRP	Node Routing Problem
NRPCL	Node Routing Problem with Capacity and Length Constraints
NRPSC	Node Routing Problem—Set Covering
NRPSP	Node Routing Problem—Set Partitioning
NRSPTW	Node Routing and Scheduling Problem With Time Windows
PCB	Printed Circuit Board
POPITT	Points Of Presence In The Territory
RDC	Regional Distribution Centre
RPP	Rural Postman Problem
RTSP	Road Travelling Salesman Problem
S/R	Storage And Retrieval
SC	Set Covering
SCOR	Supply Chain Operations References
SESC	Single-Echelon Single-Commodity
SKU	Stock Keeping Unit
SPL	Simple Plant Location
STSP	Symmetric Travelling Salesman Problem
TAP	Traffic Assignment Problem

TEMC	Two-Echelon Multicommodity
TEU	Twenty-foot Equivalent Unit
TL	Truckload
TS	Tabu Search
TSP	Travelling Salesman Problem
UB	Upper Bound
VAP	Vehicle Allocation Problem
VMR	Vendor-Managed Resupplying
VRDP	Vehicle Routing and Dispatching Problem
VRP	Vehicle Routing Problem
VRSP	Vehicle Routing and Scheduling Problem
W/RPS	Walk/Ride and Pick Systems
ZIO	Zero Inventory Ordering

Problems and Website

This textbook contains questions and problems at the end of every chapter. Some are discussion questions while others focus on modelling or algorithmic issues. The answers to these problems are available on the book's website

http://wileylogisticsbook.dii.unile.it,

which also contains additional material (FAQs, software, further modelling exercises, links to other websites, etc.).

Acknowledgements

We thank all the individuals and organizations who helped in one way or another to produce this textbook. First and most of all, we would like to thank Professor Lucio Grandinetti (University of Calabria) for his encouragement and support. We are grateful to the reviewers whose comments were invaluable in improving the organization and presentation of the book. We are also indebted to Fabio Fiscaletti (Pfizer Pharmaceuticals Group) and Luca Lenzi (ExxonMobil Chemical), who provided several helpful ideas. In addition, we thank HEC Montréal for its financial support. Our thanks also go to Maria Teresa Guaglianone, Francesca Vocaturo and Sandro Zacchino for their technical assistance, and to Nicole Paradis for carefully editing and proofreading the material. Finally, the book would not have taken shape without the very capable assistance of Rob Calver, our editor at Wiley.

About the Authors

Gianpaolo Ghiani is Associate Professor of Operations Research at the University of Lecce, Italy. His main research interests lie in the field of combinatorial optimization, particularly in vehicle routing, location and layout problems. He has published in a variety of journals, including *Mathematical Programming, Operations Research Letters, Networks, Transportation Science, Optimization Methods and Software, Computers and Operations Research, International Transactions in Operational Research, European Journal of Operational Research, Journal of the Operational Research Society, Parallel Computing* and *Journal of Intelligent Manufacturing Systems*. His doctoral thesis was awarded the Transportation Science Dissertation Award from INFORMS in 1998. He is an editorial board member of *Computers & Operations Research*.

Gilbert Laporte obtained his PhD in Operations Research at the London School of Economics in 1975. He is Professor of Operations Research at HEC Montréal, Director of the Canada Research Chair in Distribution Management, and Adjunct Professor at the University of Alberta. He is also a member of GERAD, of the Centre for Research on Transportation (serving as director from 1987 to 1991), and Fellow of the Center for Management of Operations and Logistics, University of Texas at Austin. He has authored or coauthored several books, as well as more than 225 scientific articles in combinatorial optimization, mostly in the areas of vehicle routing, location, districting and timetabling. He is the current editor of *Computers & Operations Research* and served as editor of *Transportation Science* from 1995 to 2002. He has received many scientific awards including the Pergamon Prize (United Kingdom), the Merit Award of the Canadian Operational Research Society, the CORS Practice Prize on two occasions, the Jacques-Rousseau Prize for Interdisciplinarity, as well as the President's medal of the Operational Research Society (United Kingdom). In 1998 he became a member of the Royal Society of Canada.

Roberto Musmanno is Professor of Operations Research at the University of Calabria, Italy. His major research interests lie in logistics, network optimization and parallel computing. He has published in a variety of journals, including *Operations Research, Transportation Science, Computational Optimization and Applications, Optimization Methods & Software, Journal of Optimization Theory and Applications, Optimization* and *Parallel Computing*. He is also a member of the Scientific

Committee of the Italian Center of Excellence on High Performance Computing, and an editorial board member of *Computers & Operations Research*.

1

Introducing Logistics Systems

1.1 Introduction

Logistics deals with the planning and control of material flows and related information in organizations, both in the public and private sectors. Broadly speaking, its mission is to get the right materials to the right place at the right time, while optimizing a given performance measure (e.g. minimizing total operating costs) and satisfying a given set of constraints (e.g. a budget constraint). In the military context, logistics is concerned with the supply of troops with food, armaments, ammunitions and spare parts, as well as the transport of troops themselves. In civil organizations, logistics issues are encountered in firms producing and distributing physical goods. The key issue is to decide how and when raw materials, semi-finished and finished goods should be acquired, moved and stored. Logistics problems also arise in firms and public organizations producing services. This is the case of garbage collection, mail delivery, public utilities and after-sales service.

Significance of logistics. Logistics is one of the most important activities in modern societies. A few figures can be used to illustrate this assertion. It has been estimated that the total logistics cost incurred by USA organizations in 1997 was 862 billion dollars, corresponding to approximately 11% of the USA Gross Domestic Product (GDP). This cost is higher than the combined annual USA government expenditure in social security, health services and defence. These figures are similar to those observed for the other North America Free Trade Agreement (NAFTA) countries and for the European Union (EU) countries. Furthermore, logistics costs represent a significant part of a company's sales, as shown in Table 1.1 for EU firms in 1993.

Logistics systems. A logistics system is made up of a set of *facilities* linked by *transportation services*. Facilities are sites where materials are processed, e.g. manufactured, stored, sorted, sold or consumed. They include manufacturing and assembly centres, warehouses, distribution centres (DCs), transshipment points, transportation terminals, retail outlets, mail sorting centres, garbage incinerators, dump sites, etc.

Introduction to Logistics Systems Planning and Control   G. Ghiani, G. Laporte and R. Musmanno
© 2004 John Wiley & Sons, Ltd   ISBN: 0-470-84916-9 (HB) 0-470-84917-7 (PB)

Table 1.1   Logistics costs (as a percentage of GDP) in EU countries (T, transportation; W, warehousing; I, inventory; A, administration).

Sector	T	W	I	A	Total
Food/beverage	3.7	2.2	2.8	1.7	10.4
Electronics	2.0	2.0	3.8	2.5	10.3
Chemical	3.8	2.3	2.6	1.5	10.2
Automotive	2.7	2.3	2.7	1.2	8.9
Pharmaceutical	2.2	2.0	2.5	2.1	8.8
Newspapers	4.7	3.0	3.6	2.1	13.4

Transportation services move materials between facilities using vehicles and equipment such as trucks, tractors, trailers, crews, pallets, containers, cars and trains. A few examples will help clarify these concepts.

ExxonMobil Chemical is one of the largest petrochemical companies in the world. Its products include olefins, aromatics, synthetic rubber, polyethylene, polypropylene and oriented polypropylene packaging films. The company operates its 54 manufacturing plants in more than 20 countries and markets its products in more than 130 countries.

The plant located in Brindisi (Italy) is devoted to the manufacturing of oriented polypropylene packaging films for the European market. Films manufactured in Brindisi that need to be metallized are sent to third-party plants located in Italy and in Luxembourg, where a very thin coating of aluminium is applied to one side. As a rule, Italian end-users are supplied directly by the Brindisi plant while customers and third-party plants outside Italy are replenished through the DC located in Milan (Italy). In particular, this warehouse supplies three DCs located in Herstal, Athus and Zeebrugge (Belgium), which in turn replenish customers situated in Eastern Europe, Central Europe and Great Britain, respectively. Further details on the ExxonMobil supply chain can be found in Section 8.2.

The Pfizer Pharmaceuticals Group is the largest pharmaceutical corporation in the world. The company manufactures and distributes a broad assortment of pharmaceutical products meeting essential medical needs, a wide range of consumer products for self-care and well-being, and health products for livestock and pets. The Pfizer logistics system comprises 58 manufacturing sites in five continents producing medicines for more than 150 countries. Because manufacturing pharmaceutical products requires highly specialized and costly machines, each Pfizer plant produces a large amount of a limited number of pharmaceutical ingredients or medicines for an international market. For example, ALFA10, a cardiovascular product, is produced in a unique plant for an international market including 90 countries. For this reason, freight transportation plays a key role in the Pfizer supply chain. A more detailed description of the Pfizer logistics system is given in Section 8.3.

Railion is an international carrier, based in Mainz (Germany), whose core business is rail transport. Railion transports a vast range of products, such as steel, coal, iron ore, paper, timber, cars, washing machines, computers as well as chemical products. In 2001 the company moved about 500 000 containers. Besides offering high-quality rail transport, Railion is also engaged in the development of integrated logistics systems. This involves close cooperation with third parties, such as road haulage, waterborne transport, forwarding and transshipment companies. More details on the freight rail transportation system at Railion can be found in Section 8.4.

The Gioia Tauro marine terminal is the largest container transshipment hub on the Mediterranean Sea and one of the largest in the world. In 1999, its traffic amounted to 2253 million Twenty-foot Equivalent Units (TEUs). The terminal is linked to nearly 50 end-of-line ports on the Mediterranean Sea. Inside the terminal is a railway station where cars can be loaded or unloaded and convoys can be formed. Section 8.5 is devoted to an in-depth description of the Gioia Tauro terminal.

The waste management system of the regional municipality of Hamilton-Wentworth (Canada) is divided into two major subsystems: the solid waste collection system and the regional disposal system. Each city or town is in charge of its own kerbside garbage collection, using either its own workforce or a contracted service. On the other hand, the regional municipality is responsible for the treatment and disposal of the collected wastes. For the purposes of municipal solid waste planning, the region is divided into 17 districts. The regional management is made up of a waste-to-energy facility, a recycling facility, a 550 acre landfill, a hazardous waste depot and three transfer stations. Section 8.6 contains a more detailed description of this logistics system.

Supply chains. A *supply chain* is a complex logistics system in which raw materials are converted into finished products and then distributed to the final users (consumers or companies). It includes suppliers, manufacturing centres, warehouses, DCs and retail outlets. Figure 1.1 shows a typical supply chain in which the production and distribution systems are made up of two stages each. In the production system, components and semi-finished parts are produced in two manufacturing centres while finished goods are assembled at a different plant. The distribution system consists

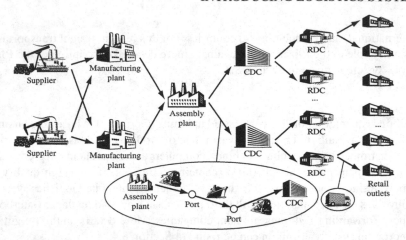

Figure 1.1   A supply chain.

of two *central distribution centres* (CDCs) supplied directly by the assembly centre, which in turn replenish two *regional distribution centres* (RDCs) each. Of course,
depending on product and demand characteristics it may be more appropriate to design
a supply chain without separate manufacturing and assembly centres (or even without
an assembly phase), without RDCs or with different kinds of facilities (e.g. crossdocks, see Section 1.2.2). Each of the transportation links in Figure 1.1 could be
a simple transportation line (e.g. a truck line) or of a more complex transportation
process involving additional facilities (e.g. port terminals) and companies (e.g. truck
carriers). Similarly, each facility in Figure 1.1 comprises several devices and subsystems. For example, manufacturing plants contain machines, buffers, belt conveyors or
other material handling equipment, while DCs include shelves, forklifts or automatic
storage and retrieval systems. Logistics is not normally associated with the detailed
planning of material flows inside manufacturing and assembly plants. Strictly speaking, topics like aggregate production planning and machine scheduling are beyond
the scope of logistics and are not examined in this textbook. The core logistics issues
described in this book are the design and operations of DCs and transportation terminals.

Push versus pull supply chains.   Supply chains are often classified as push or pull
systems. In a *pull* (or *make-to-order* (MTO)) system, finished products are manufactured only when customers require them. Hence, in principle, no inventories are
needed at the manufacturer. In a *push* (or *make-to-stock* (MTS)) system, production
and distribution decisions are based on forecasts. As a result, production anticipates
effective demand, and inventories are held in warehouses and at the retailers. Whether
a push system is more appropriate than a pull system depends on product features,
manufacturing process characteristics, as well as demand volume and variability.
MTO systems are more suitable whenever lead times are short, products are costly,
and demand is low and highly variable. In some cases, a mixed approach can be used.

For example, in *make-to-assembly* (MTA) systems components and semi-finished products are manufactured in a push-based manner while the final assembly stage is pull-based. Hence, the work-in-process inventory at the end of the first stage is used to assemble the finished product as demand arises. These parts are then assembled as soon as customer orders are received.

Product and information flows in a supply chain. Products flow through the supply chain from raw material sources to customers, except for obsolete, damaged and nonfunctioning products which have to be returned to their sources for repair or disposal. Information follows a reverse path. It traverses the supply chain backward from customers to raw material suppliers. In an MTO system, end-user orders are collected by salesmen and then transmitted to manufacturers who in turn order the required components and semi-finished products from their suppliers. Similarly, in an MTS system, past sales are used to forecast future product demand and associated material requirements.

Product and information flows cannot move instantaneously through the supply channel. First, freight transportation between raw material sources, production plants and consumption sites is usually time consuming. Second, manufacturing can take a long time, not only because of processing itself, but also because of the limited plant capacity (not all products in demand can be manufactured at once). Finally, information can flow slowly because order collection, transmission and processing take time, or because retailers place their orders periodically (e.g. once a week), and distributors make their replenishment decisions on a periodic basis (e.g. twice a week).

Degree of vertical integration and third-party logistics. According to a classical economic concept, a supply chain is said to be *vertically integrated* if its components (raw material sources, plants, transportation system, etc.) belong to a single firm. Fully vertically integrated systems are quite rare. More frequently the supply chain is operated by several independent companies. This is the case of manufacturers buying raw materials from outside suppliers, or using contractors to perform particular services, such as container transportation and warehousing. The relationships between the companies of a supply chain may be *transaction based* and *function specific* (as those illustrated in the previous example), or they can be *strategic alliances*. Strategic alliances include *third-party logistics* (3PL) and *vendor-managed resupply*. 3PL is a long-term commitment to use an outside company to perform all or part of a company's product distribution. It allows the company to focus on its core business while leaving distribution to a logistics outsourcer. 3PL is suitable whenever the company is not willing to invest much in transportation and warehousing infrastructures, or whenever the company is unable to take advantage of economies of scale because of low demand. On the other hand, 3PL causes the company to lose control of distribution and may possibly generate higher logistics costs.

Retailer-managed versus vendor-managed resupply. Traditionally, customers (both retailers or final consumers) have been in charge of monitoring their inventory levels and place purchase orders to vendors (*retailer-managed* systems). In recent years, there has been a growth in *vendor-managed systems*, in which vendors monitor customer sales (or consumption) and inventories through *electronic data interchange* (EDI), and decide when and how to replenish their customers. Vendors are thus able to achieve cost savings through a better coordination of customer deliveries while customers do not need to allocate costly resources to inventory management. Vendor-managed resupply is popular in the gas and soft drink industries, although it is gaining in popularity in other sectors. In some vendor-managed systems, the retailer owns the goods sitting on the shelves, while in others the inventory belongs to the vendor. In the first case, the retailer is billed only at the time where it makes a sale to a customer.

1.2   How Logistics Systems Work

Logistics systems are made up of three main activities: order processing, inventory management and freight transportation.

1.2.1   Order processing

Order processing is strictly related to information flows in the logistics system and includes a number of operations. Customers may have to request the products by filling out an order form. These orders are transmitted and checked. The availability of the requested items and customer's credit status are then verified. Later on, items are retrieved from the stock (or produced), packed and delivered along with their shipping documentation. Finally, customers have to be kept informed about the status of their orders.

Traditionally, order processing has been a very time-consuming activity (up to 70% of the total order-cycle time). However, in recent years it has benefited greatly from advances in electronics and information technology. *Bar code* scanning allows retailers to rapidly identify the required products and update inventory level records. *Laptop computers* and *modems* allow salespeople to check in real time whether a product is available in stock and to enter orders instantaneously. EDI allows companies to enter orders for industrial goods directly in the seller's computer without any paperwork.

1.2.2   Inventory management

Inventory management is a key issue in logistics system planning and operations. *Inventories* are stockpiles of goods waiting to be manufactured, transported or sold. Typical examples are

- components and semi-finished products (*work-in-process*) waiting to be manufactured or assembled in a plant;

- merchandise (raw material, components, finished products) transported through the supply chain (*in-transit inventory*);

- finished products stocked in a DC prior to being sold;

- finished products stored by end-users (consumers or industrial users) to satisfy future needs.

There are several reasons why a logistician may wish to hold inventories in some facilities of the supply chain.

Improving service level. Having a stock of finished goods in warehouses close to customers yields shorter lead times.

Reducing overall logistics cost. Freight transportation is characterized by economies of scale because of high fixed costs. As a result, rather than frequently delivering small orders over long distances, a company may find it more convenient to satisfy customer demand from local warehouses (replenished at low frequency).

Coping with randomness in customer demand and lead times. Inventories of finished goods (*safety stocks*) help satisfy customer demand even if unexpected peaks of demand or delivery delays occur (due, for example, to unfavourable weather or traffic conditions).

Making seasonal items available throughout the year. Seasonal products can be stored in warehouses at production time and sold in subsequent months.

Speculating on price patterns. Merchandise whose price varies greatly during the year can be purchased when prices are low, then stored and finally sold when prices go up.

Overcoming inefficiencies in managing the logistics system. Inventories may be used to overcome inefficiencies in managing the logistics system (e.g. a distribution company may hold a stock because it is unable to coordinate supply and demand).

Holding an inventory can, however, be very expensive for a number of reasons (see Table 1.1). First, a company that keeps stocks incurs an *opportunity* (or *capital*) cost represented by the *return on investment* the firm would have realized if money had been better invested. Second, warehousing costs must be incurred, whether the warehouse is privately owned, leased or public (see Chapter 4 for a more detailed analysis of inventory costs).

The aim of *inventory management* is to determine stock levels in order to minimize total operating cost while satisfying customer service requirements. In practice, a good inventory management policy should take into account five issues: (1) the relative importance of customers; (2) the economic significance of the different products; (3) transportation policies; (4) production process flexibility; (5) competitors' policies.

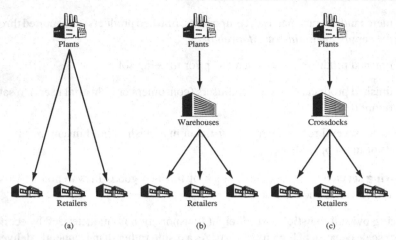

Figure 1.2 Distribution strategies: (a) direct shipment; (b) warehousing; (c) crossdocking.

Inventory and transportation strategies.   Inventory and transportation policies are intertwined. When distributing a product, three main strategies can be used: direct shipment, warehousing, crossdocking.

If a *direct shipment* strategy is used, goods are shipped directly from the manufacturer to the end-user (the retailers in the case of retail goods) (see Figure 1.2a). Direct shipments eliminate the expenses of operating a DC and reduce lead times. On the other hand, if a typical customer shipment size is small and customers are dispersed over a wide geographic area, a large fleet of small trucks may be required. As a result, direct shipment is common when fully loaded trucks are required by customers or when perishable goods have to be delivered timely.

Warehousing is a traditional approach in which goods are received by warehouses and stored in tanks, pallet racks or on shelves (see Figure 1.2b). When an order arrives, items are retrieved, packed and shipped to the customer. Warehousing consists of four major functions: reception of the incoming goods, storage, order picking and shipping. Out of these four functions, storage and order picking are the most expensive because of inventory holding costs and labour costs, respectively.

Crossdocking (also referred to as *just-in-time distribution*) is a relatively new logistics technique that has been successfully applied by several retail chains (see Figure 1.2c). A *crossdock* is a transshipment facility in which incoming shipments (possibly originating from several manufacturers) are sorted, consolidated with other products and transferred directly to outgoing trailers without intermediate storage or order picking. As a result, shipments spend just a few hours at the facility. In *pre-distribution crossdocking*, goods are assigned to a retail outlet before the shipment leaves the vendor. In *post-distribution crossdocking*, the crossdock itself allocates goods to the retail outlets. In order to work properly, crossdocking requires high volume and low variability of demand (otherwise it is difficult to match supply and demand) as well as easy-to-handle products. Moreover, a suitable information system is needed to coordinate inbound and outbound flows.

Centralized versus decentralized warehousing. If a warehousing strategy is used, one has to decide whether to select a centralized or a decentralized system. In *centralized warehousing*, a single warehouse serves the whole market, while in *decentralized warehousing* the market is divided into different zones, each of which is served by a different (smaller) warehouse. Decentralized warehousing leads to reduced lead times since warehouses are much closer to customers. On the other hand, centralized warehousing is characterized by lower facility costs because of larger economies of scale. In addition, if customers' demands are uncorrelated, the aggregate safety stock required by a centralized system is significantly smaller than the sum of the safety stocks in a decentralized system. This phenomenon (known as *risk pooling*) can be explained qualitatively as follows: under the above hypotheses, if the demand from a customer zone is higher than the average, then there will probably be a customer zone whose demand is below average. Hence, demand originally allocated to a zone can be reallocated to the other and, as a result, lower safety stocks are required. A more quantitative explanation of risk pooling will be given in Section 2.2. Finally, *inbound* transportation costs (the costs of shipping the goods from manufacturing plants to warehouses) are lower in a centralized system while *outbound* transportation costs (the costs of delivering the goods from the warehouses to the customers) are lower in a decentralized system.

1.2.3 Freight transportation

Freight transportation plays a key role in today's economies as it allows production and consumption to take place at locations that are several hundreds or thousands of kilometres away from each other. As a result, markets are wider, thus stimulating direct competition among manufacturers from different countries and encouraging companies to exploit economies of scale. Moreover, companies in developed countries can take advantage of lower manufacturing wages in developing countries. Finally, perishable goods can be made available in the worldwide market.

Freight transportation often accounts for even two-thirds of the total logistics cost (see Table 1.1) and has a major impact on the level of customer service. It is therefore not surprising that transportation planning plays a key role in logistics system management.

A manufacturer or a distributor can choose among three alternatives to transport its materials. First, the company may operate a private fleet of owned or rented vehicles (*private transportation*). Second, a carrier may be in charge of transporting materials through direct shipments regulated by a contract (*contract transportation*). Third, the company can resort to a carrier that uses common resources (vehicles, crews, terminals) to fulfil several client transportation needs (*common transportation*).

In the remainder of this section, we will illustrate the main features of freight transportation from a logistician's perspective. A more detailed analysis is provided in Chapters 6 and 7.

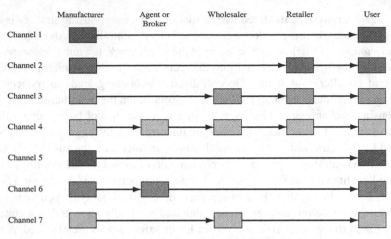

Figure 1.3   Channels of distribution.

Distribution channels.   Bringing products to end-users or into retail stores may be a complex process. While a few manufacturing firms sell their own products to end-users directly, in most cases *intermediaries* participate in product distribution. These can be *sales agents* or *brokers*, who act for the manufacturer, or *wholesalers*, who purchase products from manufacturers and resell them to retailers, who in turn sell them to end-users. Intermediaries add a markup to the cost of a product but on the whole they benefit consumers because they provide lower transportation unit costs than manufacturers would be able to achieve. A *distribution channel* is a path followed by a product from the manufacturer to the end-user. A relevant marketing decision is to select an appropriate combination of distribution channels for each product. Figure 1.3 illustrates the main distribution channels. Channels 1–4 correspond to consumer goods while channels 5–7 correspond to industrial goods. In channel 1, there are no intermediaries. This approach is suitable for a restricted number of products (cosmetics and encyclopaedias sold door-to-door, handicraft sold at local flea markets, etc.). In channel 2, producers distribute their products through retailers (e.g. in the tyre industry). Channel 3 is popular whenever manufacturers distribute their products only in large quantities and retailers cannot afford to purchase large quantities of goods (e.g. in the food industry). Channel 4 is similar to channel 3 except that manufacturers are represented by sales agents or brokers (e.g. in the clothing industry). Channel 5 is used for most industrial goods (raw material, equipment, etc.). Goods are sold in large quantities so that wholesalers are useless. Channel 6 is the same as channel 5, except that manufacturers are represented by sales agents or brokers. Finally, channel 7 is used for small accessories (paper clips, etc.).

Freight consolidation.   A common way to achieve considerable logistics cost savings is to take advantage of economies of scale in transportation by consolidating small shipments into larger ones. *Consolidation* can be achieved in three ways. First, small shipments that have to be transported over long distances may be consolidated

Table 1.2 Main features of the most common containers used for transporting solid goods.

Type	Size (m^3)	Tare (kg)	Capacity (kg)	Capacity (m^3)
ISO 20	5.899 × 2.352 × 2.388	2300	21 700	33.13
ISO 40	12.069 × 2.373 × 2.405	3850	26 630	67.80

so as to transport large shipments over long distances and small shipments over short distances (*facility consolidation*). Second, less-than-truckload pick-up and deliveries associated with different locations may be served by the same vehicle on a multi-stop route (*multi-stop consolidation*). Third, shipment schedules may be adjusted forward or backward so as to make a single large shipment rather than several small ones (*temporal consolidation*).

Modes of transportation. Transportation services come in a large number of variants. There are five basic modes (ship, rail, truck, air and pipeline), which can be combined in several ways in order to obtain door-to-door services such as those provided, for example, by intermodal carriers and small shipment carriers.

Merchandise is often consolidated into *pallets* or *containers* in order to protect it and facilitate handling at terminals. Common pallet sizes are 100×120 cm^2, 80×100 cm^2, 90×110 cm^2 and 120×120 cm^2. Containers may be refrigerated, ventilated, closed or with upper openings, etc. Containers for transporting liquids have capacities between 14 000 and 20 000 l. The features of the most common containers for transporting solid goods are given in Table 1.2.

When selecting a carrier, a *shipper* must take two fundamental parameters into account: price (or cost) and transit time.

The cost of a shipper's operated transportation service is the sum of all costs associated with operating terminals and vehicles. The price of a *transportation service* is simply the rate charged by the carrier to the shipper. A more detailed analysis of such costs is reported in Chapters 6 and 7. Air is the most expensive mode of transportation, followed by truck, rail, pipeline and ship. According to recent surveys, transportation by truck is approximately seven times more expensive than by train, which is four times more costly than by ship.

Transit time is the time a shipment takes to move between its origin to its destination. It is a random variable influenced by weather and traffic conditions. A comparison between the average transit times of the five basic modes is provided in Figure 1.4. One must bear in mind that some modes (e.g. air) have to be used jointly with other modes (e.g. truck) to provide door-to-door transportation. The standard deviation and the coefficient of variation (standard deviation over average transit time) of the transit time are two measures of the reliability of a transportation service (see Table 1.3).

Rail. Rail transportation is inexpensive (especially for long-distance movements), relatively slow and quite unreliable. As a result, the railroad is a slow mover of raw

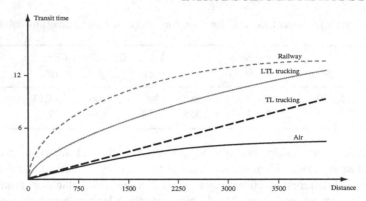

Figure 1.4   Average transit time (in days) as a function of distance (in kilometres)
between origin and destination.

Table 1.3   Reliability of the five basic modes of transportation expressed by the standard
deviation and the coefficient of variation of the transit time.

Ranking	Standard deviation	Coefficient of variation
1	Pipeline	Pipeline
2	Airplane	Airplane
3	Truck	Train
4	Train	Truck
5	Ship	Ship

materials (coal, chemicals, etc.) and of low-value finished products (paper, tinned
food, etc.). This is due mainly to three reasons:

- convoys transporting freight have low priority compared to trains transporting
  passengers;

- direct train connections are quite rare;

- a convoy must include tens of cars in order to be worth operating.

Truck.   Trucks are used mainly for moving semi-finished and finished products.
Road transportation can be *truckload* (TL) or *less-than-truckload* (LTL). A TL service moves a full load directly from its origin to its destination in a single trip (see
Figure 1.5). If shipments add up to much less than the vehicle capacity (LTL loads), it
is more convenient to resort to several trucking services in conjunction with consolidation terminals rather than use direct shipments (see Figure 1.6). As a result, LTL
trucking is slower than TL trucking.

Air.   Air transportation is often used along with road transportation in order to provide door-to-door services. While air transportation is in principle very fast, it is

Figure 1.5 Example of TL transportation.

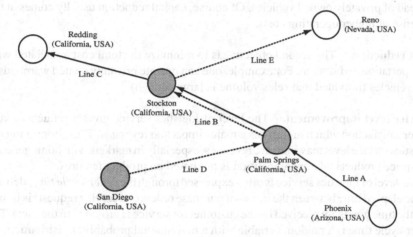

Figure 1.6 Example of LTL transportation.

slowed down in practice by freight handling at airports. Consequently, air transportation is not competitive for short and medium haul shipments. In contrast, it is quite popular for the transportation of high-value products over long distances.

Intermodal transportation. Using more than one mode of transportation can lead to transportation services having a reasonable trade-off between cost and transit time. Although there are in principle several combinations of the five basic modes of transportation, in practice only a few of them turn out to be convenient. The most frequent intermodal services are air–truck (*birdyback*) transportation, train–truck (*piggyback*) transportation, ship–truck (*fishyback*) transportation. Containers are the most common load units in intermodal transportation and can be moved in two ways:

- containers are loaded on a truck and the truck is then loaded onto a train, a ship or an airplane (*trailer on flatcar*);

- containers are loaded directly on a train, a ship or an airplane (*container on flatcar*).

1.3   Logistics Managerial Issues

When devising a logistics strategy, managers aim at achieving a suitable compromise between three main objectives: capital reduction, cost reduction and service level improvement.

Capital reduction.   The first objective is to reduce as much as possible the level of investment in the logistics system (which depends on owned equipment and inventories). This can be accomplished in a number of ways, for example, by choosing public warehouses instead of privately owned warehouses, and by using common carriers instead of privately owned vehicles. Of course, capital reduction usually comes at the expense of higher operating costs.

Cost reduction.   The second objective is to minimize the total cost associated with transportation and storage. For example, one can operate privately owned warehouses and vehicles (provided that sales volume is large enough).

Service level improvement.   The level of logistics service greatly influences customer satisfaction which in turn has a major impact on revenues. Thus, improving the logistics service level may increase revenues, especially in markets with homogeneous low-price products where competition is not based on product features.

The level of logistics service is often expressed through the *order-cycle time*, defined as the elapsed time between the instant a purchase order (or a service request) is issued and the time goods are received by the customer (or service is provided to the user). The order-cycle time is a random variable with a multinomial probability distribution. To illustrate, the probability density function of the supply chain of Figure 1.1 is depicted in Figure 1.7. When a retailer outlet issues an order, the following events may occur:

(a) if the goods required by the outlet are available at the associated RDC, the merchandise will be delivered shortly;

(b) otherwise, the RDC has to resupply its stocks by placing an order to the CDC, in which case the shipment to the retailer will be further delayed;

(c) if the goods are not available even at the CDC, the plants will be requested to produce them.

Let p_a, p_b and p_c be the probabilities of events a, b and c, and let $f_a(t)$, $f_b(t)$, $f_c(t)$ be the (conditional) probability density functions of the order-cycle time in case events a, b and c occur, respectively. The probability density function of the order-cycle time is then

$$f(t) = p_a f_a(t) + p_b f_b(t) + p_c f_c(t).$$

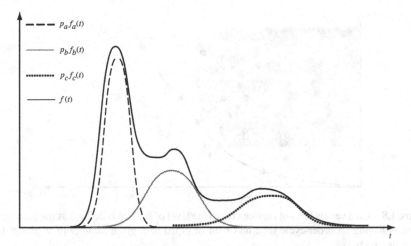

Figure 1.7 Probability density function of the order-cycle time.

Cost versus level of service relationship. Different logistics systems can be classified on the basis of classical multi-objective analysis concepts. Each logistics system is characterized by a level of investment, a cost and a level of service. For example, a system with privately owned warehouses and fleets can be characterized by a high level of investment, a relatively low cost and a high level of logistics service. In what follows the focus will be on the cost–service relationship. System A is said to be *dominated* by a system B (see Figure 1.8)) if the cost of A is higher or equal to the cost of B, the level of service of A is less or equal to the level of service of B and at least one of these two inequalities holds strictly. For example, in Figure 1.8, alternative configurations 2, 3, 4 and 5 are dominated by system 1, while 3, 4, 5 and 7 are dominated by 6. The undominated alternatives are called *efficient* (or *Pareto optimal*) and define the *cost versus level of service curve*.

Sales versus service relationship. The level of logistics service greatly influences sales volume (see Figure 1.9). If service is poor, few sales are generated. As service approaches that of the competition, the sales volume grows. As service is further improved, sales are captured from competing suppliers (provided that other companies do not change their logistics system). Finally, if service improvements are carried too far, sales continue to increase but at a much slower rate. The sales versus service relationship can be estimated by means of buyer surveys and computer simulations.

Determining the optimal service level. The cost versus level of service and sales versus level of service relationships can be used to determine the level of service that maximizes the profit contribution to the firm, as shown in Figure 1.9. The optimal service level usually lies between the low and high extremes. In practice, a slightly different approach is often used: first, a customer service level is set; then the logistics system is designed in order to meet that service level at minimum cost.

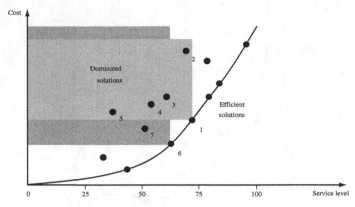

Figure 1.8 Cost versus level of service curve (the level of service is defined as the percentage of orders having an order-cycle time less than or equal to a given number of working days (e.g. four days)).

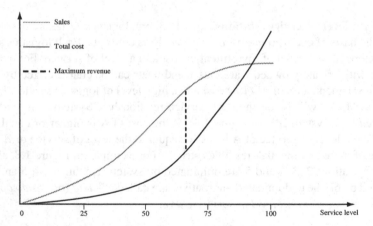

Figure 1.9   Determination of the optimum service level.

1.4   Emerging Trends in Logistics

In recent years, several strategic and technological changes have had a marked impact on logistics. Among these, three are worthy of mention: globalization, new information technologies and e-commerce.

Globalization.   An increasing number of companies operate at the world level in order to take advantage of lower manufacturing costs or cheap raw materials available in some countries. This is sometimes achieved through acquisitions or strategic alliances with other firms. As a result of globalization, transportation needs have increased. More parts and semi-finished products have to be moved between production sites, and transportation to markets tends to be more complex and costly. The

Table 1.4 Main differences between traditional logistics and e-logistics.

	Traditional logistics	E-logistics
Type of load	High volumes	Parcels
Customer	Known	Unknown
Average order value	>$1000	<$100
Destinations	Concentrated	Highly scattered
Demand trend	Regular	Lumpy

increase in multimodal container transportation is a direct consequence of globalization. Also, as a result of globalization, more emphasis must be put on the efficient design and management of supply chains, sometimes at the world level.

Information technologies. Suppliers and manufacturers make use of EDI. This enables them to share data on stock levels, timing of deliveries, positioning of in-transit goods in the supply chain, etc. At the operational level, *geographic information systems* (GISs), *global positioning systems* (GPSs) and on-board computers allow dispatchers to keep track of the current position of vehicles and to communicate with drivers. Such technologies are essential to firms engaged in express pick-up and delivery operations, and to long-haul trucking companies.

E-commerce. An increasing number of companies make commercial transactions through the internet. It is common to distinguish between *business-to-business* (B2B) and *business-to-consumers* (B2C) transactions. The growth of e-commerce parallels that of globalization and information technologies. As a result of e-commerce the volume of goods between producers and retailers should go down while more direct deliveries should be expected between manufacturers and end-users.

E-commerce leads to a more complex organization of the entire logistics system (*e-logistics*), which should be able to manage small- and medium-size shipments to a large number of customers, sometimes scattered around the world. Furthermore, the return flow of defective (or rejected) goods becomes a major issue (*reverse logistics*). Table 1.4 reports the main differences between traditional logistics and e-logistics.

In an e-logistics system different approaches for operating warehouses and distribution are generally adopted. The *virtual warehouse* and the *Points Of Presence In The Territory* (POPITT) are just a few examples. A virtual warehouse is a facility where suppliers and distributors keep their goods in stock in such a way that the e-commerce company can fulfil its orders. A POPITT is a company-owned facility where customers may go either for purchasing and fetching the ordered goods, or for returning defective products. Unlike traditional shops, a POPITT only stores already sold goods waiting to be picked up by customers and defective products waiting to be returned to the manufacturers. This solution simplifies distribution management but reduces customer service level since it does not allow for home deliveries.

1.5   Logistics Decisions

When designing and operating a logistics system, one needs to address several fundamental issues. For example, should new facilities (manufacturing and assembly centres, CDCs, RDCs, etc.) be opened? What are their best configuration, size and location? Should any existing facility be divested, displaced or sized down? Where should materials and components be acquired and stored? Where should manufacturing and assembly take place? Where should finished goods be stored? Should warehouses be company-owned or leased? Where should spare parts be stocked? How should production be planned? How should warehouses operate? (Should goods be stored in racks or should they be stacked? Should goods be retrieved by a team of human order pickers or by automated devices?) When and how should each stocking point be resupplied? What mode of transportation should be used to transport products? Should vehicles be company-owned or leased? What is the best fleet size? How should shipment be scheduled? How should vehicles be routed? Should some transportation be carried out by common carriers?

Logistics decisions are traditionally classified as strategic, tactical and operational, according to the planning horizon.

Strategic decisions.   *Strategic decisions* have long-lasting effects (usually over many years). They include logistics systems design and the acquisition of costly resources (facility location, capacity sizing, plant and warehouse layout, fleet sizing). Because data are often incomplete and imprecise, strategic decisions generally use forecasts based on aggregated data (obtained, for example, by grouping individual products into *product families* and aggregating individual customers into customer zones).

Tactical decisions.   *Tactical decisions* are made on a medium-term basis (e.g. monthly or quarterly) and include production and distribution planning, as well as resource allocation (storage allocation, order picking strategies, transportation mode selection, consolidation strategy). Tactical decisions often use forecasts based on disaggregated data.

Operational decisions.   *Operational decisions* are made on a daily basis or in real-time and have a narrow scope. They include warehouse order picking as well as shipment and vehicle dispatching. Operational decisions are customarily based on very detailed data.

1.5.1   Decision support methods

Quantitative analysis is essential for intelligent logistics decision-making. *Operations research* offers a variety of planning tools.

There are three basic situations in which quantitative analysis may be helpful.

- If a logistics system already exists, one may wish to compare the current system design (or a current operating policy) to an industry standard.

- One may wish to evaluate specified alternatives. In particular, one may wish to answer a number of *what-if* questions regarding specified alternatives to the existing system.

- One may wish to generate a configuration (or a policy) which is *optimal* (or at least *good*) with respect to a given performance measure.

Benchmarking. Benchmarking consists of comparing the performance of a logistics system to a 'best-practice' standard, i.e. the performance of an industry leader in logistics operations. The most popular logistics benchmarking is based on the *supply chain operations references* (SCOR) model. The SCOR model makes use of several performance parameters that range from highly aggregated indicators (named *key performance indicators* (KPIs)) to indicators describing a specific operational issue.

Simulation. *Simulation* enables the evaluation of the behaviour of a particular configuration or policy by considering the dynamics of the system. For instance, a simulation model can be used to estimate the average order retrieval time in a given warehouse when a specific storage policy is used. Whenever a different alternative has to be evaluated, a new simulation is run. For instance, if the number of order pickers is increased by one, a new simulation is required. Simulation models can easily incorporate a large amount of details, such as individual customer ordering patterns. However, detailed simulations are time consuming and can be heavy when a large number of alternatives are considered.

Optimization. The decision-making process can sometimes be cast as a *mathematical optimization problem*. 'Easy' (*polynomial*) optimization problems can be consistently solved within a reasonable amount of time even if instance size is large. This is the case, for example, in *linear programming* (LP) problems and, in particular, of *linear network flow* (NF) *problems* (linear programs with tens of thousands of variables and constraints can be optimized quickly on a personal computer). NP-hard optimization problems can be solved consistently within a reasonable amount of time *only if* instance size is sufficiently small. Most *integer programming* (IP), *mixed-integer programming* (MIP), and *nonlinear programming* (NLP) problems are difficult to optimize. Unfortunately, several classes of logistics decisions (production planning, location decisions, vehicle routing and scheduling, etc.) can only be modelled as IP or MIP problems. This has motivated the development of fast *heuristic* algorithms that search for good but not necessarily the best solutions. Popular examples of heuristics include rounding the solution of the continuous relaxation of an IP or MIP model, *local search*, *simulated annealing* and *tabu search*. In order to work properly, such procedures must be tailored to the problem at hand. As a result, a slight change to problem features may entail a significant modification to the heuristic.

Table 1.5   Annual sales forecast and total cost (in millions of dollars)
for different service levels.

	Percentage of orders filled within three working days						
	70%	75%	80%	85%	90%	95%	100%
Annual sales	4.48	6.67	8.17	9.34	9.87	10.56	11.52
Annual cost	4.41	5.55	5.99	6.22	6.87	7.44	12.84

When using an optimization model, a key aspect is to keep model size as small as possible. As a result, unlike simulation models, optimization models do not customarily consider systems dynamics issues.

Continuous approximation methods.   *Continuous approximation methods* can be used whenever customers are so numerous that demand can be seen as a continuous spatial function. Approximation often yields closed-form solutions and can be used as a simple heuristic.

This textbook presents the main mathematical optimization and simulation methods used for decision-making in logistics management. Other approaches such as the SCOR model and the continuous approximation method are described in the references listed at the end of the chapter.

1.5.2   Outline of the book

The remainder of this textbook describes the main quantitative methods used for the planning, organizing and controlling of logistics systems. The material is divided into five major streams: forecasting logistics requirements (Chapter 2); designing the logistics network (Chapter 3); managing inventories (Chapter 4); designing and operating warehouses and crossdocks (Chapter 5); planning and controlling long-haul and short-haul freight transportation (Chapters 6 and 7). Finally, Chapter 8 provides supplementary material as well as some case studies that show how efficient logistics plans can be devised by applying or adapting the quantitative methods presented in Chapters 2 to 7.

1.6   Questions and Problems

1.1 Why does a push-based supply chain react more slowly to changing demand than a pull-based system?

1.2 Discuss the impact of product diversification (the increase in the number of product variants) on logistics systems planning and control.

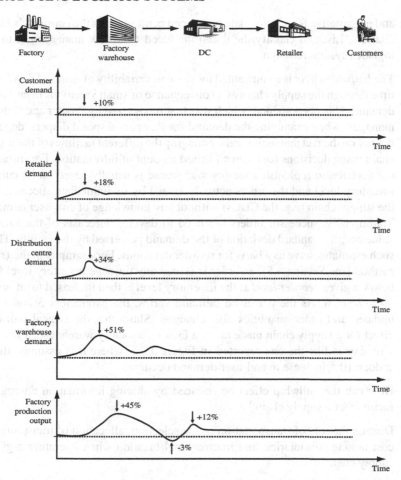

Figure 1.10 The bullwhip effect.

1.3 CalFruit is an emerging Californian distributor of high-quality fresh fruits and vegetables, and packaged food. Because the company operates in a very competitive market, the crucial factor influencing sales volume is the time required to meet orders. On the basis of the historical data, the logistician of the company has estimated that the service level (expressed as the percentage of orders filled within three working days) influences annual sales volume and total cost as reported in Table 1.5. Determine the service level that maximizes revenue.

1.4 Norsk is a Danish producer of dairy products with five subsidiaries in the EU countries and a large network of distributors in North America. The company has recently decided to redesign its Scandinavian distribution network where 140 warehouses have been transformed into pure stocking points, while administrative activities have been concentrated in 14 new regional logistics centres and quantitative forecasting data have been centralized at the company's head-quarters. List and classify the decisions faced by Norsk management in the logistics system redesign.

1.5 The *bullwhip effect* is an unwanted increase in variability of material flows over time through the supply chain as a consequence of small variations in customer demand. This phenomenon, which was first recognized by Procter and Gamble managers when examining the demand for Pampers disposal diapers, depends mainly on the fact that individuals managing the different facilities of the supply chain make decisions based on a limited amount of information. For instance, the decision to replenish a factory warehouse is usually based on its current inventory level and the orders actually issued by its immediate successors in the supply chain (e.g. the CDCs) without any knowledge of end-user demand. Traditionally, successor orders are used to develop forecasts of the average value and the standard deviation of the demand perceived by the facility. Then, such estimates serve as a basis for reorder decisions. For example, in the (s, S) method (see Section 4.8), an order is issued any time the inventory level falls below a given *reorder level s*; the inventory level is then increased to an *order-up-to-level S*. As the perceived demand varies, the parameters S and s are updated and order quantities also changed. Show that the typical bullwhip effect for a supply chain made up of a factory, a factory warehouse, a DC and a retailer is like the one reported in Figure 1.10 (where it is assumed that a sudden 10% increase in end-user demand occurs).

1.6 How can the bullwhip effect be reduced by sharing information among the facilities of a supply chain?

1.7 Discuss the role of transportation mode selection, allocation of transportation cost among subsidiaries, and international taxation when operating a global supply chain.

1.8 Illustrate how a distribution company can take advantage of on-vehicle GPSs.

1.9 Which are the most relevant issues when selecting a company supplier?

1.10 What are the main issues in reverse logistics?

1.7 Annotated Bibliography

A detailed introduction to business logistics is:

1. Ballou R 1998 *Business Logistics Management: Planning, Organizing, and Controlling the Supply Chain*. Prentice Hall, New York.

Statistics reported in Table 1.1 are derived from the following survey:

2. Kearney AT 1993 *Logistics Excellence in Europe*. European Logistics Association.

Table 1.4 is taken from:

3. Bayles DL 2000 *E-commerce Logistics and Fulfillment: Delivering the Goods*. Prentice Hall, New York.

The reference manual to the SCOR approach can be found on the website:

4. Supply Chain Council homepage, http://www.supply-chain.org.

An introductory text to exact and heuristic algorithms for IP and MIP problems is:

5. Wolsey LA 1998 *Integer Programming*. Wiley, New York.

Continuous approximation methods are surveyed in the book:

6. Daganzo CF 1996 *Logistics System Analysis*. Springer, Berlin.

2

Forecasting Logistics Requirements

2.1 Introduction

Forecasting is an attempt to determine in advance the most likely outcome of an uncertain variable. Planning and controlling logistics systems need predictions for the level of future economic activities because of the time lag in matching supply to demand. Typical decisions that must be made before some data are known consider virtually every aspect of the network planning process (including facility location and capacity purchasing) as well as production scheduling, inventory management and transportation planning.

Logistics requirements to be predicted include customer demand, raw material prices, labour costs and lead times. In this chapter forecasting techniques are described with respect to demand although they are equally applicable to other kinds of data.

Forecasting methods are equally relevant to every kind of logistics system, although they are crucial for MTS systems (see Section 1.1), where inventory levels have to be set in every facility.

Forecasting is based on some hypotheses. No forecasting method can be deemed to be superior to others in every respect. As a matter of fact, in order to generate a forecast the demand must show some degree of regularity. For instance, the demand pattern must remain nearly the same in the future or the demand entries must depend to some extent on the past values of a set of variables. Items for which these hypotheses hold are said to have a *regular* demand. This is often the case when there are many customers that individually purchase a small fraction of the whole sales volume.

Lumpy demand. When demand is *lumpy* or *irregular* (see Figure 2.1), there is so much randomness in the demand pattern that no reliable prediction can be made. This is typically the case when large and rare customer orders dominate the demand pattern or when the volume of each item is low (this often happens when the degree of

Introduction to Logistics Systems Planning and Control   G. Ghiani, G. Laporte and R. Musmanno
© 2004 John Wiley & Sons, Ltd   ISBN: 0-470-84916-9 (HB) 0-470-84917-7 (PB)

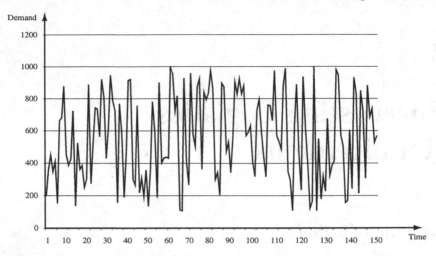

Figure 2.1   Irregular demand pattern of an item. Although the demand pattern is discrete we use a continuous graph representation. This convention will be used throughout the book.

product diversification is high). When dealing with such items, two alternatives should be explored. If demand is low, accuracy is not usually a key issue and an overestimate can be used (this could lead, for example, to a higher safety stock). As an alternative, the processes of the supply chain (namely, manufacturing and transportation) could be made more flexible in order to obtain a *quick response*. If this is feasible, an MTO system is able to satisfy promptly each customer request.

Long-term, medium-term and short-term forecasts.   Demand forecasts are organized by periods of time into three general categories. *Long-term* forecasts span a time horizon from one to five years. Predictions for longer periods are very unreliable, since political and technological issues come into play. Long-term forecasts are used for deciding whether a new item should be put on the market, or whether an old one should be withdrawn, as well as in designing a logistics network. Such forecasts are often generated for a whole group of commodities (or services) rather than for a single item (or service). Moreover, in the long term, sector forecasts are more common than corporate ones. *Medium-term* forecasts extend over a period from a few months to one year. They are used for tactical logistical decisions, such as setting annual production and distribution plans, inventory management and slot allocation in warehouses. *Short-term forecasts* cover a time interval from a few days to several weeks. They are employed to schedule and re-schedule resources in order to meet medium-term production and distribution targets. As service requests are received, there is less need for forecasts. Consequently, forecasts for a shorter time interval (a few hours or a single day) are quite uncommon (see Figure 2.2).

The role of the logistician in generating forecasts.   Medium- and long-term demand forecasts are hardly ever left to the logistician. More frequently, this task is

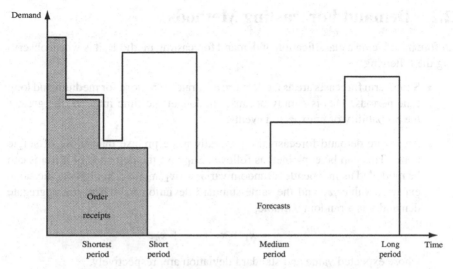

Figure 2.2 Demand pattern of an item. In the shortest period the forecasts are replaced by the order receipts.

assigned to marketing managers who try to influence demand, for example, by launching an advertising campaign for those items whose sales are in decline. On the other hand, the logistician will often produce short-term demand forecasts.

Spatial location of demand. Since in most cases customers are geographically dispersed, it is worth estimating not only when, but also where demand volume will occur. This is because decisions such as warehouse location and inventory level setting are affected by the spatial location of demand. To this end, a *top–down* or a *bottom–up* approach can be utilized. In the top–down method, the entire demand is globally forecasted and then divided heuristically among geographic areas (e.g. on the basis of the most recent sales quotas). On the contrary, in the bottom–up technique, the demand pattern of an item is estimated in each geographical area, and then aggregated if necessary.

Derived versus independent demand. The demand for certain items (e.g. the finished goods of a manufacturing firm) cannot be related to the demand of some other commodities. However, there are some products (like the raw materials and the components required by a production schedule) whose demand can be derived deterministically from the requirements of some other items (e.g. finished goods). For example, the number of loudspeakers needed when assembling a TV set can be easily calculated as a multiple of the number of finished items. Since even moderately complex products can contain several hundreds of different components, such calculations are often performed through computerized procedures, such as *manufacturing resource planning* (MRP).

2.2   Demand Forecasting Methods

Before introducing a classification of demand forecasting methods, it is worth observing the following.

- Short-term forecasts are as a rule more accurate than those for medium and long time periods. This is simply because the longer the time interval, the greater the probability of unexpected events.

- Aggregate demand forecasts are generally more precise than those of single items. This can be explained as follows. Suppose the demands of n items can be modelled as independent random variables $x_1, x_2, \ldots, x_n$, having the same expected value μ_x and the same standard deviation σ_x. Then, the aggregate demand y is a random variable,

$$y = x_1 + x_2 + \cdots + x_n,$$

whose expected value and standard deviation are, respectively,

$$\mu_y = n\mu_x \quad \text{and} \quad \sigma_y^2 = n\sigma_x^2.$$

It follows that the ratio between the standard deviation σ_y and the average μ_y is

$$\frac{\sigma_y}{\mu_y} = \frac{1}{\sqrt{n}} \frac{\sigma_x}{\mu_x}. \tag{2.1}$$

Equation (2.1) indicates that the relative dispersion of the aggregate demand around the correspondent expected value μ_y is less than the relative dispersion of a single item demand.

Forecasting approaches can be classified in two main categories: *qualitative* and *quantitative* methods.

2.2.1   Qualitative methods

Qualitative methods are mainly based on workforce experience or on surveys, although they can also make use of simple mathematical tools to combine different forecasts. Qualitative methods are usually employed for long- and medium-term forecasts, when there is insufficient history to use a quantitative approach. This is the case, for example, when a new product or service is launched on the market, when a product packaging is changed, or when the future demand pattern is expected to be affected by political changeovers or by technological advances.

The most widely used qualitative methods are *sales force assessment*, *market research* and the *Delphi method*. In the first approach, a forecast is developed by company salesmen. As a rule, the workforce can provide accurate estimates since it is close to customers. Market research is based on interviews with potential consumers or users. It is time consuming and requires a deep knowledge of sampling theory.

For these reasons it is used only occasionally, for example, when deciding whether a new product should be launched. In the Delphi method, a series of questionnaires is submitted to a panel of experts. Every time a group of questions is answered, new sets of information become available. Then a new questionnaire is prepared in such a way that every expert is faced with the new findings. This procedure eliminates the bandwagon effect of majority opinion. The Delphi method terminates as soon as all experts share the same viewpoint. This technique is mainly used to estimate the influence of political or macro-economical changes on an item demand.

2.2.2 Quantitative methods

Quantitative methods can be used every time there is sufficient demand history. Such techniques belong to two main groups: *causal methods* and *time series extrapolation*. Causal methods are based on the hypothesis that future demand depends on the past or current values of some variables. They include *regression, econometric models, input–output models, life-cycle analysis, computer simulation models* and *neural networks*. Most of these approaches are difficult to implement, even for larger companies. In practice, only single or multiple regression is used for logistics planning and control. Time series extrapolation presupposes that some features of the past demand time pattern will remain the same. The demand pattern is then projected in the future. This can be done in a number of ways, including the *elementary technique, moving averages, exponential smoothing techniques*, the *decomposition approach* and the *Box–Jenkins method*. The choice of the most suitable quantitative forecasting technique (see also Section 2.9) depends on the kind of historical data available and the type of product (or service). However, as a rule, it is best to select the simplest possible approach. This principle is based on the following observations.

- Forecasts obtained by using simple techniques are easier to understand and explain. This is a fundamental aspect when large sums of money are involved in the decision-making process.

- In a business context, complex forecasting procedures seldom yield better results than simple ones.

This rule is often kept in mind by logisticians, as confirmed by several surveys carried out in North America and in the EU (see, for example, Table 2.1).

The usage frequencies reported in columns 2 and 3 of Table 2.1 should be adjusted in order to take into account the variable levels of familiarity of the decision makers with different forecasting methods (column 4). For example, when comparing the decomposition technique and the more complex Box–Jenkins method in the medium term, one should consider the different level of familiarity that the decision makers have (57% and 37%, respectively) with such approaches. This can be done by computing the values that the quotas of use would likely have if all the decision makers knew both techniques ($12/0.57 = 21\%$ and $5/0.37 = 13.5\%$, respectively).

Table 2.1 Quota of use of the main quantitative forecasting methods in USA (1994). Reprinted from *Interfaces* **24**(2), 92–100, Sanders NR and Manrodt KB 1994 Forecasting practices in US corporations: survey results, ©1994, with the permission of INFORMS.

Forecasting method	Use (%) in short term	Use (%) in medium term	Level (%) of familiarity
Decomposition	7	12	57
Elementary technique	19	14	84
Moving average	33	28	96
Exponential smoothing	20	17	83
Regression	25	26	83
Box–Jenkins	2	5	37

2.2.3 Notation

In the remainder of this chapter, we will assume, as usual, that the time horizon has been divided into a finite number of *time periods* and that all periods have the same duration. Moreover, we will use the following notation. Let d_t, $t = 1, \ldots, T$, be the demand of a given product (or a service) at time period t, where T indicates the time period in correspondence of the latest demand entry available. Moreover, let

$$p_t(\tau), \quad \tau = 1, 2, \ldots,$$

be the τ periods ahead forecast made at time t (i.e. the forecast of demand $d_{t+\tau}$ generated at time t). If $\tau = 1$, a one-period-ahead forecast has to be generated and the notation can be simplified to

$$p_t(1) = p_{t+1}.$$

As explained later, it is worth defining a forecast error in order to evaluate, *a posteriori* (i.e. once the forecasted demands become known), the deviation of the demand from its forecast. The error made by using forecasting $p_i(\tau)$ instead of demand d_t is given by

$$e_i(\tau) = d_t - p_i(\tau), \quad i + \tau = t.$$

As before, the notation can be simplified if $\tau = 1$:

$$e_{t-1}(1) = e_t.$$

2.3 Causal Methods

Causal methods exploit the strong correlation between the future demand of some items (or services) and the past (or current) values of some *causal variables*. For example, the demand for economy cars depends on the level of economic activity and, therefore, can be related to the GDP. Similarly, the demand for spare parts can be associated with the number of installed devices using them.

Table 2.2 Monthly exports of Italian avicultural meat (in hundreds of kilograms) to Germany during 1994 and 1995.

Month	Quantity	Month	Quantity
Jan 94	8257	Jan 95	9 443
Feb 94	8659	Feb 95	9 671
Mar 94	8906	Mar 95	11 624
Apr 94	8601	Apr 95	11 371
May 94	8084	May 95	10 627
Jun 94	8669	Jun 95	11 141
Jul 94	8608	Jul 95	10 993
Aug 94	9186	Aug 95	10 572
Sep 94	9162	Sep 95	10 817
Oct 94	9475	Oct 95	11 133
Nov 94	9196	Nov 95	10 761
Dec 94	9283	Dec 95	10 560

The major advantage of causal methods is their ability to anticipate variations in demand. As such, they are very effective for medium- and long-term forecasts. Unfortunately, in several cases, it is difficult to identify any causal variable having a strong correlation with future demands. Moreover, it is even more difficult to find a causal variable that leads the forecasted variable in time. For these reasons, causal methods are less popular than those based on the time series extrapolation.

As explained in the previous section, a number of different techniques can be classified as causal method although only regression is widely used by logisticians. In this section regression-based forecasting is described while the other causal methods are outlined in Section 2.8.

Regression is a statistical method that relates a dependent variable y (representing, for example, future demand d_{T+1}) to some causal variables $x_1, x_2, \ldots, x_n$ whose value is known or can be predicted:

$$y = f(x_1, x_2, \ldots, x_n).$$

Such a relation can be linear,

$$y = a_0 + a_1 x_1 + a_2 x_2 + \cdots + a_n x_n,$$

or even nonlinear. It is assumed that a set of observed values of the causal variables and the corresponding values of the dependent variable are available. A function $f(\cdot)$ is then selected as the one that interpolates best such observations (if $f(\cdot)$ is linear, this amounts to applying the least squares method in order to estimate $a_0, a_1, \ldots, a_n$).

PAI is an association of Italian farmers which, at the end of 1995, ordered marketing research to estimate future export levels of some Italian goods to the other EU

Table 2.3   Average monthly LIT/DM exchange rate
from November 1993 to December 1995.

Month	Exchange rate	Month	Exchange rate
Nov 93	980.619 52	Dec 94	1039.120 00
Dec 93	987.036 82	Jan 95	1051.678 10
Jan 94	975.963 50	Feb 95	1078.278 00
Feb 94	971.129 00	Mar 95	1201.790 43
Mar 94	986.006 96	Apr 95	1239.064 44
Apr 94	957.251 05	May 95	1172.814 09
May 94	961.928 64	Jun 95	1170.013 64
Jun 94	977.984 09	Jul 95	1158.960 95
Jul 94	996.441 43	Aug 95	1111.316 36
Aug 94	1011.230 45	Sep 95	1104.762 86
Sep 94	1010.387 27	Oct 95	1135.130 00
Oct 94	1018.733 33	Nov 95	1124.658 10
Nov 94	1028.189 05	Dec 95	1106.748 89

countries. The estimate of Italian avicultural meat demand in Germany in January
1996 was based on the data reported in Table 2.2.

It was assumed that the export volume was affected by the exchange rate between
Italian lira and German mark (LIT/DM) (see Table 2.3).

More specifically, it was assumed that the exports in a given month depended on
the exchange rate in the two months before. This hypothesis was confirmed by the
correlation indices (ρ_1 and ρ_2) between the corresponding time series (see Table 2.4):

$$\rho_1 = 0.881 \quad \text{and} \quad \rho_2 = 0.822.$$

Therefore, the required demand was estimated as

$$y = f(x_1, x_2) = a_0 + a_1 x_1^2 + a_2 x_2^2 + a_3 x_1 x_2 + a_4 x_1 + a_5 x_2,$$

where x_1 and x_2 are the average monthly exchange LIT/DM rates in December 1995
and November 1995, respectively ($x_1 = 1106.748\,89$ and $x_2 = 1124.658\,10$). The
coefficients $a_0, a_1, \ldots, a_5$ were estimated through multiple regression applied to data
in Table 2.4. The results were as follows:

$$a_0 = -75\,811.605, \qquad a_1 = -0.014, \qquad a_2 = 0.038,$$
$$a_3 = -0.088, \qquad a_4 = 135.354, \qquad a_5 = 13.435.$$

As a consequence, the estimated exports in January 1996 were

$$y = 10\,682.11 \text{ hundred kg.}$$

Table 2.4 Monthly exports of avicultural meat (in hundreds of kilograms) and average monthly LIT/DM exchange rate during the two previous months.

Month	Quantity	Exchange rate one month before	Exchange rate two months before
Jan 94	8 257	987.036 82	980.619 52
Feb 94	8 659	975.963 50	987.036 82
Mar 94	8 906	971.129 00	975.963 50
Apr 94	8 601	986.006 96	971.129 00
May 94	8 084	957.251 05	986.006 96
Jun 94	8 669	961.928 64	957.251 05
Jul 94	8 608	977.984 09	961.928 64
Aug 94	9 186	996.441 43	977.984 09
Sep 94	9 162	1011.230 45	996.441 43
Oct 94	9 475	1010.387 27	1011.230 45
Nov 94	9 196	1018.733 33	1010.387 27
Dec 94	9 283	1028.189 05	1018.733 33
Jan 95	9 443	1039.120 00	1028.189 05
Feb 95	9 671	1051.678 10	1039.120 00
Mar 95	11 624	1078.278 00	1051.678 10
Apr 95	11 371	1201.790 43	1078.278 00
May 95	10 627	1239.064 44	1201.790 43
Jun 95	11 141	1172.814 09	1239.064 44
Jul 95	10 993	1170.013 64	1172.814 09
Aug 95	10 572	1158.960 95	1170.013 64
Sep 95	10 817	1111.316 36	1158.960 95
Oct 95	11 133	1104.762 86	1111.316 36
Nov 95	10 761	1135.130 00	1104.762 86
Dec 95	10 560	1124.658 10	1135.130 00

2.4 Time Series Extrapolation

Time series extrapolation methods assume that the main features of past demand pattern will be replicated in the future. A forecast is then obtained by extrapolating (projecting) the demand pattern. Such techniques are suitable for short- and medium-term predictions, where the probability of a changeovers is low.

As explained in Section 2.2, time series extrapolation can be done in a number of ways. The classical decomposition method is depicted in this section, while the elementary technique, moving averages and exponential smoothing techniques are described in Sections 2.5, 2.6 and 2.7. Moreover, the Box–Jenkins method is outlined in Section 2.8.

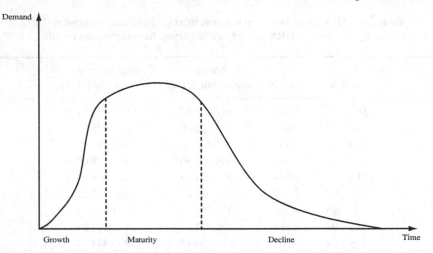

Figure 2.3   Life cycle of a product or service.

2.4.1   Time series decomposition method

The time series decomposition method is based on the assumption that the demand pattern of a product (or a service) can be decomposed into the following four effects: *trend*, *cyclical variation*, *seasonal variation* and *residual variation*.

Trend.   The trend is the long-term modification of demand over time; it may depend on changes in population and on the product (or service) life cycle (see Figure 2.3).

Cyclical variation.   Cyclical variation is caused by the so-called business cycle, which depends on macro-economic issues. It is quite irregular, but its pattern is roughly periodic.

Seasonal variation.   Seasonal variation is caused by the periodicity of several human activities. Typical examples are the ups and downs in the demand of some items over the year. This type of effect can also be observed on a weekly basis (e.g. some product sales are higher on weekends than on working days).

Residual variation.   Residual variation is the portion of demand that cannot be interpreted as trend, cyclical or seasonal variation. It is often the result of numerous causes, each of which has a small impact. If there are no other predictable variations in the demand, the residual effect is a random variable with unit expected value (assuming that demand is modelled as the product of the four effects).

In the sequel we assume that the way the four components are combined together is multiplicative,

$$d_t = q_t v_t s_t r_t, \quad t = 1, \ldots, T,$$

where q_t represents the trend at time period t (expressed in the same units as the demand), v_t is the cyclical effect at time period t, s_t is the seasonal variation at time period t, and r_t is the residual variation at time period t. It is worth noting that all factors are greater than or equal to 0. Also note that if M is the periodicity of the seasonal variation, then the average of the seasonal effects over M consecutive time periods is equal to 1:

$$\frac{\sum_{t=j+1}^{j+M} s_t}{M} = 1, \quad j = 0, 1, \ldots, T - M. \tag{2.2}$$

In Figure 2.4 a typical demand pattern is reported. The decomposition method is made up of three steps: in the first phase, the demand time series d_t, $t = 1, \ldots, T$, is decomposed into the four components q_t, v_t, s_t, r_t, $t = 1, \ldots, T$; in the second phase, the time series of q, v and s are projected into one or more future time periods (it is worth noting that the residual variation cannot be predicted); finally, in the third phase the projected values are combined,

$$p_T(\tau) = q_T(\tau)v_T(\tau)s_T(\tau), \quad \tau = 1, 2, \ldots, \tag{2.3}$$

to obtain the required demand forecasts. The decomposition phase is carried out as follows.

Evaluation of the product $(qv)_t$

The product $(qv)_t$ is obtained by removing from the time series d_t, $t = 1, \ldots, T$, the seasonal effect and the random fluctuation. This can be done by observing that the average value of the demand over M consecutive time periods is not affected by the seasonal fluctuations. Furthermore, by so doing we also remarkably reduce the influence of the random fluctuations, especially if M is relatively high (see also the next section). Therefore, the computation of the following quantities,

$$\frac{d_1 + \cdots + d_M}{M},$$

$$\frac{d_2 + \cdots + d_{M+1}}{M},$$

$$\vdots$$

$$\frac{d_{T-M+1} + \cdots + d_T}{M},$$

allows us to determine a series of demand entries without the seasonal and residual effects.

If M is odd, each average can be associated with the central period of the corresponding time interval. Thus,

$$(qv)_{\lceil M/2 \rceil} = \frac{d_1 + \cdots + d_M}{M},$$

$$(qv)_{\lceil M/2 \rceil + 1} = \frac{d_2 + \cdots + d_{M+1}}{M},$$

$$(qv)_{T - \lceil M/2 \rceil + 1} = \frac{d_{T-M+1} + \cdots + d_T}{M}.$$

Hence the required time series is $(qv)_t$, $t = \lceil \frac{1}{2}M \rceil, \ldots, T - \lceil \frac{1}{2}M \rceil + 1$.

If M is even, one can use a weighted average of $M + 1$ demand entries in which the first and the last ones have a weight of $\frac{1}{2}$ and all other values have a unit weight. Then, the time series $(qv)_t$, $t = \frac{1}{2}M + 1, \ldots, T - \frac{1}{2}M$ is given by

$$(qv)_t = \frac{\frac{1}{2}d_{t-M/2} + d_{t-M/2+1} + \cdots + \frac{1}{2}d_{t+M/2}}{M}, \quad t = \frac{1}{2}M + 1, \ldots, T - \frac{1}{2}M.$$

Evaluation of q_t and v_t

In most cases, it can be assumed that the trend is described by a simple functional relation, such as a linear or quadratic function. Then the trend is obtained by applying a simple regression to the time series $(qv)_t$, $t = 1, \ldots, T$ (for the sake of simplicity, we assume that $(qv)_t$ spans $t = 1, \ldots, T$, although, as we have seen previously, it is defined over a shorter time interval). Once q_t, $t = 1, \ldots, T$, is determined, the cyclical effect v_t, $t = 1, \ldots, T$ can be computed for each $t = 1, \ldots, T$ as follows:

$$v_t = \frac{(qv)_t}{q_t}.$$

Evaluation of s_t and r_t

The time series $(sr)_t$, $t = 1, \ldots, T$, which includes both the seasonal variation and the random fluctuation, can be computed for each $t = 1, \ldots, T$ as follows:

$$(sr)_t = \frac{d_t}{(qv)_t}.$$

The seasonal effect can then be expressed by means of M indices $\bar{s}_1, \ldots, \bar{s}_M$, defined as

$$s_{kM+t} = \bar{s}_t, \quad t = 1, \ldots, M, \ k = 0, 1, \ldots.$$

Each index $\bar{s}_t$, $t = 1, \ldots, M$, represents the average of the values $(sr)_t$, $t = 1, \ldots, T$, associated with homologous time periods (i.e. $\bar{s}_t$, $t = 1, \ldots, M$, is the average of $(sr)_t$, $(sr)_{M+t}$, $(sr)_{2M+t}$, $\ldots$). This procedure is correct because, as explained previously, the average calculation reduces greatly the random fluctuation. Finally, we observe that, on the basis of the definition of seasonal index, we have

$$\frac{\sum_{t=1}^{M} \bar{s}_t}{M} = 1. \tag{2.4}$$

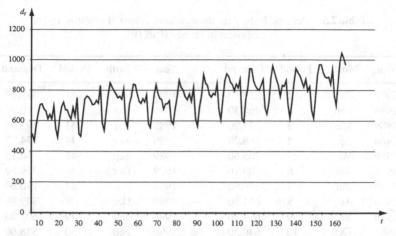

Figure 2.4 Demand pattern of electrosurgical equipment in France.

If Equation (2.4) is not satisfied, the following normalized indices $\tilde{s}_t$, $t = 1, \ldots, M$, are used:

$$\tilde{s}_t = \frac{M\bar{s}_t}{\sum_{t'=1}^{M} \bar{s}_{t'}}, \qquad t = 1, \ldots, M.$$

It is easy to show that the indices $\tilde{s}_t$, $t = 1, \ldots, M$, verify the relation,

$$\frac{\sum_{t=1}^{M} \tilde{s}_t}{M} = 1.$$

The random values r_t, $t = 1, \ldots, T$, can be obtained by dividing each term of the time series $(sr)_t$, $t = 1, \ldots, T$, by the correspondent seasonal index s_t, $t = 1, \ldots, T$, that is

$$r_t = \frac{(sr)_t}{s_t}.$$

If the decomposition has been executed correctly, the time series r_t, $t = 1, \ldots, T$, has an expected value close to 1.

The second phase of the decomposition method amounts to projecting the effects q, v and s previously determined over one or more future time periods; this is very easy to accomplish for the trend and the seasonal effect. However, the cyclical trend is much harder to extrapolate in a quantitative fashion. As a result, it is often estimated qualitatively, on the basis of the macro-economic forecasts. If no such information is available, it can be assumed that

$$v_T(\tau) = v_T, \qquad \tau = 1, 2, \ldots.$$

Finally, in the third phase, a forecast is generated by combining the projections obtained in step two, according to Equation (2.3).

Table 2.5   Demand history (in thousands of euros) of electrosurgical
equipment in France (Part I).

Year	Month	Period	Demand	Year	Month	Period	Demand
1986	Jan	1	511.70	1989	May	41	848.40
1986	Feb	2	468.30	1989	Jun	42	820.40
1986	Mar	3	571.90	1989	Jul	43	795.90
1986	Apr	4	648.20	1989	Aug	44	774.90
1986	May	5	705.60	1989	Sep	45	750.40
1986	Jun	6	709.10	1989	Oct	46	759.50
1986	Jul	7	676.90	1989	Nov	47	740.60
1986	Aug	8	661.50	1989	Dec	48	809.90
1986	Sep	9	611.80	1990	Jan	49	603.40
1986	Oct	10	640.50	1990	Feb	50	558.60
1986	Nov	11	611.10	1990	Mar	51	711.20
1986	Dec	12	697.20	1990	Apr	52	760.90
1987	Jan	13	548.80	1990	May	53	840.00
1987	Feb	14	492.10	1990	Jun	54	835.80
1987	Mar	15	613.20	1990	Jul	55	777.00
1987	Apr	16	692.30	1990	Aug	56	727.30
1987	May	17	721.70	1990	Sep	57	714.00
1987	Jun	18	672.00	1990	Oct	58	744.80
1987	Jul	19	670.60	1990	Nov	59	723.10
1987	Aug	20	635.60	1990	Dec	60	770.70
1987	Sep	21	611.80	1991	Jan	61	581.00
1987	Oct	22	686.00	1991	Feb	62	555.80
1987	Nov	23	630.70	1991	Mar	63	665.70
1987	Dec	24	750.40	1991	Apr	64	770.70
1988	Jan	25	515.20	1991	May	65	836.50
1988	Feb	26	498.40	1991	Jun	66	779.10
1988	Mar	27	627.20	1991	Jul	67	745.50
1988	Apr	28	741.30	1991	Aug	68	739.20
1988	May	29	760.90	1991	Sep	69	676.20
1988	Jun	30	754.60	1991	Oct	70	710.50
1988	Jul	31	733.60	1991	Nov	71	711.90
1988	Aug	32	704.90	1991	Dec	72	731.50
1988	Sep	33	709.80	1992	Jan	73	598.50
1988	Oct	34	733.60	1992	Feb	74	578.90
1988	Nov	35	714.70	1992	Mar	75	675.50
1988	Dec	36	831.60	1992	Apr	76	756.00
1989	Jan	37	586.60	1992	May	77	865.20
1989	Feb	38	536.90	1992	Jun	78	819.00
1989	Mar	39	654.50	1992	Jul	79	800.80
1989	Apr	40	767.90	1992	Aug	80	758.10

Table 2.6 Demand history (in thousands of euros) of electrosurgical equipment in France (Part II).

Year	Month	Period	Demand	Year	Month	Period	Demand
1992	Sep	81	737.80	1996	Mar	123	721.00
1992	Oct	82	774.90	1996	Apr	124	877.10
1992	Nov	83	728.00	1996	May	125	959.70
1992	Dec	84	817.60	1996	Jun	126	916.30
1993	Jan	85	618.10	1996	Jul	127	870.80
1993	Feb	86	565.60	1996	Aug	128	832.30
1993	Mar	87	691.60	1996	Sep	129	760.20
1993	Apr	88	768.60	1996	Oct	130	833.70
1993	May	89	903.00	1996	Nov	131	827.40
1993	Jun	90	847.70	1996	Dec	132	864.50
1993	Jul	91	830.90	1997	Jan	133	705.60
1993	Aug	92	772.10	1997	Feb	134	619.50
1993	Sep	93	755.30	1997	Mar	135	723.10
1993	Oct	94	779.10	1997	Apr	136	847.70
1993	Nov	95	770.00	1997	May	137	942.90
1993	Dec	96	844.20	1997	Jun	138	917.00
1994	Jan	97	671.30	1997	Jul	139	897.40
1994	Feb	98	607.60	1997	Aug	140	859.60
1994	Mar	99	737.80	1997	Sep	141	821.80
1994	Apr	100	863.10	1997	Oct	142	872.20
1994	May	101	908.60	1997	Nov	143	795.90
1994	Jun	102	891.10	1997	Dec	144	824.60
1994	Jul	103	853.30	1998	Jan	145	669.90
1994	Aug	104	836.50	1998	Feb	146	618.10
1994	Sep	105	797.30	1998	Mar	147	756.00
1994	Oct	106	840.70	1998	Apr	148	901.60
1994	Nov	107	816.90	1998	May	149	968.80
1994	Dec	108	872.20	1998	Jun	150	968.80
1995	Jan	109	613.90	1998	Jul	151	921.20
1995	Feb	110	595.00	1998	Aug	152	891.10
1995	Mar	111	744.10	1998	Sep	153	882.00
1995	Apr	112	812.00	1998	Oct	154	887.60
1995	May	113	941.50	1998	Nov	155	840.00
1995	Jun	114	940.10	1998	Dec	156	935.90
1995	Jul	115	863.10	1999	Jan	157	763.70
1995	Aug	116	829.50	1999	Feb	158	700.00
1995	Sep	117	808.50	1999	Mar	159	844.20
1995	Oct	118	800.10	1999	Apr	160	989.10
1995	Nov	119	836.50	1999	May	161	1045.80
1995	Dec	120	870.80	1999	Jun	162	1012.90
1996	Jan	121	684.60	1999	Jul	163	970.90
1996	Feb	122	644.70				

Table 2.7   Computation of combined trend and cyclical effects $(qv)_t$, $t = 1, \ldots, 163$, in the P&D problem.

t	$(qv)_t$	t	$(qv)_t$
1		152	864.65
2		153	871.73
3		154	879.05
4		155	885.91
5		156	890.95
6		157	894.86
7	627.70	158	
8	630.23	159	
9	632.95	160	
10	636.50	161	
11	639.01	162	
12	638.14	163	
...	...		

P&D is a French consulting firm which was entrusted in July 1999 to estimate the future demand of electrosurgical equipment in France for the subsequent six months. The sales over the past 13 years and seven months are available (Tables 2.5 and 2.6, Figure 2.4).

The duration M of the seasonal cycle was assumed to be equal to 12 and the trend was assumed to be linear. Then the decomposition method was applied. The intermediate and final results are summarized in Tables 2.7–2.12 and in Figures 2.5–2.12. The trend equation is $q_t = 638.51 + 1.43t$. The seasonal indices $\bar{s}_1, \ldots, \bar{s}_{12}$ (see Table 2.10 and Figure 2.9) satisfy Equation (2.4). As the expected value of the random variation is approximately 1, the demand decomposition can be deemed to be satisfactory. The demand forecasts from August 1999 to January 2000 (the first six months ahead) were obtained by combining the projections of the trend with the seasonal and cyclical effects. The latter was estimated (see Figure 2.11) by using a quadratic regression curve defined on the basis of v_t, $t = 150, \ldots, 157$. In particular, it was assumed that

$$v_t = f(t - 149)^2 + g(t - 149) + h, \quad t = 150, 151, \ldots.$$

The values of the coefficients f, g and h that best fit the cyclical effect for $t = 150, \ldots, 157$ are

$$f = -0.0004, \qquad g = 0.0094, \qquad h = 0.9856.$$

Table 2.8 Trend q_t, $t = 7, \ldots, 157$, and cyclical effect v_t, $t = 7, \ldots, 157$, in the P&D problem.

t	$(qv)_t$	q_t	v_t
7	627.70	648.52	0.97
8	630.23	649.95	0.97
9	632.95	651.39	0.97
10	636.50	652.82	0.98
11	639.01	654.25	0.98
12	638.14	655.68	0.97
...	...	...	...
152	864.65	856.02	1.01
153	871.73	857.45	1.02
154	879.05	858.88	1.02
155	885.91	860.31	1.03
156	890.95	861.74	1.03
157	894.86	863.17	1.04

Table 2.9 Evaluation of combined seasonal and residual effects $(sr)_t$, $t = 1, \ldots, 157$, in the P&D problem.

t	d_t	$(qv)_t$	$(sr)_t$
7	676.90	627.70	1.08
8	661.50	630.23	1.05
9	611.80	632.95	0.97
10	640.50	636.50	1.01
11	611.10	639.01	0.96
12	697.20	638.14	1.09
...	...	...	...
152	891.10	864.65	1.03
153	882.00	871.73	1.01
154	887.60	879.05	1.01
155	840.00	885.91	0.95
156	935.90	890.95	1.05
157	763.70	894.86	0.85

2.5 Further Time Series Extrapolation Methods: the Constant Trend Case

We first analyse the case in which the past demand pattern does not show any relevant cyclical and seasonal effects, and the trend is constant. We suppose initially that a forecast must be generated only for the next period ahead.

Table 2.10   Computation of the seasonal indices $\bar{s}_t$, $t = 1, \ldots, 12$, in the P&D problem.

t	$\bar{s}_t$	t	$\bar{s}_t$
1	0.82	7	1.07
2	0.76	8	1.02
3	0.92	9	0.98
4	1.05	10	1.02
5	1.15	11	0.99
6	1.11	12	1.08

Table 2.11   Computation of the residual variations r_t, $t = 7, \ldots, 157$, in the P&D problem.

t	$(sr)_t$	s_t	r_t
7	1.08	1.07	1.01
8	1.05	1.02	1.02
9	0.97	0.98	0.98
10	1.01	1.02	0.98
11	0.96	0.99	0.97
12	1.09	1.08	1.01
...	...	...	...
152	1.03	1.02	1.01
153	1.01	0.98	1.03
154	1.01	1.02	0.99
155	0.95	0.99	0.96
156	1.05	1.08	0.98
157	0.85	0.82	1.04

Table 2.12   Demand forecasts $p_T(\tau)$, $T = 163$, $\tau = 1, \ldots, 6$, in the P&D problem.

Month	$T + \tau$	$q_T(\tau)$	$v_T(\tau)$	$s_T(\tau)$	$p_T(\tau)$
Aug 99	164	873.19	1.04	1.02	933.84
Sep 99	165	874.62	1.04	0.98	895.28
Oct 99	166	876.05	1.04	1.02	932.52
Nov 99	167	877.48	1.04	0.99	898.46
Dec 99	168	878.92	1.03	1.08	976.03
Jan 00	169	880.35	1.03	0.82	743.76

2.5.1   Elementary technique

The forecast for the first time period ahead is simply given by

$$p_{T+1} = d_T.$$

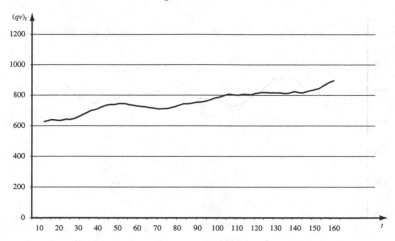

Figure 2.5 Combined trend and cyclical effects $(qv)_t$, $t = 7, \ldots, 157$, in the P&D problem.

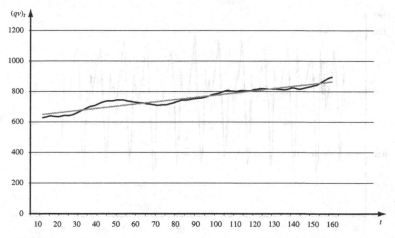

Figure 2.6 Linear trend (in grey) q_t, $t = 7, \ldots, 157$, in the P&D problem.

The method is straightforward. The forecast time series reproduces the demand pattern with one period delay. Consequently, it usually produces rather poor predictions.

Sarath is a Malaysia-based distributor of Korean appliances. The sales volume of portable TV sets during the last 12 weeks in Kuala Lumpur is shown in Table 2.13.

The demand pattern is depicted in Figure 2.13. It can be seen that the trend is constant. By using the elementary technique, we obtain

$$p_{13} = d_{12} = 1177.$$

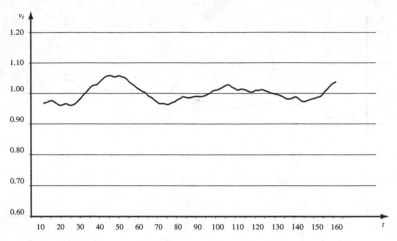

Figure 2.7 Cyclical effect v_t, $t = 7, \ldots, 157$, in the P&D problem.

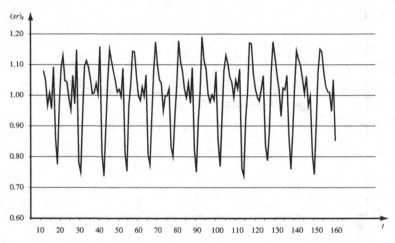

Figure 2.8 Combined seasonal effect and random fluctuation $(sr)_t$, $t = 7, \ldots, 157$, in the P&D problem.

2.5.2 Moving average method

The moving average method uses the average of the r most recent demand entries as the forecast for first period ahead ($r \geqslant 1$):

$$p_{T+1} = \sum_{k=0}^{r-1} \frac{d_{T-k}}{r}.$$

If r is chosen equal to 1, the moving average method reduces to the elementary technique.

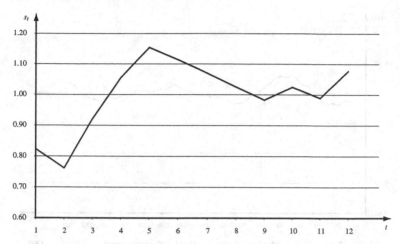

Figure 2.9 Seasonal effect $s_t = \bar{s}_t$, $t = 1, \ldots, 12$, in the P&D problem.

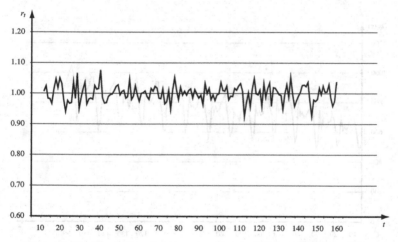

Figure 2.10 Random variation r_t, $t = 7, \ldots, 157$, in the P&D problem.

Table 2.13 Number of portable TV sets sold by Sarath company in the last 12 weeks.

Time period	Quantity	Time period	Quantity
1	1180	7	1162
2	1176	8	1163
3	1185	9	1180
4	1163	10	1170
5	1188	11	1161
6	1172	12	1177

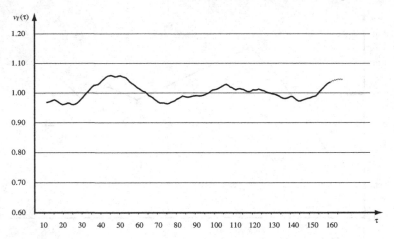

Figure 2.11   Extrapolation of the cyclical effect (in grey) $v_{\bar{t}}(\tau)$, $\bar{t} = 157$, $\tau = 1, \ldots, 12$, in the P&D problem.

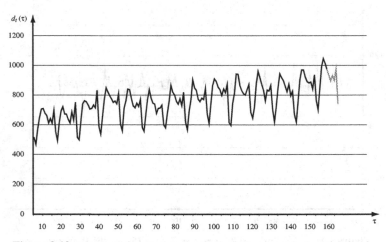

Figure 2.12   Demand forecasting (in grey) of electrosurgical equipment in France for the subsequent six months.

Using the moving average method for solving the Sarath problem above, we obtain the forecasts,

$$p_{13} = \frac{d_{12} + d_{11}}{2} = 1169,$$

$$p_{13} = \frac{d_{12} + d_{11} + d_{10}}{3} = 1169.33,$$

with $r = 2$ and $r = 3$, respectively.

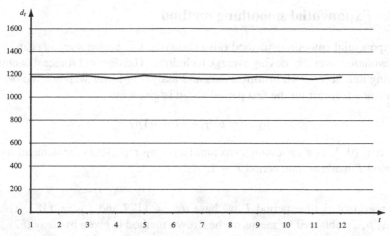

Figure 2.13 Demand pattern of portable TV sets sold by the Sarath company in the last 12 weeks.

When using the moving average method, one should wait for the first r demand data to be available before producing a forecast. To overcome this drawback, the forecasts for time periods $T < r$ are obtained by using the average of the data available in the first T periods, i.e.

$$p_{T+1} = \sum_{k=0}^{T-1} \frac{d_{T-k}}{r}, \quad T < r.$$

For example, if $T = 1$, then $p_2 = d_1$, whereas if $T = 2$, then $p_3 = (d_1 + d_2)/2$.

A key aspect of the moving average method is the choice of parameter r. A small value of r allows a rapid adjustment of the forecast to demand fluctuations but, at the same time, increases the influence of random perturbations. In contrast, a high value of r effectively filters the random effect, but produces a slow adaptation to demand variations. This phenomenon can be explained as follows. Let $d_1, \ldots, d_T$ be independent random variables with expected value μ and standard deviation σ. The random variable p_{T+1} will have an expected value $\mu_{p_{T+1}}$ defined as

$$\mu_{p_{T+1}} = \sum_{k=0}^{r-1} \frac{\mu_{d_{T-k}}}{r} = \sum_{k=0}^{r-1} \frac{\mu}{r} = \mu$$

(i.e. the same expected value as $d_1, \ldots, d_T$), and a standard deviation

$$\sigma_{p_{T+1}} = \frac{1}{\sqrt{r}} \sigma.$$

It follows that the dispersion of p_{T+1} is less than that of $d_1, \ldots, d_T$ and decreases as r increases.

2.5.3   Exponential smoothing method

The exponential smoothing method (also known as the *Brown method*) can be seen as an evolution over the moving average technique. The demand forecast is obtained by taking into account all historical data and assigning lower weights to older data. The demand forecast for the first period ahead is given by

$$p_{T+1} = \alpha d_T + (1 - \alpha)p_T, \tag{2.5}$$

where $\alpha \in (0, 1)$ is a *smoothing constant*. Here, p_T represents the demand forecast for period T made at time period $T - 1$.

Suppose that at time period T we have $d_T = 1177$ and $p_T = 1182$. Then the forecast p_{T+1} obtained by means of the Brown method is given by ($\alpha = 0.2$)

$$p_{T+1} = 0.2 \times 1177 + (1 - 0.2) \times 1182 = 1181.$$

Rewriting Equation (2.5) as

$$p_{T+1} = p_T + \alpha(d_T - p_T) = p_T + \alpha e_T,$$

we obtain the following interpretation: the demand forecast at time period $T + 1$ corresponds to the sum of the demand value estimated at time period $T - 1$ and a fraction of the forecasting error at time period T. This means that if the value of p_T is overestimated with respect to d_T, the forecasting value p_{T+1} is lower than p_T. Vice versa, if p_T is an underestimate of d_T, then p_{T+1} is increased.

The demand history is embedded into p_T, and hence does not appear explicitly in the previous formula. Applying Equation (2.5) recursively, all the demand history appears explicitly:

$$p_T = \alpha d_{T-1} + (1 - \alpha)p_{T-1}.$$

From Equation (2.5), we obtain

$$p_{T+1} = \alpha d_T + (1 - \alpha)[\alpha d_{T-1} + (1 - \alpha)p_{T-1}].$$

Iterating these substitutions (and taking into account that p_2 can be assumed equal to d_1), the following relation is obtained:

$$p_{T+1} = \alpha \sum_{k=0}^{T-2} (1 - \alpha)^k d_{T-k} + (1 - \alpha)^{T-1}d_1.$$

In this relation past demand entries are multiplied by exponentially decreasing weights (this is where the name of the method comes from). Finally, we observe that the sum of all weights $\alpha \sum_{k=0}^{T-2} (1 - \alpha)^k + (1 - \alpha)^{T-1}$ is equal to 1.

Table 2.14 Demand forecasts of portable TV sets in the Sarath problem.

t	p_t	t	p_t
2	1180.00	8	1172.51
3	1178.80	9	1169.65
4	1180.66	10	1172.76
5	1175.36	11	1171.93
6	1179.15	12	1168.65
7	1177.01	13	1171.16

If we use the exponential smoothing method (with $\alpha = 0.3$) for solving the forecasting problem of Sarath company, we get

$$p_2 = d_1 = 1180.$$

For $t = 3$,
$$p_3 = \alpha d_2 + (1 - \alpha)p_2 = 1178.80.$$

Proceeding recursively up to $t = 12$, we obtain the results reported in Table 2.14.

2.5.4 Choice of the smoothing constant

The choice of a value for α plays an important role in the exponential smoothing method. High values of α give a larger weight to the most recent historical data and therefore allow us to follow rapidly the demand variations. On the other hand, lower values of α yield a forecasting method less dependent on the random fluctuation but, at the same time, cannot take quickly into account the most recent variations of the demand.

In practice, the value of α is frequently chosen between 0.01 and 0.3. However, a larger value may be preferable if rapid demand changes are anticipated.

In order to estimate the best value of α, it is worth evaluating *a posteriori* the errors that would have been made in the past if the Brown method had been applied with different values of α (e.g. the values between 0.1 and 0.5, with a step length equal to 0.05). A more detailed treatment of this topic will be given in Section 2.9.

2.5.5 The demand forecasts for the subsequent time periods

The methods just illustrated can be used to forecast demand one period ahead. In order to predict demand for the subsequent time periods, it is sufficient to recall that the trend is assumed to be constant. Consequently,

$$p_T(\tau) = p_{T+1}, \quad \tau = 2, 3, \ldots, \bar{\tau},$$

where the forecasting value p_{T+1} is obtained by using any technique described above and $\bar{\tau}$ represents the duration of the forecasting time horizon. In such a context, the forecasting time horizon is said to be *rolling*, because, once a new demand value becomes available, the time horizon shifts one time period ahead.

The four-weeks-ahead forecasts are needed for the Sarath problem. By using the moving average method with $r = 2$, the predictions are

$$p_{13}[= p_{12}(1)] = p_{12}(2) = p_{12}(3) = \frac{d_{12} + d_{11}}{2} = 1169.$$

At time period $t = 13$, the rolling forecasting horizon would cover the time periods $t = 14, 15, 16$. For example, if $d_{13} = 1173$, then

$$p_{13}(1) = p_{13}(2) = p_{13}(3) = \frac{d_{13} + d_{12}}{2} = 1175.$$

Thus, for $t = 14$ a new updated value is available ($p_{13}(1) = 1175$), which substitutes the previous one ($p_{12}(2) = 1169$). Similar considerations are valid for $t = 15$.

2.6 Further Time Series Extrapolation Methods: the Linear Trend Case

If the trend is linear and no cyclical or seasonal effect is displayed, the forecasting methods are based on the following computational scheme:

$$p_T(\tau) = a_T + b_T\tau, \quad \tau = 1, 2, \ldots.$$

For estimating a_T and b_T, we can use the techniques illustrated below.

2.6.1 Elementary technique

This is the simplest technique:

$$a_T = d_T \quad \text{and} \quad b_T = d_T - d_{T-1}.$$

Sarath company also distributes satellite receivers. The items sold to the stores located in Kuala Lumpur district during the last 12 weeks are reported in Table 2.15. As shown in Figure 2.14, the trend is linear. By using the elementary technique to forecast the demand in τth periods ahead, we get

$$p_{12}(\tau) = a_{12} + b_{12}\tau = d_{12} + (d_{12} - d_{11})\tau = 1230 + 100\tau, \quad \tau = 1, 2, \ldots.$$

In particular, for the first time period ahead ($\tau = 1$):

$$p_{13} = 1330.$$

Table 2.15 Number of kits for satellite equipment furnished by Sarath company in the last 12 weeks.

Time period	Quantity	Time period	Quantity
1	630	7	895
2	730	8	1010
3	880	9	1030
4	850	10	1150
5	910	11	1130
6	890	12	1230

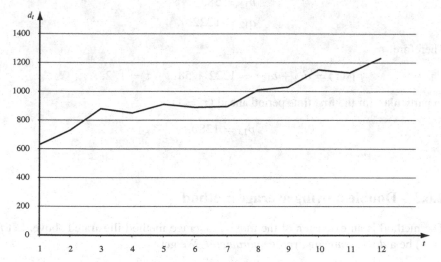

Figure 2.14 Demand pattern of kit for satellite equipment furnished by Sarath company in the last 12 weeks.

2.6.2 Linear regression method

In order to estimate a_T and b_T, this method determines the regression line which best interpolates the r most recent demand entries (i.e. $d_{T-r+1}, \ldots, d_{T-1}, d_T$):

$$b_T = \frac{-\frac{1}{2}(r-1)\sum_{k=0}^{r-1} d_{T-k} + \sum_{k=0}^{r-1} k d_{T-k}}{\frac{1}{4}r(r-1)^2 - \frac{1}{6}r(r-1)(2r-1)},$$

$$a_T = \frac{\sum_{k=0}^{r-1} d_{T-k} + \frac{1}{2}b_T r(r-1)}{r}.$$

The satellite receiver demand of Sarath company can be predicted by means of the linear regression method. If $r = 4$, then

$$\sum_{k=0}^{r-1} d_{T-k} = 4540,$$

$$\sum_{k=0}^{r-1} k d_{T-k} = 6520,$$

and

$$b_{12} = 58,$$
$$a_{12} = 1222.$$

Therefore,

$$p_{12}(\tau) = a_{12} + b_{12}\tau = 1222 + 58\tau, \quad \tau = 1, 2, \ldots.$$

In particular, for the first time period ahead ($\tau = 1$):

$$p_{13} = 1280.$$

2.6.3   Double moving average method

The method is an extension of the moving average method illustrated above. Let r (> 1) be a *double moving average parameter*. We get

$$a_T = 2\gamma_T - \eta_T,$$

$$b_T = \frac{2}{r-1}(\gamma_T - \eta_T),$$

where γ_T is the average of the r most recent demand entries,

$$\gamma_T = \sum_{k=0}^{r-1} \frac{d_{T-k}}{r},$$

and η_T represents the average of the r most recent average demand entries, i.e.

$$\eta_T = \sum_{k=0}^{r-1} \frac{\gamma_{T-k}}{r}.$$

Whenever r past demand data are not available ($T < r$), the computation of γ_T and η_T can be executed along the guidelines illustrated for the moving average method.

We will now solve the satellite receiver problem of Sarath company by means of the double moving average method (with $r = 3$). We get

$$\gamma_{12} = \frac{d_{12} + d_{11} + d_{10}}{3} = 1170,$$

$$\eta_{12} = \frac{\gamma_{12} + \gamma_{11} + \gamma_{10}}{3},$$

where

$$\gamma_{11} = \frac{d_{11} + d_{10} + d_9}{3} = 1103.33,$$

$$\gamma_{10} = \frac{d_{10} + d_9 + d_8}{3} = 1063.33.$$

Therefore,

$$\eta_{12} = 1112.22,$$

from which

$$a_{12} = 2\gamma_{12} - \eta_{12} = 1227.78,$$

$$b_{12} = \gamma_{12} - \eta_{12} = 57.78.$$

Therefore,

$$p_{12}(\tau) = a_{12} + b_{12}\tau = 1227.78 + 57.78\tau, \quad \tau = 1, 2, \ldots.$$

In particular, for the first period ahead ($\tau = 1$):

$$p_{13} = 1285.56.$$

2.6.4 The Holt method

The exponential smoothing method, introduced in Section 2.5.3, is unable to deal with a linear trend. The Holt method is a modification of the exponential smoothing method and is based on the following two relations:

$$a_T = \alpha d_T + (1 - \alpha)(a_{T-1} + b_{T-1}), \quad (2.6)$$

$$b_T = \beta(a_T - a_{T-1}) + (1 - \beta)b_{T-1}. \quad (2.7)$$

Applying recursively Equations (2.6) and (2.7), it is possible to express a_T and b_T as a function of the past demand entries $d_1, \ldots, d_T$.

Table 2.16 Computation of a_t and b_t, $t = 1, \ldots, 12$, in the Sarath satellite receiver forecasting problem.

t	a_t	b_t
1	630.00	0.00
2	660.00	9.00
3	732.30	27.99
4	787.20	36.06
5	849.29	43.87
6	892.21	43.59
7	923.56	39.91
8	977.43	44.10
9	1024.07	44.86
10	1093.26	52.16
11	1140.79	50.77
12	1203.09	54.23

In order to start the procedure, a_1 and b_1 must be specified. They can be chosen in the following way:

$$a_1 = d_1 \quad \text{and} \quad b_1 = 0.$$

In this way, we have $p_2 = p_1(1) = a_1 + b_1 = d_1$, as in the exponential smoothing method. The choice of parameters α and β is conducted according to the same criteria illustrated for the exponential smoothing method.

Applying the Holt method (with $\alpha = \beta = 0.3$) to the Sarath satellite receiver forecasting problem, the values of a_t and b_t, $t = 1, \ldots, 12$, in Table 2.16 are generated. As a result,

$$p_{12}(\tau) = a_{12} + b_{12}\tau = 1203.09 + 54.23\tau, \quad \tau = 1, 2, \ldots.$$

In particular, for the first period ahead ($\tau = 1$):

$$p_{13} = 1257.32.$$

2.7 Further Time Series Extrapolation Methods: the Seasonal Effect Case

This section describes the main forecasting method when the demand pattern displays a constant or linear trend, and a seasonal effect.

Table 2.17 Number of air conditioners sold in the last 24 months by Sarath company.

Time period	Quantity	Time period	Quantity
1	915	13	815
2	815	14	1015
3	1015	15	915
4	1115	16	1315
5	1415	17	1215
6	1615	18	1615
7	1515	19	1315
8	1415	20	1115
9	815	21	1115
10	615	22	915
11	315	23	715
12	815	24	615

2.7.1 Elementary technique

If the trend is constant, then

$$p_T(\tau) = d_{T+\tau-M}, \quad \tau = 1, \ldots, M. \tag{2.8}$$

On the basis of Equation (2.8), the forecast related to the time period $T + \tau$ corresponds to the demand value M time periods back. More generally, for a temporal horizon whose length is superior to one cycle, we have

$$p_T(kM + \tau) = d_{T+\tau-M}, \quad \tau = 1, \ldots, M, \ k = 1, 2, \ldots.$$

The air conditioners supplied by Sarath company during the last 24 months are reported in Table 2.17. The demand pattern (Figure 2.15) displays a linear trend and a seasonal effect with a cycle duration $M = 12$. By using the elementary technique, we get

$$p_{24}(\tau) = d_{24+\tau-12}, \quad \tau = 1, \ldots, 12.$$

In particular,

$$p_{25} = d_{25-12} = d_{13} = 815,$$

$$p_{24}(2) = d_{26-12} = d_{14} = 1015.$$

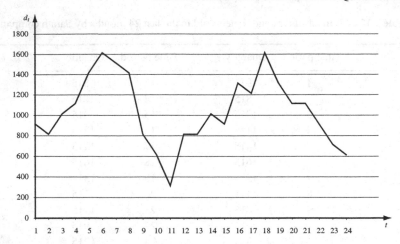

Figure 2.15   Demand pattern of Sarath air conditioners for the last 24 months.

2.7.2   Revised exponential smoothing method

This method can be used whenever the trend is constant. It is based on the following computational scheme,

$$p_T(\tau) = a_T s_{T+\tau}, \quad \tau = 1, \ldots, M, \tag{2.9}$$

where a_T takes into account the constant trend (and can be interpreted as the forecasted demand without the seasonal effect), whereas $s_{T+\tau}$ is the seasonal index (see Section 2.4.1) for period $T + \tau$. More generally, for a time horizon whose duration is greater than one cycle time, we get

$$p_T(kM + \tau) = a_T s_{T+\tau}, \quad \tau = 1, \ldots, M, \ k = 1, 2, \ldots.$$

Assuming, without loss of generality, that the available historical data are sufficient to cover an integer number $K = T/M$ of cycles, the parameters a_T and $s_{T+\tau}$, $\tau = 1, \ldots, M$, can be computed by the following relations,

$$a_T = \alpha \frac{d_T}{s_T} + (1 - \alpha)a_{T-1}, \tag{2.10}$$

$$s_{T+\tau} = s_{kM+\tau} = \beta \frac{d_{(K-1)M+\tau}}{a_{(K-1)M+\tau}} + (1 - \beta)s_{(K-1)M+\tau}, \quad \tau = 1, \ldots, M, \tag{2.11}$$

where α and β are smoothing constants ($0 \leqslant \alpha, \beta \leqslant 1$). Equation (2.10) expresses a_T as the weighted sum of two components: the first, d_T/s_T, represents the value of the demand at time period T without the seasonal effect, while the second represents the forecast, without the seasonal effect, at time period $T - 1$. A similar interpretation can be given to Equation (2.11). However, in this case, it is necessary to take into account the periodicity of the seasonal effect. Using the reasoning of Sections 2.5.3 and 2.6.4, it is possible to develop recursively Equations (2.10) and (2.11) in such a way that all demand entries $d_1, \ldots, d_T$ appear explicitly. To start the procedure we set a_0 equal to the average demand in the first time cycle, i.e.

$$a_0 = \bar{d}_{(1)} = \frac{d_1 + \cdots + d_M}{M},$$

whereas we can select the following initial estimate of s_t, $t = 1, \ldots, M$:

$$s_t = \frac{d_t/\bar{d}_{(1)} + d_{t+M}/\bar{d}_{(2)} + \cdots + d_{t+(K-1)M}/\bar{d}_{(K)}}{K}. \tag{2.12}$$

It is worth noting that the numerator is the sum of the demand entries of the tth time period for each cycle $(d_t, d_{t+M}, \ldots)$ divided by the average demand of the corresponding cycles $(\bar{d}_{(1)}, \bar{d}_{(2)}, \ldots)$. Equation (2.12) implies

$$\sum_{t=1}^{M} s_t = \frac{T}{K} = M,$$

i.e. the average seasonal index for the first cycle is equal to 1. However, this condition cannot be satisfied for the subsequent cycles, and for this reason, in order to respect Equation (2.2), it is necessary to normalize the indices s_t, $t = (k-1)M + 1, \ldots, kM$, $k = 2, 3, \ldots$.

To solve the air conditioner forecasting problem of Sarath company, we use the revised exponential smoothing method with $\alpha = \beta = 0.3$. To this end, we compute the mean value of the demand during the two time cycles,

$$\bar{d}_{(1)} = \frac{d_1 + \cdots + d_{12}}{12} = 1031.67,$$

$$\bar{d}_{(2)} = \frac{d_{13} + \cdots + d_{24}}{12} = 1056.67,$$

from which we obtain

$$a_0 = \bar{d}_{(1)} = 1031.67.$$

Then we compute the seasonal indices s_t, $t = 1, \ldots, 12$ (see Table 2.18), using Equation (2.12). We observe that

$$\bar{s} = \frac{1}{12} \sum_{t=1}^{12} s_t = 1.$$

Using Equations (2.10) and (2.11), we obtain the results reported in Tables 2.19 and 2.20, where the values of s_t, $t = 13, \ldots, 36$, are already normalized. From Equation (2.9), we get

$$p_{25} = p_{24}(1) = 959.20,$$
$$p_{24}(2) = 1016.66,$$

Table 2.18   Time series of s_t, $t = 1, \ldots, 12$, in the Sarath air conditioner forecasting problem.

t	s_t	t	s_t
1	0.83	7	1.36
2	0.88	8	1.21
3	0.92	9	0.92
4	1.16	10	0.73
5	1.26	11	0.49
6	1.55	12	0.69

Table 2.19   Time series of a_t, $t = 1, \ldots, 24$, in the Sarath air conditioner forecasting problem.

t	a_t	t	a_t
1	1053.25	13	969.16
2	1016.61	14	1035.30
3	1040.86	15	1016.74
4	1016.31	16	1056.87
5	1048.13	17	1022.92
6	1046.90	18	1029.51
7	1067.89	19	1007.53
8	1097.37	20	975.86
9	1033.17	21	1062.15
10	975.60	22	1135.18
11	875.39	23	1269.56
12	969.19	24	1140.58

and so on. Figure 2.16 shows both the demand pattern during the first 24 months and the forecasting of the subsequent 12 months.

2.7.3   The Winters method

The Winters method can be used whenever there is a linear trend and a seasonal effect:

$$p_T(kM + \tau) = [a_T + b_T(kM + \tau)]s_{T+\tau}, \qquad \tau = 1, \ldots, M, \ k = 1, 2, \ldots . \quad (2.13)$$

As in the revised exponential smoothing method, we assume that the historical data available are enough to have an integer number $K = T/M$ of cycles. The Winters method is based on the following relationships for the computation of a_T, b_T and

Table 2.20 Time series of s_t, $t = 13, \ldots, 36$, in the Sarath air conditioner forecasting problem.

t	s_t	t	s_t
13	0.84	25	0.84
14	0.85	26	0.89
15	0.94	27	0.93
16	1.14	28	1.17
17	1.29	29	1.26
18	1.55	30	1.55
19	1.38	31	1.35
20	1.24	32	1.21
21	0.88	33	0.93
22	0.70	34	0.73
23	0.45	35	0.49
24	0.73	36	0.67

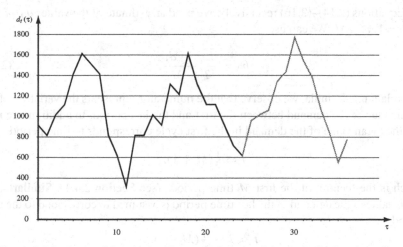

Figure 2.16 Demand forecasting (in grey) of air conditioners in the Sarath problem.

$s_{T+\tau}$, $\tau = 1, \ldots, M$:

$$a_T = \alpha \left(\frac{d_T}{s_T} \right) + (1 - \alpha)(a_{T-1} + b_{T-1}), \tag{2.14}$$

$$b_T = \eta(a_T - a_{T-1}) + (1 - \eta)b_{T-1}, \tag{2.15}$$

$$s_{T+\tau} = s_{KM+\tau} = \beta \left(\frac{d_{(K-1)M+\tau}}{a_{(K-1)M+\tau}} \right) + (1 - \beta)s_{(K-1)M+\tau}, \quad \tau = 1, \ldots, M, \tag{2.16}$$

where α, η and β are smoothing constants chosen in the interval $(0, 1)$. In order to

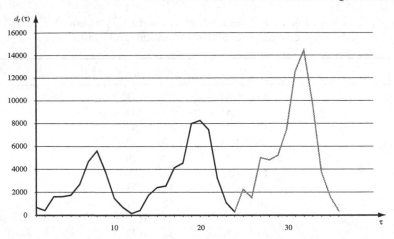

Figure 2.17 Demand pattern and demand forecasting (in grey) of microwave ovens in the Sarath problem.

use Equations (2.14)–(2.16) recursively, we need an estimate of the values a_0, b_0 and s_t, $t = 1, \ldots, M$. We can use

$$b_0 = \frac{\bar{d}_{(K)} - \bar{d}_1}{T - M}. \tag{2.17}$$

To explain this formula, we observe that the numerator represents the variation of the mean value of the demand between the first and the last period. In addition, we note that the mean value of the demand in the first cycle corresponds to time period

$$t = \tfrac{1}{2}(M + 1),$$

which is the 'centre' of the first M time periods (see Section 2.4.1). Similarly, the mean value of the demand of the last time period is assumed to correspond to the time period

$$t = T - \tfrac{1}{2}(M - 1).$$

Hence, there are $T - M$ time periods between the centres of the first and the last period. The parameter a_0 can be determined as follows:

$$a_0 = \bar{d}_{(1)} - \tfrac{1}{2}(M + 1)b_0. \tag{2.18}$$

An estimate of the seasonal indices during the first period is given by

$$s_t = \frac{d_t}{a_0 + b_0 t}, \quad t = 1, \ldots, M, \tag{2.19}$$

subject to a normalization, if necessary.

Table 2.21 reports the number of microwave ovens sold by Sarath company in the last 24 months in Southern Malaysia. The demand pattern (see Figure 2.17) has both a linear trend and a seasonal effect ($M = 12$). In order to forecast the demand for the subsequent 12 months, we use the Winters method. First, we compute the mean demand value in the $K = 2$ time cycles, that is

$$\bar{d}_{(1)} = \frac{d_1 + \cdots + d_{12}}{12} = 2089.58,$$

$$\bar{d}_{(2)} = \frac{d_{13} + \cdots + d_{24}}{12} = 3674.50.$$

Then, we determine the value b_0 through Equation (2.17),

$$b_0 = \frac{\bar{d}_{(1)} + \bar{d}_{(2)}}{12} = 132.08,$$

and the value a_0 through Equation (2.18),

$$a_0 = \bar{d}_{(1)} - 6.5 b_0 = 1231.09.$$

The seasonal indices s_t, $t = 1, \ldots, M$ (already normalized), determined by Equation (2.19), are reported in Table 2.22.

Using Equations (2.14)–(2.16) with $\alpha = \eta = \beta = 0.1$, we obtain the results reported in Tables 2.23 and 2.24.

The s_t values, $t = 13, \ldots, 36$, are already in normalized form. From Equation (2.13) we obtain the forecasting values for the subsequent 12 months, reported in Table 2.25. Figure 2.17 shows both the demand pattern in the first 24 months and the forecasts in the subsequent 12 months.

2.8 Advanced Forecasting Methods

For the sake of completeness, six advanced forecasting techniques are outlined in this section. As stated in Section 2.2, the first five approaches can be classified as casual methods, whereas the sixth one is a complex time series extrapolation technique.

Econometric models. Econometric models consist of several interrelated regression equations linking the demand to be forecasted and its main determinants. The parameters of such relations have to be estimated simultaneously.

Input–output models. Input–output analysis, introduced by Wassily Leontief, is concerned with the interdependence among the various industries and sectors of the economy. Once such a dependence has been established, the variation in demand of a commodity can be forecasted given the forecasts for the other commodities (obtained, for example, through econometric models).

Table 2.21   Number of microwave ovens produced in
the last 24 months in the Sarath problem.

Time period	Quantity	Time period	Quantity
1	682	13	416
2	416	14	1746
3	1613	15	2411
4	1613	16	2544
5	1746	17	4140
6	2677	18	4539
7	4672	19	7997
8	5603	20	8263
9	3741	21	7465
10	1480	22	3209
11	682	23	1081
12	150	24	283

Table 2.22   Time series of s_t, $t = 1, \ldots, 12$, in
the Sarath microwave ovens forecasting problem.

t	s_t	t	s_t
1	0.50	7	2.17
2	0.28	8	2.45
3	0.99	9	1.55
4	0.92	10	0.58
5	0.92	11	0.25
6	1.32	12	0.05

Life-cycle analysis.   Most items (and services) pass through the usual stages of
introduction, growth, maturity and decline, as shown by the 'S curve' of Figure 2.3.
In each stage the product (or service) is demanded by a particular subset of potential
customers. Life-cycle analysis attempts to predict clients' demand through an analysis
of their behaviour.

Computer simulation models.   Computer simulation can be used to estimate the
impact of changes of policy (e.g. inventory policies, production schedules) on the
demand for finished goods.

Neural networks.   Neural networks are made up of a set of elementary nonlinear
systems reproducing the behaviour of biological neurons. If properly trained by means
of the past demand entries, they can be used to make a forecast.

Table 2.23 Time series of a_t and b_t, $t = 1, \ldots, 24$, in the Sarath microwave ovens forecasting problem.

t	a_t	b_t	t	a_t	b_t
1	1362.96	132.06	13	2733.31	110.74
2	1494.82	132.04	14	3186.32	144.96
3	1626.67	132.02	15	3241.05	135.94
4	1758.50	132.00	16	3316.38	129.88
5	1890.32	131.98	17	3549.47	140.20
6	2022.13	131.96	18	3663.31	137.57
7	2153.93	131.95	19	3789.22	136.40
8	2285.72	131.93	20	3869.94	130.83
9	2417.51	131.92	21	4082.83	139.04
10	2549.30	131.90	22	4352.16	152.07
11	2681.08	131.89	23	4478.58	149.50
12	2812.87	131.88	24	4695.76	156.27

Table 2.24 Time series of s_t, $t = 13, \ldots, 36$, in the Sarath microwave ovens forecasting problem.

t	s_t	t	s_t
13	0.50	25	0.47
14	0.28	26	0.31
15	0.99	27	0.97
16	0.92	28	0.91
17	0.92	29	0.95
18	1.32	30	1.32
19	2.17	31	2.17
20	2.45	32	2.43
21	1.55	33	1.58
22	0.58	34	0.60
23	0.25	35	0.25
24	0.05	36	0.05

Box–Jenkins method. The Box–Jenkins method is made up of three steps (identification, parameter evaluation and diagnostic check). In the first phase, the most appropriate forecasting method is selected from a set of techniques. To this end, the past demand entries are used to generate a set of autocorrelation functions, which are then compared. In the second phase, the coefficients of the forecasting method are selected so as to minimize the mean squared error. Finally, an autocorrelation function of the error is determined to verify the adequacy of the method chosen and its corresponding parameters. In case of negative result, the entire procedure is executed

Table 2.25   Demand forecast of Sarath microwave ovens in the subsequent 12 months.

τ	$p_{24}(\tau)$	τ	$p_{24}(\tau)$
1	2266.79	7	12 560.23
2	1533.73	8	14 427.88
3	5009.31	9	9 640.73
4	4815.66	10	3 741.53
5	5207.48	11	1 627.72
6	7431.86	12	355.89

again by discarding the forecasting method previously chosen. Of course, when new demand entries become available, the whole procedure is run again.

2.9   Selection and Control of Forecasting Methods

Forecasting methods can be evaluated through accuracy measures calculated on the basis of errors made in the past. Such measures can be employed to select the most precise approach. Moreover, in the case of periodic predictions (like those required by inventory management), forecasting errors should be monitored in order to adjust parameters if needed. For the sake of brevity, we examine these issues for the case where a one-period-ahead forecast has to be generated.

2.9.1   Accuracy measures

To evaluate the accuracy of a forecasting method, the errors made in the past have to be computed. Then a number of indices (the *mean absolute deviation* (MAD), the *mean absolute percentage deviation* (MAPD) and the *mean squared error* (MSE)) at time period t can be defined:

$$\text{MAD}_t = \frac{\sum_{k=2}^{t} |e_k|}{t - 1}, \tag{2.20}$$

$$\text{MAPD}_t = 100 \frac{\sum_{k=2}^{t} |e_k|/d_k}{t - 1}, \tag{2.21}$$

$$\text{MSE}_t = \frac{\sum_{k=2}^{t} e_k^2}{t - 2}, \tag{2.22}$$

where $1 < t \leqslant T$ for Equations (2.20) and (2.21), and $2 < t \leqslant T$ for Equation (2.22). These three accuracy measures can be used at time period $t = T$ to establish a comparison between different forecasting methods. In particular, MAPD_T can be used to evaluate the quality of a forecasting method (see Table 2.26).

Table 2.26 Evaluation of the forecasting accuracy through $MAPD_T$.

$MAPD_T$	Quality of forecast
$\leqslant 10\%$	Very good
$> 10\%, \leqslant 20\%$	Good
$> 20\%, \leqslant 30\%$	Moderate
$> 30\%$	Poor

Table 2.27 Mean absolute deviation in the Sarath microwave ovens forecasting problem.

α	MAD_{12}
0.05	9.30
0.10	9.27
0.15	9.28
0.20	9.33
0.25	9.40
0.30	9.50
0.35	9.63
0.40	9.79
0.45	9.96

The accuracy of the exponential smoothing method will be evaluated for different values of the smoothing constant α for the TV sets forecasting problem of Sarath company. By using MAD_{12} (see Table 2.27), $\alpha = 0.1$ comes out to give the most precise forecast. With this value of the smoothing constant we obtain $MAPD_{12} = 0.79\%$, which corresponds to a very good accuracy.

2.9.2 Forecast control

A forecasting method works correctly if the errors are random and not systematic. Typical systematic errors occur when the demand value is constantly underestimated or overestimated, and a seasonal variation is not taken into account. Forecasting control can be done through a *tracking signal* or a *control chart*. The tracking signal S_t, $1 < t \leqslant T$, is defined as the ratio between the cumulative error and the MAD_t,

$$S_t = \frac{E_t}{MAD_t},$$

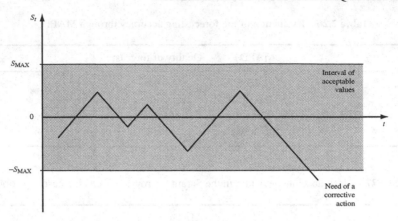

Figure 2.18   Use of a tracking signal for a forecasting control.

where

$$E_t = \sum_{k=2}^{t} e_k.$$

The tracking signal is greater than zero if the forecast systematically underestimates the demand; vice versa, a negative value of S_t indicates a systematic overestimate of the demand. For this reason, a forecast is assumed to be unbiased if the tracking signal falls in the range $\pm S_{\max}$. The value of $S_{\max}$ is established heuristically, and usually varies between 3 and 8. If the tracking signal is outside this interval, the parameters of the forecasting method should be modified or a different forecasting method should be selected (see Figure 2.18).

Unlike tracking signals, control charts are based on the plot of single errors e_t. Under the hypothesis that the expected value of the errors is zero, a forecast is effective if each error $e_k, k = 2, \ldots, t$, is in the confidence interval $\pm m\sigma_t$, where σ_t is the standard deviation of the errors. An estimate of σ_t can be obtained as

$$\sigma_t = \sqrt{\mathrm{MSE}_t}.$$

The parameter m can be related to the probability that the error be in the interval $\pm m\sigma_t$. If the errors are normally distributed with zero mean, the error belongs to the interval $\pm 2\sigma_t$ with a probability around 97.7%, and to the interval $\pm 3\sigma_t$ with a probability around 99.8%. Finally, it is worth observing that the interval $\pm 3\sigma_t$ corresponds approximately to the band ± 4 of the tracking signal.

In addition to the previous analytical check, it is useful to verify visually whether the error pattern reveals the possibility of improving the forecast by introducing suitable modifications. Here are three of the most common pathological situations.

- The errors have an expected value different from zero; this means that the forecast is biased (see Figure 2.19).

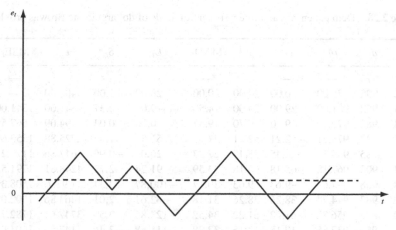

Figure 2.19 Nonzero mean error.

- The error pattern shows a positive or negative trend; in this case, the accuracy of the forecasting method is progressively diminishing.

- The error pattern is periodic; this can happen if an existing seasonal effect has not been identified.

By using the tracking signal (± 4 band) and the control chart ($\pm 3\sigma_T$), we can monitor the demand forecast of sports goods for Browns supermarkets. The forecasting technique is the exponential smoothing method with $\alpha = 0.3$. The demand history for the last few months, and the required forecasts (both in hundreds of dollars), are reported in Table 2.28. The tracking signal S_t, $t = 2, \ldots, T$, is always in the interval ± 4. On the basis of this preliminary evaluation, we can state that the forecast is under control. To make a further check, the expected value and the standard deviation of the error at time period $t = T$ are estimated. The results are -3.02 and 45.36, respectively. We observe that, since the average error is much less than the average demand, we can consider the forecast unbiased. Furthermore, since all errors are in the interval $\pm 3\sigma_T$, even this test suggests that the forecast is under control. Finally, the examination of the control chart (see Figure 2.20) does not show any systematic error.

2.10 Questions and Problems

2.1 In containerized freight transportation, empty containers have to be periodically allocated to depots in order to satisfy future customer demands. How would you forecast the demand for carrier ISO 20 refrigerated containers?

Table 2.28   Demand entries and forecasts (in hundreds of dollars) in the Browns problem.

t	d_t	p_t	e_t	$|e_t|$	MAD_t	E_t	S_t	e_t^2	MSE_t
1	975	—	—	—	—	—	—	—	—
2	995	975.00	20.00	20.00	20.00	20.00	1.00	400.00	—
3	952	981.00	−29.00	29.00	24.50	−9.00	−0.37	841.00	1241.00
4	982	972.30	9.70	9.70	19.57	0.70	0.04	94.09	667.55
5	923	975.21	−52.21	52.21	27.73	−51.51	−1.86	2725.88	1353.66
6	985	959.55	25.45	25.45	27.27	−26.06	−0.96	647.86	1177.21
7	902	967.18	−65.18	65.18	33.59	−91.24	−2.72	4248.81	1791.53
8	938	947.63	−9.63	9.63	30.17	−100.87	−3.34	92.7	1508.39
9	983	944.74	38.26	38.26	31.18	−62.61	−2.01	1463.86	1502.03
10	895	956.22	−61.22	61.22	34.52	−123.83	−3.59	3747.61	1782.73
11	950	937.85	12.15	12.15	32.28	−111.68	−3.46	147.56	1601.04
12	1020	941.50	78.50	78.50	36.48	−33.17	−0.91	6162.77	2057.21
13	—	965.05	—	—	—	—	—	—	—

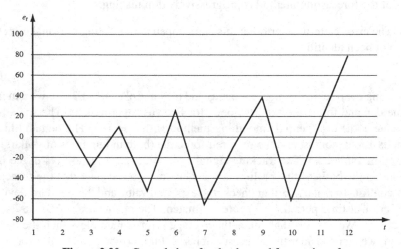

Figure 2.20   Control chart for the demand forecasting of
sports goods in the Browns problem.

2.2 Illustrate a logistics system where there is little need for forecasting.

2.3 To what extent are the forecasting practices different in an MTS and in an MTA
system?

2.4 How would you predict the future demand of a new product?

2.5 Your company is planning to add extra capacity to a plant currently manufacturing 110 000 items per year. You are asked to suggest how much capacity should
be added to the factory. After an accurate sales forecast over the next few years

Table 2.29 Number of installed heaters by Hot Spot and service requests.

| | Installed heaters | | | |
Year	Less than two years ago	More than two years ago	Total	Service requests
1995	260 000	69 500	329 500	18 672
1996	265 000	74 200	339 200	19 076
1997	287 800	82 850	370 650	20 994
1998	313 750	90 550	404 300	23 249
1999	345 350	97 150	442 500	25 025
2000	379 050	105 950	485 000	28 111
2001	416 950	111 550	528 500	30 985
2002	459 100	117 000	576 100	33 397
2003	502 550	123 200	625 750	

you are quite sure that the most likely value of the annual demand is 140 000 items and that the MSE is equal to 10^8. You also know that your company loses \$3 for each unit of unused capacity and \$7 for each unit of unsatisfied demand. How much capacity should your company buy? (Hint: suppose that the forecasting error can be assumed to be normally distributed.)

2.6 Hot Spot is a firm based in the USA whose core business is the maintenance of home heaters. The company usually forecasts service requests on the basis of the number of installed heaters. Make a forecast of the service requests in 2003 in New Jersey by using data in Table 2.29.

To this purpose use

(a) a single regression analysis (service requests versus total installed heaters);

(b) a multiple regression analysis (service requests versus the number of heaters installed less than two years ago and at least two years).

Which technique is the most accurate? Why?

2.7 Sunshine Ltd is one of the world's leading suppliers of fast-moving goods in household care and personal product categories. According to management, its facial soap sales depend mainly on the promotion expenditure that the company and its competitor make. Table 2.30 reports facial soap sales in Canada versus (a) Sunshine promotion expenditure, and (b) the competitor's promotion expenditure divided by the Sunshine promotion expenditure. Make a sales forecast for the next two periods under the hypotheses that Sunshine increases its promotion expenditure to 14.5 millions of Canadian dollars and the com-

Table 2.30 Sunshine Ltd facial soap sales (in millions of Canadian dollars) in Canada.

Period (trimester, year)	Sunshine promotion expenditure	Competitor's promotion expenditure/Sunshine promotion expenditure	Sales of facial soaps in Canada
I 2001	6.0	1.2	46.8
II 2001	6.8	1.2	52.7
III 2001	7.5	1.4	60.5
IV 2001	7.5	1.5	56.6
I 2002	9.0	1.5	64.4
II 2002	10.5	1.7	74.1
III 2002	12.0	1.8	72.2
IV 2002	12.0	1.5	78.0
I 2003	13.5	1.4	87.8
II 2003	13.5	1.5	95.6

Table 2.31 Number of light trucks sold by Mitsumishi.

Month	Year			
	2000	2001	2002	2003
January	22 882	23 478	24 768	24 765
February	19 981	17 019	19 351	21 739
March	18 811	20 967	23 953	25 153
April	19 352	19 759	18 855	20 515
May	27 226	22 200	28 414	24 038
June	14 932	24 162	18 537	26 151
July	18 531	20 275	22 845	
August	8 523	7 949	9 451	
September	13 064	14 328	15 842	
October	13 733	16 691	16 409	
November	12 597	13 784	13 881	
December	7 645	10 986	11 230	

petitor's promotion expenditure remains the same as in the second trimester of 2003.

2.8 Mitsumishi is a Korean company whose number of light trucks sold between January 2000 and June 2003 is reported in Table 2.31.

The company usually carries out a promotion in May and/or in June. The sales improvement achieved during the last few promotions are shown in Table 2.32.

Table 2.32 Light trucks sales improvement of Mitsumishi.

Period		Sales improvement (%)
May	2000	+30
May	2001	+20
June	2001	+15
May	2002	+25
June	2003	+30

Table 2.33 Sales of the new spiced food (in hundreds of kilograms) produced by Mare Nostrum.

Month	Year		
	2001	2002	2003
January	130 000	141 988	156 467
February	129 720	142 376	158 137
March	129 703	143 636	159 140
April	129 633	144 543	161 156
May	129 632	147 534	162 835
June	129 854	148 919	165 479
July	130 436	150 961	
August	132 751	152 748	
September	133 334	152 977	
October	133 761	154 387	
November	135 286	156 856	
December	136 800	157 349	

(a) Using a classical decomposition method, forecast the sales for the next six months.

(b) Plot a control chart of the error over the last three months. Are you able to detect any anomaly?

2.9 Mare Nostrum, a canned tuna manufacturer based in Sicily (Italy), has marketed a new spiced food line in June 2002. Now the management wishes to project sales for improved planning of production and logistics operations. Sales data for one and a half years are reported in Table 2.33.

(a) Plot the data on a graph. What important observations can you make about the demand pattern? Which data are relevant and should be used for forecasting purposes?

Table 2.34   Demand of refrigerated trucks (in europallets) between Antwerp and Brussels over the last ten weeks.

						Week				
Day	1	2	3	4	5	6	7	8	9	10
Monday	67	68	76	75	75	82	77	88	84	84
Tuesday	54	57	59	57	58	69	65	57	72	56
Wednesday	47	49	49	52	57	59	52	54	68	59
Thursday	40	45	46	43	48	49	55	50	59	52
Friday	60	63	68	69	72	69	68	66	63	69

(b) Using the classical time series decomposition analysis, predict the expected sales over the next six months.

(c) Repeat the forecast by applying the Holt method.

(d) Estimate the MAPD of both methods using the last six months. Which approach seems to work best?

2.10 The Belgian Trucking Company needs to determine the number of refrigerated trucks to satisfy the transportation demand between Antwerp and Brussels on a daily basis. The volume of the demand for the last weeks is given in Table 2.34.

(a) Using the Winters method, predict the expected number of pallets to be transported for the next week.

(b) Estimate the error in the above forecast using the last three weeks.

(c) Construct a 95% confidence interval on the forecasting. (Hint: assume a normal distribution of demand.)

2.11   Annotated Bibliography

An in-depth treatment of the forecasting methods is reported in:

1. Montgomery DC, Johnson LA and Gradiner JS 1990 *Forecasting and Time Series Analysis*. McGraw-Hill, New York.

For a detailed description of the statistical methods, the reader can refer to:

2. Sandy R 1990 *Statistics for Business and Economics*. McGraw-Hill, New York.

The results of the survey reported in Table 2.1 are taken from:

3. Sanders NR and Manrodt KB 1994 Forecasting practices in US corporations: survey results. *Interfaces* **24**(2), 92–100.

3
Designing the Logistics Network

3.1 Introduction

In business logistics the network planning process consists of designing the system through which commodities flow from suppliers to demand points, while in the public sector it consists of determining the set of facilities from which users are serviced. In both cases the main issues are to determine the number, location, equipment and size of new facilities, as well as the divestment, displacement or downsizing of facilities. Of course, the objectives and constraints vary depending on the sector (private or public) and on the type of facilities (plants, CDCs, RDCs, regional and field warehouses, retail outlets, dumpsites, incinerators, ambulance parking places, fire stations, etc.). The aim generally pursued in business logistics is the minimization of the annual total logistics cost subject to side constraints related to facility capacity and required customer service level (recall the discussion in Chapter 1). As a rule, the cost to be minimized is associated with facility operations (manufacturing, storage, sorting, consolidation, selling, incineration, parking, etc.), and to transportation between facilities, or between facilities and users. Also, when designing the logistics network for a utility company, different objectives, such as achieving equity in servicing users, may have to be considered.

Research in logistics network design dates back to the early location theories of the 19th century. Since then a variety of models and solution methodologies has been proposed and analysed. In this chapter some of the most important facility location problems are examined. To put this analysis in the right perspective, a number of relevant issues are first introduced and discussed.

When location decisions are needed. Facility location decisions must obviously be made when a logistics system is started from scratch. They are also required as a consequence of variations in the demand pattern or spatial distribution, or following modifications of materials, energy or labour cost. In particular, location decisions are often made when new products or services are launched, or outdated products are withdrawn from the market.

Introduction to Logistics Systems Planning and Control   G. Ghiani, G. Laporte and R. Musmanno
© 2004 John Wiley & Sons, Ltd   ISBN: 0-470-84916-9 (HB) 0-470-84917-7 (PB)

Location decisions may be strategic or tactical.   Whereas facilities are purchased or built, location decisions involve sizeable investments. In this case, changing sites or equipment is unlikely in the short or medium term. This may be true even if facilities are leased. On the other hand, if space and equipment are rented (e.g. from a public warehouses) or operations are subcontracted, location decisions can be reversible in the medium term.

Location and allocation decisions are intertwined.   Location decisions are strictly related to those of defining facility area boundaries (i.e. allocating demand to facilities). For example, in a two-echelon distribution system (see Figure 3.1), opening a new RDC must be accompanied by a redefinition of the sales districts along with a different allocation of the RDCs to the CDCs and of the CDCs to the production plants. For this reason location problems are sometimes referred to as *location–allocation* problems.

Location decisions may affect demand.   Facility location may affect the demand volume. For example, opening a new RDC may lead to the acquisition of customers who previously could not be served at a satisfactory level of service because they lived too far away.

3.2   Classification of Location Problems

Location problems come in a variety of forms, which can be classified with respect to a number of criteria. The classification proposed below is *logistics-oriented*.

Time horizon.   In *single-period problems*, facility location decisions must be made at the beginning of the planning horizon on the basis of the forecasted logistics requirements. In *multi-period problems* one has to decide, *at the beginning* of the planning horizon, a sequence of changes to be made at given time instants *within* the planning horizon.

Facility typology.   In *single-type* location problems, a single type of facility (e.g. only RDCs) are located. Instead, in *multi-type* problems several kinds of facility (e.g. both CDCs and RDCs) are located.

Material flows.   In *single-commodity problems* it can be assumed that a single homogeneous flow of materials exists in the logistics system, while in *multicommodity problems* there are several items, each with different characteristics. In the latter case each commodity is associated with a specific flow pattern.

Interaction among facilities.   In complex logistics systems there can be material flows among facilities of the same kind (e.g. component flows among plants). In this case, optimal facility locations depend not only on the spatial distribution of finished

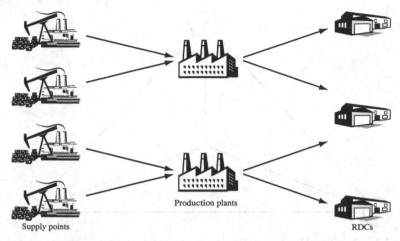

Supply points   Production plants   RDCs

Figure 3.1 A two-echelon single-type location problem.

product demand but also on the mutual position of the facilities (*location problems with interaction*).

Dominant material flows. *Single-echelon* location problems are single-type problems such that *either* the material flow coming out *or* the material flow entering the facilities to be located is negligible. In *multiple-echelon* problems, both inbound and outbound commodities are relevant. This is the case, for example, when DCs have to be located taking into account both the transportation cost from plants to DCs and the transportation cost from DCs to customers. In multiple-echelon problems, constraints aiming at balancing inbound and outbound flows have to be considered.

Demand divisibility. In some distribution systems it is required, for administrative or book-keeping reasons, that each facility or customer be supplied by a single centre, while in others a facility or a customer may be served by two or more centres. In the former case demand is said to be *divisible* while in the latter it is *indivisible*.

Influence of transportation on location decisions. Most location models assume that transportation cost between two facilities, or between a facility and an user, is computed as a suitable transportation rate multiplied by the freight volume and the distance between the two points. Such an approach is appropriate if vehicles travel by means of a direct route. However, if each vehicle makes collections or deliveries to several points, then a transportation rate cannot easily be established. In such cases the routes followed by the vehicles should be taken explicitly into account when locating the facilities (*location-routing* models). To illustrate this concept, consider Figure 3.2, where a warehouse serves three sales districts located at the vertices of triangle ABC. Under the hypothesis that the facility fixed cost is independent of the site, there can be two extreme cases.

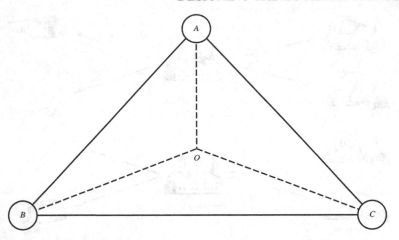

Figure 3.2   The optimal location of a warehouse depends on the way customers are serviced.

- Each customer requires a full-load supply and, therefore, the optimal location of the DC is equal to the Steiner point O.

- A single vehicle can service all points and hence the facility can be located at any point of the triangle ABC perimeter.

The interdependence between facility locations and vehicle routes is particularly strong when dealing with mail distribution, solid waste collection or road mainte-nance, as the users are located almost continuously on the road network.

Retail location.   When planning a store network, the main issue is to optimally locate a set of retail outlets that *compete* with other stores for customers. In such a context, predicting the expected revenues of a new site is difficult since it depends on a number of factors such as location, sales area and level of competition. Retail location problems can be modelled as *competitive location models*, the analysis of which is also beyond the scope of this textbook. The reader should again consult the references quoted in the last section of this chapter for further details.

Modelling and solving location problems

In the remainder of this chapter, some selected facility location problems are modelled and solved as MIP problems. An optimal solution can be determined in principle by means of a general-purpose or tailored *branch-and-bound* algorithm. Such an approach only works for relatively simple problems (such as the *single-echelon single-commodity* (SESC) problem) whenever instance size is small. For *multiple-echelon multiple-commodity* problems, determining an optimal solution can be prohibitive even if the number of potential facilities is relatively small (less than 100). As a result, heuristic procedures capable of determining a 'good' feasible solution in a reasonable amount of time can be very useful. To evaluate whether a heuristic solution provides a tight *upper bound* (UB) on the optimal solution value, it is useful to determine a *lower bound* (LB) on the optimal solution value. This yields a ratio (UB − LB)/LB which represents an overestimate of relative deviation of the heuristic solution value from the optimum.

3.3 Single-Echelon Single-Commodity Location Models

The SESC location problem is based on the following assumptions:

- the facilities to be located are homogeneous (e.g. they are all regional ware-houses);

- either the material flow coming out *or* the material flow entering such facilities is negligible;

- all material flows are homogeneous and can therefore be considered as a single commodity;

- transportation cost is linear or piecewise linear and concave;

- facility operating cost is piecewise linear and concave (or, in particular, constant).

The second assumption is the most restrictive. It holds in contexts where locating production plants whose finished product (e.g. steel) weighs much less than the raw materials (iron and coal in the example) used in the manufacturing process. Another application arises in warehouse location for a distribution company (see Figure 3.3), whereas the goods are purchased at a price inclusive of transportation cost up to the warehouses. We examine the case where inbound flows are negligible, although the same methodology can be applied without any change to the case where they are important and outbound flows are negligible. Moreover, to simplify the exposition, it is assumed that the facilities to be located are warehouses and the demand points are customers.

The problem can be modelled through a bipartite complete directed graph $G(V_1 \cup V_2, A)$, where the vertices in V_1 stand for the potential facilities, the vertices in V_2 represent the customers, and the arcs in $A = V_1 \times V_2$ are associated with the material flows between the potential facilities and the demand points.

In what follows, we further assume that

- the demand is divisible (see Section 3.2).

Let d_j, $j \in V_2$, be the demand of customer j; q_i, $i \in V_1$, the capacity of the potential facility i; u_i, $i \in V_1$, a decision variable that accounts for operations in potential facility i; s_{ij}, $i \in V_1, j \in V_2$, a decision variable representing the amount of product sent from site i to demand point j; $C_{ij}(s_{ij})$, $i \in V_1, j \in V_2$, the cost of transporting s_{ij} units of product from site i to customer j; $F_i(u_i)$, $i \in V_1$, the cost for operating

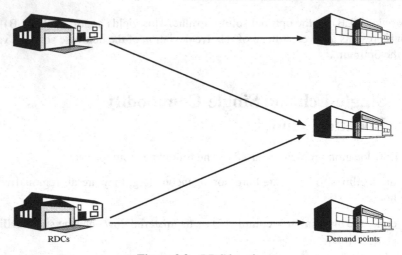

Figure 3.3   RDC location.

potential facility i at level u_i. Then the problem can be expressed in the following way.

Minimize

$$\sum_{i \in V_1} \sum_{j \in V_2} C_{ij}(s_{ij}) + \sum_{i \in V_1} F_i(u_i) \tag{3.1}$$

subject to

$$\sum_{j \in V_2} s_{ij} = u_i, \quad i \in V_1, \tag{3.2}$$

$$\sum_{i \in V_1} s_{ij} = d_j, \quad j \in V_2, \tag{3.3}$$

$$u_i \leqslant q_i, \quad i \in V_1, \tag{3.4}$$

$$s_{ij} \geqslant 0, \quad i \in V_1, j \in V_2, \tag{3.5}$$

$$u_i \geqslant 0, \quad i \in V_1. \tag{3.6}$$

Variables u_i, $i \in V_1$, implicitly define a location decision since a facility $i \in V_1$ is open if only if u_i is strictly positive. Variables s_{ij}, $i \in V_1, j \in V_2$, determine customer allocations to facilities. The objective function (3.1) is the sum of the facility operating costs plus the transportation cost between facilities and users. Constraints (3.2) state that the sum of the flows outgoing a facility equals its activity level. Constraints (3.3) ensure that each customer demand is satisfied, while constraints (3.4) force the activity level of a facility not to exceed the corresponding capacity.

Model (3.1)–(3.6) is quite general and can be easily adapted to the case where, in order to have an acceptable service level, some arcs $(i, j) \in A$ having a travel time larger than a given threshold cannot be used (we remove the corresponding decision variables s_{ij}, for the appropriate $i \in V_1$ and $j \in V_2$ (see Figure 3.4)). In the remainder

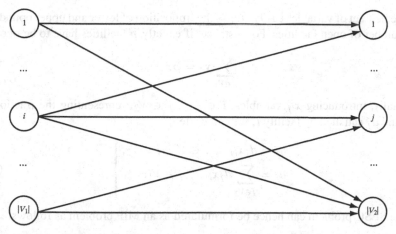

Figure 3.4 Graph representation of the single-echelon location problem (note that the arcs $(1, j)$ and $(|V_1|, 1)$ are absent, given that the corresponding travel times are longer than the given threshold).

of this section two particular cases of the SESC location problem are examined in detail. In the first case, the transportation cost per unit of commodity is constant, and facility operating costs consist of fixed costs. In the second case, transportation costs are still linear but facilities operating costs are piecewise linear and concave. Both problems are NP-hard and can be modelled as MIP problems. Hence they can be solved through general purpose (or tailored) branch-and-bound algorithms.

3.3.1 Linear transportation costs and facility fixed costs

If the transportation costs per unit of flow are constant, then

$$C_{ij}(s_{ij}) = \bar{c}_{ij}s_{ij}, \quad i \in V_1, \ j \in V_2.$$

Moreover, if facility costs $F_i(u_i)$ are described by a *fixed cost* f_i and a constant *marginal cost* g_i, then

$$F_i(u_i) = \begin{cases} f_i + g_i u_i, & \text{if } u_i > 0, \\ 0, & \text{if } u_i = 0, \end{cases} \quad i \in V_1. \tag{3.7}$$

Equation (3.7) leads to the introduction in problem (3.1)–(3.6) of a binary variable y_i replacing u_i, for each $i \in V_1$, whose value is equal to 1 if potential facility i is open, and 0 otherwise. If g_i is negligible, then Equation (3.7) is replaced by

$$F_i(y_i) = f_i y_i, \quad i \in V_1,$$

and constraints (3.2) and (3.4) become

$$\sum_{j \in V_2} s_{ij} \leqslant q_i y_i, \quad i \in V_1.$$

The new set of variables easily allows the imposition of lower and upper bounds on the number of open facilities. For instance, if exactly p facilities have to be opened, then

$$\sum_{i \in V_1} y_i = p.$$

Finally, introducing x_{ij} variables, $i \in V_1$, $j \in V_2$, representing the fraction of demand d_j satisfied by facility i, we can write

$$\left. \begin{array}{l} s_{ij} = d_j x_{ij}, \quad i \in V_1, \ j \in V_2, \\ u_i = \sum_{j \in V_2} d_j x_{ij}, \quad i \in V_1. \end{array} \right\} \tag{3.8}$$

The SESC problem can hence be formulated as an MIP problem as follows.

Minimize

$$\sum_{i \in V_1} \sum_{j \in V_2} c_{ij} x_{ij} + \sum_{i \in V_1} f_i y_i \tag{3.9}$$

subject to

$$\sum_{i \in V_1} x_{ij} = 1, \quad j \in V_2, \tag{3.10}$$

$$\sum_{j \in V_2} d_j x_{ij} \leqslant q_i y_i, \quad i \in V_1, \tag{3.11}$$

$$\sum_{i \in V_1} y_i = p, \tag{3.12}$$

$$0 \leqslant x_{ij} \leqslant 1, \quad i \in V_1, \ j \in V_2, \tag{3.13}$$

$$y_i \in \{0, 1\}, \quad i \in V_1, \tag{3.14}$$

where

$$c_{ij} = \bar{c}_{ij} d_j, \quad i \in V_1, \ j \in V_2, \tag{3.15}$$

is the transportation cost incurred for satisfying the entire demand d_j of customer $j \in V_2$ from facility $i \in V_1$. It is worth noting that on the basis of constraints (3.10), variables x_{ij}, $i \in V_1$, $j \in V_2$, cannot take a value larger than 1. Therefore, relations (3.13) can be written more simply in the form $x_{ij} \geqslant 0$, $i \in V_1$, $j \in V_2$. Relations $x_{ij} \leqslant 1$, $i \in V_1$, $j \in V_2$, are fundamental in a context where the constraints (3.10) are relaxed, as in the Lagrangian procedure described below.

The SESC formulation (3.9)–(3.14) is quite general and can sometimes be simplified. In particular, if constraint (3.12) is removed, the model is known as *capacitated plant location* (CPL). If relations (3.11) are also deleted, the so-called *simple plant location* (SPL) model is obtained. If fixed costs f_i are the same for each $i \in V_1$, $d_j = 1$ for each $j \in V_2$ and $q_i = |V_2|$ for each $i \in V_1$, then formulation (3.9)–(3.14) is known as a *p-median* model.

Koster Express is an American LTL express carrier operating in Oklahoma (USA). The carrier service is organized through a distribution subsystem, a group of terminals and a long-haul transportation subsystem. The distribution subsystem uses a set of trucks, based at the terminals, that pick up the outgoing goods from 1:00 p.m. to 7:00 p.m., and deliver the incoming goods from 9:00 a.m. to 1:00 p.m. The terminals are equipped areas where the outgoing items are collected and consolidated on pallets, the incoming pallets are opened up, and their items are classified and directed to the distribution. At present the firm has nine terminals (located in Ardmore, Bartlesville, Duncan, Enid, Lawton, Muskogee, Oklahoma City, Ponca City and Tulsa). The long-haul transportation subsystem provides the transport, generally during the night, of the consolidated loads between the origin and destination terminals. For that purpose, trucks with a capacity between 14 and 18 pallets (and a maximum weight of between 0.8 and 1 ton) are used. During the last six years, the long-haul transportation subsystem has had two hubs in Duncan and Tulsa. The Duncan hub receives the goods from Ardmore, Lawton, Oklahoma City and Duncan itself, while the Tulsa hub collects the items coming from the other terminals. In each hub the goods assigned to the terminals of the other hub are sent on large trucks, generally equipped with trailers. For example, goods coming from Enid and destined to Oklahoma City are brought to the Tulsa hub, from where they are sent to the Duncan hub together with the goods coming from Bartlesville, Muskogee, Ponca City and Tulsa itself and directed to the terminals served by the Duncan hub; finally, they are sent to Oklahoma City. Goods destined to other terminals of the same hub are stocked for a few hours until the trucks coming from the other hub arrive, and only then are they sent to their destinations. For example, goods to Lawton coming from Ardmore are sent to the Duncan hub, stocked until the arrival of the trucks from the Tulsa hub and then, sent to Lawton, together with the goods coming from all the other terminals.

Recently, the firm was offered a major growth opportunity with the opening of new terminals in Altus, Edmond and Stillwater. At the same time the management decided to relocate its two hubs. The team hired to carry out a preliminary analysis decided to consider only the transportation cost from each terminal to the hub and vice versa (neglecting, therefore, both the transportation cost between the two hubs and the cost, yet considerable, associated with the possible divestment of the pre-existing hub). Under the hypothesis that each terminal can accommodate a hub, the problem can be formulated a p-median problem in the following way.

Minimize

$$\sum_{i \in V_1} \sum_{j \in V_2} c_{ij} x_{ij}$$

subject to

$$\sum_{i \in V_1} x_{ij} = 1, \quad j \in V_2,$$

Table 3.1   Distances (in miles) between terminals in the Koster Express problem (Part I).

	Altus	Ardmore	Bartlesville	Duncan	Edmond	Enid
Altus	0.0	169.8	291.8	88.2	153.9	208.2
Ardmore	169.8	0.0	248.6	75.9	112.5	199.0
Bartlesville	291.8	248.6	0.0	231.5	146.0	132.4
Duncan	88.2	75.9	231.5	0.0	93.5	137.5
Edmond	153.9	112.5	146.0	93.5	0.0	88.8
Enid	208.2	199.0	132.4	137.5	88.8	0.0
Lawton	54.2	115.8	238.7	34.1	100.7	145.0
Muskogee	274.2	230.4	92.2	213.5	145.7	166.4
Oklahoma City	141.1	100.5	151.4	80.9	14.4	87.6
Ponca City	245.0	202.2	70.2	184.8	91.9	64.5
Stillwater	209.2	162.6	115.0	145.3	53.0	65.8
Tulsa	248.0	204.6	45.6	187.8	102.2	118.4

$$\sum_{j \in V_2} x_{ij} \leqslant |V_2| y_i, \quad i \in V_1,$$

$$\sum_{i \in V_1} y_i = 2,$$

$$x_{ij} \in \{0, 1\}, \quad i \in V_1, \ j \in V_2,$$

$$y_i \in \{0, 1\}, \quad i \in V_1,$$

where $V_1 = V_2$ represent the set of old and new terminals; y_i, $i \in V_1$, is a binary decision variable whose value is equal to 1 if terminal i accommodates a hub, 0 otherwise; x_{ij}, $i \in V_1$, $j \in V_2$, is a binary decision variable whose value is equal to 1 if the hub located in terminal i serves terminal j, 0 otherwise; however, due to the particular structure of the problem constraints, variables x_{ij} cannot be fractional and greater than 1, therefore $x_{ij} \in \{0, 1\}$, $i \in V_1$, $j \in V_2$ can be replaced with $x_{ij} \geqslant 0$, $i \in V_1$, $j \in V_2$.

Since the goods incoming and outgoing every day from each terminal can be transferred by a single truck with a capacity of 14 pallets, the daily transport cost (in dollars) between a pair of terminals $i \in V_1$ and $j \in V_2$ is given by $c_{ij} = 2 \times 0.74 \times l_{ij}$, where 0.74 is the transportation cost (in dollars per mile), and l_{ij} is the distance (in miles) between the terminals (see Tables 3.1 and 3.2).

The optimal solution leads to the opening of two hubs located in Duncan and Stillwater (the daily total cost being $1081.73). Terminals in Altus, Ardmore, Duncan and Lawton are assigned to the Duncan hub, while the Stillwater hub serves the terminals in Bartlesville, Edmond, Enid, Muskogee, Oklahoma City, Ponca City, Stillwater and Tulsa.

Table 3.2 Distances (in miles) between terminals in the Koster Express problem (Part II).

	Lawton	Muskogee	Oklahoma City	Ponca City	Stillwater	Tulsa
Altus	54.2	274.2	141.1	245.0	209.2	248.0
Ardmore	115.8	230.4	100.5	202.2	162.6	204.6
Bartlesville	238.7	92.2	151.4	70.2	115.0	45.6
Duncan	34.1	213.5	80.9	184.8	145.3	187.8
Edmond	100.7	145.7	14.4	91.9	53.0	102.2
Enid	145.0	166.4	87.6	64.5	65.8	118.4
Lawton	0.0	220.6	88.0	191.9	152.5	194.9
Muskogee	220.6	0.0	140.4	142.5	119.2	48.1
Oklahoma City	88.0	140.4	0.0	104.7	66.6	107.6
Ponca City	191.9	142.5	104.7	0.0	41.9	96.5
Stillwater	152.5	119.2	66.6	41.9	0.0	71.2
Tulsa	194.9	48.1	107.6	96.5	71.2	0.0

Demand allocation

For a given set $\bar{V}_1 \subseteq V_1$ of open facilities, an optimal demand allocation to $\bar{V}_1$ can be determined by means of the following LP model.

Minimize

$$\sum_{i \in \bar{V}_1} \sum_{j \in V_2} c_{ij} x_{ij} \qquad (3.16)$$

subject to

$$\sum_{i \in \bar{V}_1} x_{ij} = 1, \qquad (3.17)$$

$$\sum_{j \in V_2} d_j x_{ij} \leqslant q_i, \quad i \in \bar{V}_1, \qquad (3.18)$$

$$x_{ij} \geqslant 0, \quad i \in \bar{V}_1, \ j \in V_2. \qquad (3.19)$$

In an optimal solution of problem (3.16)–(3.19), some x_{ij}^* values may be fractional (i.e. the demand of a vertex $j \in V_2$ can be satisfied by more than one vertex $i \in V_1$) because of capacity constraints (3.18). However, in the absence of a capacity constraint (as in the SPL and p-median models), there exists at least one optimal solution such that the demand of each vertex $j \in V_2$ is satisfied by a single facility $i \in V_1$ (*single assignment* property). This solution can be defined as follows. Let $i_j \in \bar{V}_1$ be a facility such that

$$i_j = \arg \min_{i \in \bar{V}_1} c_{ij}.$$

Then, the allocation variables can be defined as follows:

$$x_{ij}^* = \begin{cases} 1, & \text{if } i = i_j, \\ 0, & \text{otherwise.} \end{cases}$$

A Lagrangian heuristic for the capacitated plant location problem

Several heuristic methods have been developed for the solution of SESC problem (3.9)–(3.14). Among them, Lagrangian relaxation techniques play a major role. They usually provide high-quality upper and lower bounds within a few iterations. For the sake of simplicity, in the remainder of this section, a Lagrangian procedure is illustrated for the CPL problem, although this approach may be used for the general SESC model (3.9)–(3.14). The fundamental step of the heuristic is the determination of a lower bound, obtained by relaxing demand satisfaction constraints (3.10) in a Lagrangian fashion. Let $\lambda_j \in \Re$ be the multiplier associated with the jth constraint (3.10). Then the relaxed problem is:

Minimize

$$\sum_{i \in V_1} \sum_{j \in V_2} c_{ij} x_{ij} + \sum_{i \in V_1} f_i y_i + \sum_{j \in V_2} \lambda_j \left(\sum_{i \in V_1} x_{ij} - 1 \right) \qquad (3.20)$$

subject to

$$\sum_{j \in V_2} d_j x_{ij} \leqslant q_i y_i, \qquad i \in V_1, \qquad (3.21)$$

$$0 \leqslant x_{ij} \leqslant 1, \quad i \in V_1, \ j \in V_2, \qquad (3.22)$$

$$y_1 \in \{0, 1\}, \qquad i \in V_1. \qquad (3.23)$$

It is easy to check that problem (3.20)–(3.23) can be decomposed into $|V_1|$ subproblems. Indeed, for a given vector $\lambda \in \Re^{|V_2|}$ of multipliers, the optimal objective value of problem (3.20)–(3.23), $\text{LB}_{\text{CPL}}(\lambda)$, can be determined by solving, *for each potential facility $i \in V_1$*, the following subproblem:

Minimize

$$\sum_{j \in V_2} (c_{ij} + \lambda_j) x_{ij} + f_i y_i \qquad (3.24)$$

subject to

$$\sum_{j \in V_2} d_j x_{ij} \leqslant q_i y_i, \qquad (3.25)$$

$$0 \leqslant x_{ij} \leqslant 1, \quad j \in V_2, \qquad (3.26)$$

$$y_i \in \{0, 1\}, \qquad (3.27)$$

and then by setting

$$\text{LB}_{\text{CPL}}(\lambda) = \sum_{i \in V_1} \text{LB}^i_{\text{CPL}}(\lambda) - \sum_{j \in V_2} \lambda_j,$$

where $\text{LB}^i_{\text{CPL}}(\lambda)$ is the optimal objective function value of subproblem (3.24)–(3.27). $\text{LB}^i_{\text{CPL}}(\lambda)$ can be easily determined by inspection, by observing that

- for $y_i = 0$, Equation (3.25) implies $x_{ij} = 0$, for each $j \in V_2$, and therefore $LB^i_{CPL}(\lambda) = 0$;

- for $y_i = 1$, subproblem (3.24)–(3.27) is a continuous knapsack problem (with an objective function to be minimized); it is well known that this problem can be solved in polynomial time by means of a 'greedy' procedure.

By suitably modifying the optimal solution of the Lagrangian relaxation, it is possible to construct a CPL feasible solution as follows.

Step 1. (*Finding the facilities to be activated.*) Let L be the list of potential facilities $i \in V_1$ sorted by nondecreasing values of $LB^i_{CPL}(\lambda)$, $i \in V_1$ (note that $LB^i_{CPL}(\lambda) \leqslant 0$, $i \in V_1, \lambda \in \Re^{|V_2|}$). Extract from L the minimum number of facilities capable to satisfy the total demand $\sum_{j \in V_2} d_j$. Let $\bar{V}_1$ be the set of facilities selected. Then $\bar{V}_1$ satisfies the relation:

$$\sum_{i \in \bar{V}_1} q_i \geqslant \sum_{j \in V_2} d_j.$$

Step 2. (*Customer allocation to the selected facilities.*) Solve the demand allocation problem (3.16)–(3.19) considering $\bar{V}_1$ as the set of facilities to be opened. Let $UB_{CPL}(\lambda)$ be the cost (3.24) associated with the optimal allocation.

The heuristic first selects the facilities characterized by the smallest $LB^i_{CPL}(\lambda)$ values and then optimally allocates the demand to them.

Thus for *each* set of multipliers $\lambda \in \Re^{|V_2|}$, the above procedure computes both a lower and upper bound ($LB_{CPL}(\lambda)$ and $UB_{CPL}(\lambda)$, respectively). If these bounds coincide, an optimal solution has been found. Otherwise, in order to determine the multipliers corresponding to the maximum possible lower bound $LB_{CPL}(\lambda)$ (or at least a satisfactory bound), the classical *subgradient algorithm* can be used. This method also generates, in many cases, better upper bounds, since the feasible solutions generated from improved lower bounds are generally less costly. Here is a schematic description of the subgradient algorithm.

Step 0. (*Initialization.*) Select a tolerance value $\varepsilon \geqslant 0$. Set $LB = -\infty$, $UB = \infty$, $k = 1$ and $\lambda^k_j = 0$, $j \in V_2$.

Step 1. (*Computation of a new lower bound.*) Solve the Lagrangian relaxation (3.20)–(3.23) using $\lambda^k \in \Re^{|V_2|}$ as a vector of multipliers. If $LB_{CPL}(\lambda^k) > LB$, set $LB = LB_{CPL}(\lambda^k)$.

Step 2. (*Computation of a new upper bound.*) Determine the corresponding feasible solution. Let $UB_{CPL}(\lambda^k)$ be its cost. If $UB_{CPL}(\lambda^k) < UB$, set $UB = UB_{CPL}(\lambda^k)$.

Step 3. (*Check of the stopping criterion.*) If $(UB - LB)/LB \leqslant \varepsilon$, *STOP*. LB and UB represent the best upper and lower bound available for z^*_{CPL}, respectively.

Table 3.3   Distances (in kilometres) between potential production plants
and markets in the Goutte problem.

	Brossard	Granby	Sainte-Julie	Sherbrooke	Valleyfield	Verdun
Brossard	0.0	76.1	30.4	139.4	72.6	11.7
Granby	76.1	0.0	71.0	77.2	144.5	83.7
LaSalle	20.8	92.9	47.2	156.1	47.5	11.7
Mascouche	54.7	113.3	52.9	187.2	93.0	45.2
Montréal	13.5	85.5	28.0	148.7	67.3	9.3
Sainte-Julie	30.4	71.0	0.0	138.2	94.5	38.1
Sherbrooke	139.4	77.2	138.2	0.0	207.9	146.9
Terrebonne	47.8	106.5	46.2	180.2	86.7	38.9
Valleyfield	72.6	144.5	94.5	207.9	0.0	63.4
Verdun	11.7	83.7	38.1	146.9	63.4	0.0

Step 4. (*Updating of the Lagrangian multipliers.*) Determine the subgradient of the jth relaxed constraint,

$$s_j^k = \sum_{i \in V_1} x_{ij}^k - 1, \quad j \in V_2,$$

where x_{ij}^k is the solution of the Lagrangian relaxation (3.20)–(3.23) using $\lambda^k \in \Re^{|V_2|}$ as Lagrangian multipliers. Then set

$$\lambda_j^{k+1} = \lambda_j^k + \beta^k s_j^k, \quad j \in V_2, \tag{3.28}$$

where β^k is a suitable scalar coefficient. Let $k = k + 1$ and go back to Step 1.

This algorithm attempts to determine an ε-optimal solution, i.e. a feasible solution with a maximum user-defined deviation ε from the optimal solution.

Computational experiments have shown that the initial values of the Lagrangian multipliers do not significantly affect the behaviour of the procedure. Hence multipliers are set equal to 0 in Step 0. Formula (3.28) can be explained in the following way. If, at the kth iteration, the left-hand side of constraint (3.10) is higher than the right-hand side ($\sum_{i \in V_1} x_{ij}^k > 1$) for a certain $j \in V_2$, the subgradient s_j^k is positive and the corresponding Lagrangian multiplier has to be increased in order to heavily penalize the constraint violation. Vice versa, if the left-hand side of constraint (3.10) is lower than the right-hand side ($\sum_{i \in V_1} x_{ij}^k < 1$) for a certain $j \in V_2$, the associated subgradient s_j^k is negative and the value of the associated multiplier must be decreased to make the service of the unsatisfied demand fraction $1 - \sum_{i \in V_1} x_{ij}^k$ more attractive. Finally, if the jth constraint (3.10) is satisfied ($\sum_{i \in V_1} x_{ij}^k = 1$), the corresponding multiplier is unchanged.

The term β^k in Equation (3.28) is a proportionality coefficient defined as

$$\beta^k = \frac{\alpha(\text{UB} - \text{LB}_{\text{CPL}}(\lambda^k))}{\sum_{j=1}^{|V_2|} (s_j^k)^2},$$

Table 3.4 Plant operating costs (in Canadian dollars per year) and capacity (in hectolitres per year) in the Goutte problem.

Site	Fixed cost	Capacity
Brossard	81 400	22 000
Granby	83 800	24 000
LaSalle	88 600	28 000
Mascouche	91 000	30 000
Montréal	79 000	20 000
Sainte-Julie	86 200	26 000
Sherbrooke	88 600	28 000
Terrebonne	91 000	30 000
Valleyfield	79 000	20 000
Verdun	80 200	21 000

Table 3.5 Demands (in hectolitres per year) of the sales districts in the Goutte problem.

Site	Demand
Brossard	14 000
Granby	10 000
Sainte-Julie	8 000
Sherbrooke	12 000
Valleyfield	10 000
Verdun	9 000

where α is a scalar arbitrarily chosen in the interval $(0, 2]$. The use of parameter β^k in Equation (3.28) limits the variations of the multipliers when the lower bound $\text{LB}_{\text{CPL}}(\lambda^k)$ approaches the current upper bound UB.

Finally, note that the $\{\text{LB}_{\text{CPL}}(\lambda^k)\}$ sequence produced by the subgradient method does not decrease monotonically. Therefore, there could exist iterations k for which $\text{LB}_{\text{CPL}}(\lambda^k) < \text{LB}_{\text{CPL}}(\lambda^{k-1})$ (this explains the lower bound update in Step 1). In practice, $\text{LB}_{\text{CPL}}(\lambda^k)$ values exhibit a zigzagging pattern.

The procedure is particularly efficient. Indeed, it generally requires, for $\varepsilon \approx 0.01$, only a few thousand iterations for problems with hundreds of vertices in V_1 and in V_2.

Goutte is a Canadian company manufacturing and distributing soft drinks. The firm has recently achieved an unexpected increase of its sales mostly because of the launch of a new beverage which has become very popular with young consumers. The management is now considering the opportunity of opening a new plant to which some of the production of the other factories could be moved.

The production process makes use of water, available anywhere in Canada in an unlimited quantity at a negligible cost, and of sugar extracts which compose a modest percentage of the total weight of the finished product. For this reason, supply costs are negligible compared to finished product distribution costs, and the transportation costs are product independent. Table 3.3 provides the distances between potential plants and markets.

The management has decided to undertake a preliminary analysis in order to evaluate the theoretical minimum cost plant configuration for different service levels. In this study, the logistics system is assumed to be designed from scratch.

The annual operating costs and capacities of the potential plants are shown in Table 3.4, while the annual demands of the sales districts are reported in Table 3.5.

The maximum distance between a potential plant and a market is successively set equal to ∞ and 70 km. The trucks have a capacity of 150 hectolitres and a cost (inclusive of the crew's wages) of 0.92 Canadian dollars per kilometre.

The Goutte problem is then modelled as a CPL formulation:

Minimize

$$\sum_{i \in V_1} \sum_{j \in V_2} c_{ij} x_{ij} + \sum_{i \in V_1} f_i y_i$$

subject to

$$\sum_{i \in V_1} x_{ij} = 1, \qquad j \in V_2,$$

$$\sum_{j \in V_2} d_j x_{ij} \leqslant q_i y_i, \quad i \in V_1,$$

$$0 \leqslant x_{ij} \leqslant 1, \qquad i \in V_1, j \in V_2,$$

$$y_i \in \{0, 1\}, \quad i \in V_1,$$

where $V_1 = \{$Brossard, Granby, LaSalle, Mascouche, Montréal, Sainte-Julie, Sherbrooke, Terrebonne, Valleyfield, Verdun$\}$ and $V_2 = \{$Brossard, Granby, Sainte-Julie, Sherbrooke, Valleyfield, Verdun$\}$; x_{ij}, $i \in V_1$, $j \in V_2$, is a decision variable representing the fraction of the annual demand of market j satisfied by plant i; y_i, $i \in V_1$, is a binary decision variable, whose value is equal to 1 if the potential plant i is open, 0 otherwise. Transportation costs c_{ij}, $i \in V_1$, $j \in V_2$ (see Table 3.6), are calculated through Equations (3.15), observing that trucks always travel with a full load on the outward journey and empty on the return journey. For example, since the distance between the potential plant located in Granby and the sales district of Brossard is 76.1 km, the transportation cost is $2 \times 76.1 \times 0.92 = 140.06$ Canadian dollars for transporting 150 hectolitres. Since the demand of the sales district of Brossard is 14 000 hectolitres per year, the annual transportation cost for satisfying the entire demand of the Brossard market from the potential plant of Granby is $(14\,000/150) \times 140.06 = 13\,072.32$ Canadian dollars.

If the maximum distance between a plant and a market is set equal to infinity, the optimal solution leads to the opening of three production plants (located in Brossard, Granby and Valleyfield) and the optimal x_{ij}^* values, $i \in V_1$, $j \in V_2$, are reported in Table 3.7.

The annual total logistics cost equals to 265 283.12 Canadian dollars. The optimal solution can be found by using the Lagrangian heuristic (with $\alpha = 1.5$). The results of the first two iterations can be summarized as follows:

$$\lambda_j^1 = 0, \quad j = 1, \ldots, 6;$$
$$LB_{CPL}(\lambda^1) = 0;$$
$$LB = 0;$$
$$y^1 = [1, 0, 0, 0, 1, 0, 0, 0, 0, 1]^T,$$

corresponding to the plants located in Brossard, Montréal and Verdun. This choice guarantees the minimum number of open plants capable of satisfying the whole demand (63 000 hectolitres per year) of the market at the lowest cost:

$$UB_{CPL}(\lambda^1) = 282\,537.24;$$
$$UB = 282\,537.24;$$
$$s_j^1 = -1, \quad j = 1, \ldots, 6;$$
$$\beta^1 = 70\,634.31;$$
$$\lambda_j^2 = -70\,634.31, \quad j = 1, \ldots, 6;$$
$$LB_{CPL}(\lambda^2) = -529\,969.24;$$
$$LB = 0;$$
$$y^2 = [1, 0, 0, 0, 1, 0, 0, 0, 1, 1]^T;$$
$$UB_{CPL}(\lambda^2) = 353\,542.24;$$
$$UB = 282\,537.24;$$
$$s^2 = [-1.00, 2.60, 9.00, -0.92, 3.50, 8.67]^T;$$
$$\beta^2 = 6887.15.$$

If the maximum distance between a potential plant and a market is 70 km, the optimal solution leads to the opening of four plants (located in Brossard, Granby, Sherbrooke and Valleyfield). The demand allocation is reported in Table 3.8.

The annual total logistics cost would be 342 784.87 Canadian dollars, with an increase of 77 501.75 Canadian dollars compared with the first solution found.

Table 3.6 Transportation costs (in Canadian dollars per year) c_{ij}, $i \in V_1$, $j \in V_2$, in the Goutte problem.

	Brossard	Granby	Sainte-Julie	Sherbrooke	Valleyfield	Verdun
Brossard	0.00	9 337.37	2 984.80	20 514.58	8 903.08	1 296.97
Granby	13 072.32	0.00	6 964.54	11 370.67	17 727.19	9 238.67
LaSalle	3 565.18	11 390.41	4 627.23	22 978.23	5 823.52	1 296.97
Mascouche	9 396.60	13 897.49	5 195.76	27 550.19	11 410.15	4 992.43
Montréal	2 321.51	10 482.34	2 747.91	21 888.54	8 251.63	1 030.47
Sainte-Julie	5 223.40	8 705.67	0.00	20 348.76	11 587.82	4 210.70
Sherbrooke	23 933.68	9 475.56	13 565.84	0.00	25 505.04	16 220.97
Terrebonne	8 208.20	13 068.37	4 532.48	26 531.56	10 640.26	4 299.53
Valleyfield	12 464.31	17 727.19	9 270.25	30 606.05	0.00	7 000.07
Verdun	2 017.50	10 265.19	3 742.85	21 627.96	7 777.85	0.00

Table 3.7 Fraction of the annual demand of the sales district $j \in V_2$ satisfied by the production plant $i \in V_1$ (no limit on the distance) in the Goutte problem.

	Brossard	Granby	Sainte-Julie	Sherbrooke	Valleyfield	Verdun
Brossard	1.00	0.00	0.75	0.00	0.00	0.22
Granby	0.00	1.00	0.25	1.00	0.00	0.00
LaSalle	0.00	0.00	0.00	0.00	0.00	0.00
Mascouche	0.00	0.00	0.00	0.00	0.00	0.00
Montréal	0.00	0.00	0.00	0.00	0.00	0.00
Sainte-Julie	0.00	0.00	0.00	0.00	0.00	0.00
Sherbrooke	0.00	0.00	0.00	0.00	0.00	0.00
Terrebonne	0.00	0.00	0.00	0.00	0.00	0.00
Valleyfield	0.00	0.00	0.00	0.00	1.00	0.78
Verdun	0.00	0.00	0.00	0.00	0.00	0.00

3.3.2 Linear transportation costs and concave piecewise linear facility operating costs

In this subsection we show how an SESC location problem with piecewise linear and concave facility operating costs can still be modelled as an MIP model. Indeed, such a problem can be transformed into model (3.9)–(3.14) by introducing dummy potential facilities and suitably defining costs and capacities. For the sake of simplicity, the topic will be illustrated in three steps.

Case A. The operating cost $F_i(u_i)$ of a facility $i \in V_1$ is given by Equation (3.7) (see

Table 3.8 Fraction of the annual demand of the sales district $i \in V_2$ satisfied by the production plant $i \in V_1$ (maximum distance of 70 km) in the Goutte problem.

	Brossard	Granby	Sainte-Julie	Sherbrooke	Valleyfield	Verdun
Brossard	1.00	0.00	1.00	0.00	0.00	0.00
Granby	0.00	1.00	0.00	0.00	0.00	0.00
LaSalle	0.00	0.00	0.00	0.00	0.00	0.00
Mascouche	0.00	0.00	0.00	0.00	0.00	0.00
Montréal	0.00	0.00	0.00	0.00	0.00	0.00
Sainte-Julie	0.00	0.00	0.00	0.00	0.00	0.00
Sherbrooke	0.00	0.00	0.00	1.00	0.00	0.00
Terrebonne	0.00	0.00	0.00	0.00	0.00	0.00
Valleyfield	0.00	0.00	0.00	0.00	1.00	1.00
Verdun	0.00	0.00	0.00	0.00	0.00	0.00

Figure 3.5):

$$F_i(u_i) = \begin{cases} f_i + g_i u_i, & \text{if } u_i > 0, \\ 0, & \text{if } u_i = 0, \end{cases} \quad i \in V_1,$$

where, using Equation (3.8), $u_i = \sum_{j \in V_2} d_j x_{ij}$.

Hence, this problem can be transformed into model (3.9)–(3.14) by rewriting the objective function (3.9) as

$$\sum_{i \in V_1} \sum_{j \in V_2} t_{ij} x_{ij} + \sum_{i \in V_1} f_i y_i, \tag{3.29}$$

where

$$t_{ij} = c_{ij} + g_i d_j, \quad i \in V_1, \ j \in V_2. \tag{3.30}$$

This transformation is based on the following observations. If the site $i \in V_1$ is opened, the first term of the objective function (3.29) includes not only the transportation cost $c_{ij} x_{ij}, \ j \in V_2$, but also the contribution $g_i d_j x_{ij}$ of the variable cost of the facility $i \in V_1$; if, instead, the site $i \in V_1$ is not opened, the variables $x_{ij}, \ j \in V_2$, take the value 0 and that does not generate any cost.

Case B. A potential facility cannot be run economically if its activity level is lower than a value q_i^- or higher than a threshold q_i^+. For intermediate values, the operating cost grows linearly (see Figure 3.6). This case can be modelled as the previous case, provided that in (3.29), (3.10)–(3.14), capacity constraints (3.11) are replaced with the following relations:

$$q_i^- y_i \leqslant \sum_{j \in V_2} d_j x_{ij} \leqslant q_i^+ y_i, \quad i \in V_1.$$

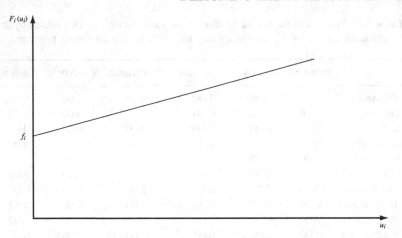

Figure 3.5   Operating cost $F_i(u_i)$ of potential facility $i \in V_1$ versus
activity level u_i (case A).

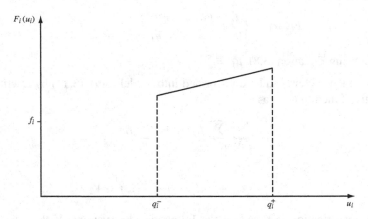

Figure 3.6   Operating cost $F_i(u_i)$ of potential facility $i \in V_1$ versus
activity level u_i (case B).

Case C. The operating cost $F_i(u_i)$ of a potential facility $i \in V_1$ is a general concave
piecewise linear function of its activity level because of economies of scale. In the
simplest case, there only are two piecewise lines (see Figure 3.7). Then

$$F_i(u_i) = \begin{cases} f_i'' + g_i'' u_i, & \text{if } u_i > u_i', \\ f_i' + g_i' u_i, & \text{if } 0 < u_i \leqslant u_i', \ i \in V_1, \\ 0, & \text{if } u_i = 0, \end{cases} \tag{3.31}$$

where $f_i' < f_i''$ and $g_i' > g_i''$. In order to model this problem as before, each potential
facility is replaced by as many artificial facilities as the piecewise lines of its cost
function. For instance, if Equation (3.31) holds, facility $i \in V_1$ is replaced by two

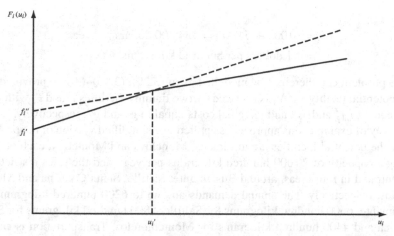

Figure 3.7 Operating cost $F_i(u_i)$ of potential facility $i \in V_1$ versus activity level u_i (case C).

artificial facilities i' and i'' whose operating costs are characterized, respectively, by fixed costs equal to f_i' and f_i'', and by marginal costs equal to g_i' and g_i''. In this way the problem belongs to case A described above since it is easy to demonstrate that in every optimal solution at most one of the artificial facilities is selected.

Logconsult is an American consulting company commissioned to propose changes to the logistics system of Gelido, a Mexican firm distributing deep-frozen food. A key aspect of the analysis is the relocation of the Gelido DCs. A preliminary examination of the problem led to the identification of about $|V_1| = 30$ potential sites where warehouses can be open or already exist. In each site several different types of warehouses can usually be installed. Here we show how the cost function of a potential facility $i \in V_1$ can be estimated. Facility fixed costs include rent, amortization of the machinery, insurance of premises and machinery, and staff wages. They add up to $80\,000$ per year. The variable costs are related to the storage and handling of goods. Logconsult has estimated the Gelido variable costs on the basis of historical data (see Table 3.9).

Facility variable costs (see Figure 3.8) are influenced by inventory costs which generally increase with the square root of the demand (see Chapter 4 for further details). This relation suggests approximating the cost function of potential facility $i \in V_1$ through Equation (3.31), where $u_i' = 3500$ hundred kilograms per year. The values of f_i', f_i'', g_i' and g_i'' can be obtained by applying linear regression (see Chapter 2) to each of the two sets of available data. This way the following relations are obtained:

$$f_i' = 80\,000 + 2252 = 82\,252 \text{ dollars per year;}$$
$$g_i' = 18.5 \text{ dollars per hundred kilograms;}$$

$$f_i'' = 80\,000 + 54\,400 = 134\,400 \text{ dollars per year};$$
$$g_i'' = 4.1 \text{ dollars per hundred kilograms.}$$

The problem can therefore be modelled as in (3.29), (3.10)–(3.14), provided that each potential facility $i \in V_1$ is replaced by two dummy facilities i' and i'' with fixed costs equal to f_i' and f_i'', and marginal costs equal to g_i' and g_i'', respectively.

By way of example, this approach is applied to a simplified version of the problem where the potential facilities are in Linares, Monclova and Monterrey, each of them having a capacity of 20 000 hundred kilograms per year, and the sales districts are concentrated in four areas, around Bustamante, Saltillo, Santa Catarina and Monte-morelos, respectively. The annual demands add up to 6200 hundred kilograms for Bustamante, 6600 hundred kilograms for Saltillo, 5800 hundred kilograms for Santa Catarina and 4400 hundred kilograms for Montemorelos. Transportation is carried out by trucks whose capacity is 10 hundred kilograms and whose cost is equal to $0.98 per mile.

The Gelido location problem can be modelled as a CPL formulation:

Minimize

$$\sum_{i \in V_1} \sum_{j \in V_2} t_{ij} x_{ij} + \sum_{i \in V_1} f_i y_i$$

subject to

$$\sum_{i \in V_1} x_{ij} = 1, \qquad j \in V_2,$$

$$\sum_{j \in V_2} d_j x_{ij} \leqslant q_i y_i, \qquad i \in V_1,$$

$$0 \leqslant x_{ij} \leqslant 1, \qquad i \in V_1, j \in V_2,$$

$$y_i \in \{0, 1\}, \qquad i \in V_1,$$

where $V_1 = \{$Linares$'$, Linares$''$, Monclova$'$, Monclova$''$, Monterrey$'$, Monterrey$''\}$, $V_2 = \{$Bustamante, Saltillo, Santa Catarina, Montemorelos$\}$. Linares$'$ and Linares$''$ represent two dummy facilities which can be opened up in Linares, with a capacity equal to

$$q_i' = 3500 \text{ hundred kilograms per year,}$$
$$q_i'' = 20\,000 \text{ hundred kilograms per year,}$$

respectively (the same goes for Monclova$'$, Monclova$''$, Monterrey$'$ and Monterrey$''$); x_{ij}, $i \in V_1$, $j \in V_2$, is a decision variable representing the fraction of the annual demand of sales district j satisfied by facility i; y_i, $i \in V_1$, is a binary decision variable, whose value is equal to 1 if the potential facility i is open, 0 otherwise. Costs t_{ij}, $i \in V_1$, $j \in V_2$, reported in Table 3.11, were obtained by means of

Table 3.9 Demand entries (in hundreds of kilograms per year) versus facility variable costs (in dollars) as reported in the past in the Logconsult problem.

	Demand	Variable cost
	1 000	17 579
	2 500	56 350
	3 500	62 208
	6 000	76 403
	8 000	85 491
	9 000	90 237
	9 500	96 251
	12 000	109 429
	13 500	107 355
	15 000	122 432
	16 000	116 816
	18 000	124 736

Table 3.10 Distances (in miles) between potential facilities and sales districts in the Logconsult problem.

	Bustamante	Saltillo	Santa Catarina	Montemorelos
Linares	165.0	132.5	92.7	32.4
Monclova	90.8	118.5	139.0	176.7
Monterrey	84.2	51.6	11.9	49.5

Equations (3.30), where the quantities c_{ij}, $i \in V_1$, $j \in V_2$, are in turn obtained through Equation (3.15). In other words, $c_{ij} = \bar{c}_{ij} d_j$, with $\bar{c}_{ij} = (0.98 \times 2 \times l_{ij})/10$, where l_{ij}, $i \in V_1$, $j \in V_2$, represents the distance (in miles) between facility i and market j (see Table 3.10).

The optimal demand allocation (see Table 3.12) leads to an optimal cost equal to $569 383.52 per year. Two facilities are located in Linares and Monterrey, with an activity level equal to 3000 hundred kilograms per year and 20 000 hundred kilograms per year, respectively (this means that the Linares' and Monterrey" dummy facilities are opened).

3.4 Two-Echelon Multicommodity Location Models

In *two-echelon multicommodity* (TEMC) location problems, homogeneous facilities have to be located as in SESC problems. However, both incoming and outgoing

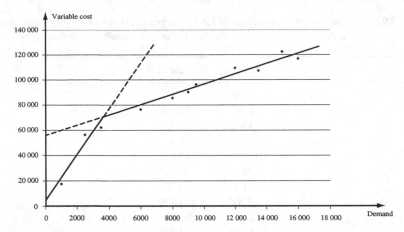

Figure 3.8 Variable cost of a facility versus demand in the Logconsult problem.

Table 3.11 Annual costs (in dollars) t_{ij}, $i \in V_1$, $j \in V_2$, in the Logconsult problem.

	Bustamante	Saltillo	Santa Catarina	Montemorelos
Linares'	315 208	293 502	212 681	109 342
Linares''	225 928	198 462	129 161	45 982
Monclova'	225 040	275 392	265 315	233 786
Monclova''	135 760	180 352	181 795	170 426
Monterrey'	217 020	188 850	120 828	124 089
Monterrey''	127 740	93 810	37 308	60 729

Table 3.12 Fraction of the annual demand of sales district $j \in V_2$ satisfied by facility $i \in V_1$, in the Logconsult problem.

	Bustamante	Saltillo	Santa Catarina	Montemorelos
Linares'	0.00	0.00	0.00	0.68
Linares''	0.00	0.00	0.00	0.00
Monclova'	0.00	0.00	0.00	0.00
Monclova''	0.00	0.00	0.00	0.00
Monterrey'	0.00	0.00	0.00	0.00
Monterrey''	1.00	1.00	1.00	0.32

commodities are relevant and material flows are not homogeneous. In what follows, transportation costs are assumed to be linear, and each facility is assumed to be characterized by a fixed and a marginal cost. If cost functions are piecewise linear and concave, transformations similar to the ones depicted in Section 3.3 can be used as a solution methodology. In the following, for the sake of clarity, it is assumed that the facilities to be located are DCs supplied directly from production plants.

TEMC problems can be modelled as MIP problems. Let V_1 be the set of production plants; V_2 the set of potential DCs, p of which are to be opened; V_3 the set of the demand points; K the set of homogeneous commodities; c_{ijr}^k, $i \in V_1, j \in V_2, r \in V_3$, $k \in K$, the unit transportation cost of commodity k from plant i to the demand point r across the DC j; d_r^k, $r \in V_3$, $k \in K$, the quantity of product k required by demand point r in a time unit (for example, one year); p_i^k, $i \in V_1, k \in K$, the maximum quantity of product k that plant i can manufacture in a time unit; q_j^- and q_j^+, $j \in V_2$, the minimum and maximum activity level of potential DC j in a time unit, respectively. Moreover, it is assumed that the operating cost of each DC $j \in V_2$ depends on the amount of commodities through a fixed cost f_j and a marginal cost g_j. Finally, it is assumed that demand is not divisible (see Section 3.2). Let $z_j, j \in V_2$, be a binary variable equal to 1 if DC j is opened, 0 otherwise; $y_{jr}, j \in V_2, r \in V_3$, a binary variable equal to 1 if demand point r is assigned to DC j, 0 otherwise; s_{ijr}^k, $i \in V_1, j \in V_2, r \in V_3, k \in K$, a continuous variable representing the quantity of item k transported from plant i to demand point r through DC j. The TEMC model is:

Minimize

$$\sum_{i \in V_1} \sum_{j \in V_2} \sum_{r \in V_3} \sum_{k \in K} c_{ijr}^k s_{ijr}^k + \sum_{j \in V_2} \left(f_j z_j + g_j \sum_{r \in V_3} \sum_{k \in K} d_r^k y_{jr} \right)$$

subject to

$$\sum_{j \in V_2} \sum_{r \in V_3} s_{ijr}^k \leqslant p_i^k, \quad i \in V_1, \, k \in K, \tag{3.32}$$

$$\sum_{i \in V_1} s_{ijr}^k = d_r^k y_{jr}, \quad j \in V_2, \, r \in V_3, \, k \in K, \tag{3.33}$$

$$\sum_{j \in V_2} y_{jr} = 1, \quad r \in V_3, \tag{3.34}$$

$$q_j^- z_j \leqslant \sum_{r \in V_3} \sum_{k \in K} d_r^k y_{jr} \leqslant q_j^+ z_j, \quad j \in V_2, \tag{3.35}$$

$$\sum_{j \in V_2} z_j = p, \tag{3.36}$$

$$z_j \in \{0, 1\}, \quad j \in V_2, \tag{3.37}$$

$$y_{jr} \in \{0, 1\}, \quad j \in V_2, \, r \in V_3, \tag{3.38}$$

$$s_{ijr}^k \geqslant 0, \quad i \in V_1, \, j \in V_2, \, r \in V_3, \, k \in K. \tag{3.39}$$

Inequalities (3.32) impose a capacity constraint for each production plant and for each product. Equations (3.33) require the satisfaction of customer demand. Equations (3.34) establish that each client must be served by a single DC. Inequalities (3.35) are DC capacity constraints. Equation (3.36) sets the number of DCs to be opened.

By removing some constraints (e.g. Equation (3.36)) or adding some new constraints (e.g. forcing the opening of a certain DC, or the assignment of some demand points to a given DC) several meaningful variants of the above model can be obtained.

- A plant $i \in V_1$ is not able to produce a certain commodity $k \in K$ (in this case we remove the decision variables s_{ijr}^k for each $r \in V_3$ and $j \in V_2$).

- A DC $j \in V_2$ cannot be serviced efficiently by a plant $i \in V_1$ because it is too far from it (in this case we remove variables s_{ijr}^k for each $r \in V_3$ and $k \in K$).

- The time required to serve a demand point $r \in V_3$ from potential DC $j \in V_2$ is too long (in this case we remove variables y_{jr} and variables s_{ijr}^k for each $i \in V_1$ and $k \in K$).

Demand allocation

When a set $\bar{z}_j$, $j \in V_2$ and $\bar{y}_{jr}$, $j \in V_2, r \in V_3$, of feasible values for the binary variables is available, determining the corresponding optimal demand allocation amounts to solving an LP problem:

Minimize

$$\sum_{i \in V_1} \sum_{j \in V_2} \sum_{r \in V_3} \sum_{k \in K} c_{ijr}^k s_{ijr}^k \tag{3.40}$$

subject to

$$\sum_{j \in V_2} \sum_{r \in V_3} s_{ijr}^k \leqslant p_i^k, \quad i \in V_1, \ k \in K, \tag{3.41}$$

$$\sum_{i \in V_1} s_{ijr}^k = d_r^k \bar{y}_{jr}, \quad j \in V_2, \ r \in V_3, \ k \in K, \tag{3.42}$$

$$x_{ijr}^k \geqslant 0, \quad i \in V_1, \ j \in V_2, \ r \in V_3, \ k \in K. \tag{3.43}$$

A Lagrangian heuristic

The Lagrangian methodology illustrated for the CPL problem can be adapted to the TEMC problem as follows. A lower bound is obtained by relaxing constraints (3.33) with multipliers θ_{jr}^k, $j \in V_2$, $r \in V_3$, $k \in K$. The relaxed problem:

Minimize

$$\sum_{i \in V_1} \sum_{j \in V_2} \sum_{r \in V_3} \sum_{k \in K} (c_{ijr}^k + \theta_{jr}^k) s_{ijr}^k + \sum_{j \in V_2} \left(f_j z_j + (g_j - \theta_{jr}^k) \sum_{r \in V_3} \sum_{k \in K} d_r^k y_{jr} \right)$$

subject to (3.32), (3.34)–(3.39), decomposes into two separate subproblems P_1 and P_2. P_1 is the LP problem:

Minimize

$$\sum_{i \in V_1} \sum_{j \in V_2} \sum_{r \in V_3} \sum_{k \in K} (c_{ijr}^k + \theta_{jr}^k) s_{ijr}^k$$

subject to (3.32) and (3.39), which can be strengthened by adding constraints:

$$\sum_{i \in V_1} \sum_{j \in V_2} s_{ijr}^k = d_r^k, \quad r \in V_3, \ k \in K,$$

$$\sum_{i \in V_1} \sum_{r \in V_3} \sum_{k \in K} s_{ijr}^k \leqslant q_j^+, \quad j \in V_2.$$

Subproblem P_2, defined as

Minimize

$$\sum_{j \in V_2} f_j z_j + \sum_{j \in V_2} \sum_{r \in V_3} \left[(g_j - \theta_{jr}^k) \left(\sum_{k \in K} d_r^k \right) \right] y_{jr}$$

subject to (3.34)–(3.38), is an SESC problem (with indivisible demand) where DCs have to be located and each customer $r \in V_3$ has a single-commodity demand equal to $\sum_{k \in K} d_r^k$. Even if this problem is still NP-hard, its solution is usually much simpler than that of the TEMC problem and can be often obtained through a general-purpose or a tailored branch-and-bound algorithm.

Let $\bar{z}_j$, $j \in V_2$ and $\bar{y}_{jr}$, $j \in V_2, r \in V_3$, be the optimal solution of P_2. If the allocation problem associated with such binary variable values is feasible, an upper bound is obtained. The above procedure is then embedded in a subgradient procedure in order to determine improved lower and upper bounds.

A Benders decomposition procedure

We now illustrate a Benders decomposition method for solving the TEMC problem. Generally speaking, a Benders decomposition method allows the determination of an optimal solution of an MIP problem by solving several MIP problems with a single continuous variable each, and several LP problems.

To simplify the exposition, the Benders decomposition method will be described assuming the problem is well posed (i.e. $\sum_{i \in V_1} p_i^k \geqslant \sum_{r \in V_3} d_r^k$, $k \in K$, and at least one solution (z_j, y_{jr}), $j \in V_2, r \in V_3$, satisfies all the constraints).

Let μ_i^k, $i \in V_1, k \in K$, and π_{jr}^k, $j \in V_2, r \in V_3, k \in K$, be the dual variables corresponding to constraints (3.41) and (3.42), respectively. The dual of problem (3.40)–(3.43) can be reformulated as

Maximize

$$\sum_{i \in V_1} \sum_{k \in K} (-p_i^k) \mu_i^k + \sum_{j \in V_2} \sum_{r \in V_3} \sum_{k \in K} (d_r^k \bar{y}_{jr}) \pi_{jr}^k \qquad (3.44)$$

subject to

$$-\mu_i^k + \pi_{jr}^k \leqslant c_{ijr}^k, \quad i \in V_1, \ j \in V_2, \ r \in V_3, \ k \in K, \qquad (3.45)$$

$$\mu_i^k \geqslant 0, \quad i \in V_1, \ k \in K. \qquad (3.46)$$

This way the optimal solution of problem (3.40)–(3.43) is determined by solving its dual. Since problem (3.40)–(3.43) always has a finite optimal solution, its dual (3.44)–(3.46) can be formulated as follows. Denote by Ω the feasible region of problem (3.44)–(3.46) and by T the set of extreme points of Ω. Moreover, let $\mu_i^k(t)$, $i \in V_1$, $k \in K$, and $\pi_{jr}^k(t)$, $j \in V_2$, $r \in V_3$, $k \in K$, be the dual solution corresponding to a generic extreme point $t \in T$. Then we can compute:

Maximize

$$\left\{ \sum_{i \in V_1} \sum_{k \in K} (-p_i^k) \mu_i^k(t) + \sum_{j \in V_2} \sum_{r \in V_3} \sum_{k \in K} (d_r^k \bar{y}_{jr}^k) \pi_{jr}^k(t) \right\}.$$

The feasible region Ω (and, consequently, also T) does not depend on the $\bar{y}_{jr}$, $j \in V_2$ and $r \in V_3$ values. This observation allows us to formulate the TEMC problem as

Minimize

$$\left\{ \sum_{j \in V_2} \left(f_j z_j + g_j \sum_{r \in V_3} \sum_{k \in K} d_r^k y_{jr} \right) \right.$$

$$\left. + \max_{t \in T} \left[\sum_{i \in V_1} \sum_{k \in K} (-p_i^k) \mu_i^k(t) + \sum_{j \in V_2} \sum_{r \in V_3} \sum_{k \in K} (d_r^k y_{jr}^k) \pi_{jr}^k(t) \right] \right\}$$

subject to (3.34)–(3.38), or equivalently (*Benders reformulation*):

Minimize

$$L \tag{3.47}$$

subject to

$$L \geqslant \sum_{j \in V_2} \left(f_j z_j + g_j \sum_{r \in V_3} \sum_{k \in K} d_r^k y_{jr} \right)$$

$$+ \sum_{i \in V_1} \sum_{k \in K} (-p_i^k) \mu_i^k(t) + \sum_{j \in V_2} \sum_{r \in V_3} \sum_{k \in K} (d_r^k y_{jr}^k) \pi_{jr}^k(t), \quad t \in T \tag{3.48}$$

and constraints (3.34)–(3.38).

Problem (3.47), (3.48), (3.34)–(3.38) becomes intractable because of the large number of constraints (3.48) (one for each extreme point of Ω). For this reason, it is convenient to resort to the relaxation of this formulation, named *master* problem, for which only a subset $T^{(h)} \subseteq T$ of constraints (3.48) is present.

Minimize

$$L \tag{3.49}$$

subject to

$$L \geqslant \sum_{j \in V_2} \left(f_j z_j + g_j \sum_{r \in V_3} \sum_{k \in K} d_r^k y_{jr} \right)$$

$$+ \sum_{i \in V_1} \sum_{k \in K} (-p_i^k) \mu_i^k(t) + \sum_{j \in V_2} \sum_{r \in V_3} \sum_{k \in K} (d_r^k y_{jr}^k) \pi_{jr}^k(t), \quad t \in T^{(h)}, \tag{3.50}$$

and constraints (3.34)–(3.38).

Let $z_j^{(h)}$, $j \in V_2$ and $y_{ij}^{(h)}$, $j \in V_2$, $r \in V_3$, be an optimal solution to the master problem, and let $L^{(h)}$ be the corresponding objective function value. Since the master problem represents a relaxation of the formulation (3.47), (3.48), (3.34)–(3.38), the value $L^{(h)}$ is a lower bound on the optimal solution value of the TEMC problem. Starting from $y_{jr}^{(h)}$, $j \in V_2$, $r \in V_3$, it is possible to determine the optimal solution $s_{ijr}^{k,(h)}$, $i \in V_1$, $j \in V_2$, $r \in V_3$, $k \in K$, of the following demand allocation problem:

Minimize

$$\sum_{i \in V_1} \sum_{j \in V_2} \sum_{r \in V_3} \sum_{k \in K} c_{ijr}^k s_{ijr}^k \tag{3.51}$$

subject to

$$\sum_{j \in V_2} \sum_{r \in V_3} s_{ijr}^k \leqslant p_i^k, \quad i \in V_1, \; k \in K, \tag{3.52}$$

$$\sum_{i \in V_1} s_{ijr}^k = d_r y_{jr}^{(h)}, \quad j \in V_2, \; r \in V_3, \; k \in K, \tag{3.53}$$

$$s_{ijr}^k \geqslant 0, \quad i \in V_1, \; j \in V_2, \; r \in V_3, \; k \in K. \tag{3.54}$$

The optimal solutions of the master and demand allocation problems, therefore, allow us to define a feasible solution $z_j^{(h)}$, $j \in V_2$, $y_{jr}^{(h)}$, $j \in V_2$, $r \in V_3$, $s_{ijr}^{k,(h)}$, $i \in V_1$, $j \in V_2$, $r \in V_3$, $k \in K$, for the TEMC problem, whose cost is equal to

$$U^{(h)} = \sum_{j \in V_2} \left(f_j z_j^{(h)} + g_j \sum_{r \in V_3} \sum_{k \in K} d_r^k y_{jr}^{(h)} \right) + \sum_{i \in V_1} \sum_{j \in V_2} \sum_{r \in V_3} \sum_{k \in K} c_{ijr}^k s_{ijr}^{k,(h)}. \tag{3.55}$$

Let $\mu_i^k(t^{(h)})$, $i \in V_1$, $k \in K$, and $\pi_{jr}^k(t^{(h)})$, $j \in V_2$, $r \in V_3$, $k \in K$, $t^{(h)} \in T$, be the dual optimal solution of the demand allocation problem (3.51)–(3.54). Since the optimum solution values of the primal and dual problems are equal, it follows that

$$\sum_{i \in V_1} \sum_{k \in K} (-p_i^k) \mu_i^k(t^{(h)}) + \sum_{j \in V_2} \sum_{r \in V_3} \sum_{k \in K} (d_r^k y_{jr}^{(h)}) \pi_i^k(t^{(h)})$$

$$= \sum_{i \in V_1} \sum_{j \in V_2} \sum_{r \in V_3} \sum_{k \in K} c_{ijr}^k s_{ijr}^{k,(h)},$$

from which

$$U^{(h)} = \sum_{j \in V_2} \left(f_j z_j^{(h)} + g_j \sum_{r \in V_3} \sum_{k \in K} d_r^k y_{jr}^{(h)} \right)$$

$$+ \sum_{i \in V_1} \sum_{k \in K} (-p_i^k) \mu_i^k (t^{(h)}) + \sum_{j \in V_2} \sum_{r \in V_3} \sum_{k \in K} (d_r^k y_{jr}^{(h)}) \pi_i^k (t^{(h)}). \qquad (3.56)$$

The implications of Equation (3.56) are clear. If $L^{(h)} < U^{(h)}$, the solution $z_j^{(h)}$, $j \in V_2$, $y_{jr}^{(h)}$, $j \in V_2$, $r \in V_3$, $s_{ijr}^{k,(h)}$, $i \in V_1$, $j \in V_2$, $r \in V_3$, $k \in K$, is not optimal for the TEMC problem. Furthermore,

$$L^{(h)} < \sum_{j \in V_2} \left(f_j z_j^{(h)} + g_j \sum_{r \in V_3} \sum_{k \in K} d_r^k y_{jr}^{(h)} \right)$$

$$+ \sum_{i \in V_1} \sum_{k \in K} (-p_i^k) \mu_i^k (t^{(h)}) + \sum_{j \in V_2} \sum_{r \in V_3} \sum_{k \in K} (d_r^k y_{jr}^{(h)}) \pi_i^k (t^{(h)}),$$

i.e. $t^{(h)} \notin T^{(h)}$. Therefore, the introduction of $t^{(h)}$ in the set $T^{(h)}$, which defines the constraints (3.50) of the master problem, yields an improvement of the relaxation and consequently the procedure can be iterated. In the case $L^{(h)} = U^{(h)}$, the optimal solution of the TEMC problem is obtained.

In a more schematic way, the Benders decomposition method can be described as follows.

Step 0. (*Initialization.*) Select a tolerance value $\varepsilon > 0$; set LB $= -\infty$, UB $= \infty$, $h = 1$, $T^{(h)} = \emptyset$.

Step 1. (*Solution of the master problem.*) Solve the master problem (3.49), (3.50), (3.34)–(3.38). Let $z_j^{(h)}$, $j \in V_2$ and $y_{jr}^{(h)}$, $j \in V_2$, $r \in V_3$, be the corresponding optimal solution and $L^{(h)}$ the corresponding objective function value; set LB $= L^{(h)}$; if (UB $-$ LB)/LB $< \varepsilon$, then *STOP*.

Step 2. (*Determination of a new feasible solution.*) Determine the optimal solution $s_{ijr}^{k,(h)}$, $i \in V_1$, $j \in V_2$, $r \in V_3$, $k \in K$, of demand allocation problem (3.51)–(3.54) and the corresponding dual solution $\mu_i^k (t^{(h)})$, $i \in V_1$, $k \in K$, and $\pi_{jr}^k (t^{(h)})$, $j \in V_2$, $r \in V_3$, $k \in K$; let $U^{(h)}$ be the objective function value obtained through Equation (3.55); if $U^{(h)} <$ UB, then set UB $= U^{(h)}$; finally, set $T^{(h)} = T^{(h)} \cup \{t^{(h)}\}$.

Step 3. Set $h = h + 1$ and go back to Step 1.

The above algorithm seeks an ε-optimal solution, i.e. a feasible solution with a maximum gap ε from the optimal solution. The efficiency of this method can be improved remarkably if some feasible solutions for the TEMC problem are available at the beginning (e.g. the solution adopted in practice and those suggested on the basis of the experience of the company's management). This yields, at Step 0 of the algorithm, a better initial choice both for the value of UB and for the set $T^{(1)}$. In particular, UB can be set equal to the best value of the objective function value

Table 3.13 Distances (in kilometres) among the production sites and the demand points through the DCs in the K9 problem.

	$i = 1$			$i = 2$		
j	$r = 1$	$r = 2$	$r = 3$	$r = 1$	$r = 2$	$r = 3$
1	400	600	1100	800	1000	1500
2	600	400	900	800	600	1100
3	1000	600	900	800	400	700
4	1600	1200	900	1200	800	500

available. Moreover, the dual solution of each feasible solution allows us to define a constraint (3.50) for the initial master problem. Moreover, because of the presence of constraints (3.34), the optimal solution of the master problem is such that each demand point $r \in V_3$ is served by one and only one DC $j_r \in V_2$, i.e.

$$y_{jr}^{(h)} = \begin{cases} 1, & \text{if } j = j_r, \\ 0, & \text{otherwise.} \end{cases}$$

Therefore, the demand allocation problem (3.51)–(3.54) in Step 2 can be further simplified as:

Minimize

$$\sum_{i \in V_1} \sum_{r \in V_3} \sum_{k \in K} c_{ij_r r}^k s_{ij_r r}^k$$

subject to

$$\sum_{r \in V_3} s_{ij_r r}^k \leqslant p_i^k, \quad i \in V_1, \ k \in K,$$

$$\sum_{i \in V_1} s_{ij_r r}^k = d_r^k, \quad r \in V_3, \ k \in K,$$

$$s_{ij_r r}^k \geqslant 0, \quad i \in V_1, \ r \in V_3, \ k \in K,$$

and can be therefore decomposed in $|K|$ *transportation problems*, the kth of which corresponds to:

Minimize

$$\sum_{i \in V_1} \sum_{r \in V_3} c_{ij_r r}^k s_{ij_r r}^k$$

subject to

$$\sum_{r \in V_3} s_{ij_r r}^k \leqslant p_i^k, \quad i \in V_1,$$

$$\sum_{i \in V_1} s_{ij_r r}^k = d_r^k, \quad r \in V_3,$$

$$s_{ij_r r}^k \geqslant 0, \quad i \in V_1, \, r \in V_3.$$

K9 is a German petrochemical company. The firm's management intends to renovate its production and distribution network, which is presently composed of two refining plants, two DCs and hundreds of sales points (gas pumps and liquefied gas retailers). After a series of meetings, it was decided to relocate the DCs, leaving the position and features of the two production plants unchanged. The products of K9 are subdivided into two homogeneous commodities (represented by the indices $k = 1, 2$): fuel for motor transportation and liquefied gas (the latter sold in cylinders). There are four potential sites suited to receive a DC and, among these, two must be selected. A DC j, $j = 1, \ldots, 4$, is economically feasible if its level of activity is higher than $q_j^- = 1\,000\,000$ hectolitres per year and lower than $q_j^+ = 2\,500\,000$ hectolitres per year; for intermediate values, the cost increases approximately with a linear trend characterized by a fixed cost of 10 million euros per year and by a marginal cost of €0.25 per hectolitre. Transportation costs c_{ijr}^k, $i \in V_1$, $j \in V_2$, $r \in V_3$, $k = 1, 2$, are equal to the cost per kilometre and hectolitre (equal to €0.67 for $k = 1$, and €0.82 for $k = 2$) multiplied by the distance between production plant $i \in V_1$ and demand point $r \in V_3$ through site $j \in V_2$ (see Table 3.13).

The market is subdivided into three districts ($r = 1, 2, 3$) characterized by demand values equal to

$$d_1^1 = 800\,000 \text{ hectolitres per year;}$$
$$d_2^1 = 600\,000 \text{ hectolitres per year;}$$
$$d_3^1 = 700\,000 \text{ hectolitres per year;}$$
$$d_1^2 = 300\,000 \text{ hectolitres per year;}$$
$$d_2^2 = 400\,000 \text{ hectolitres per year;}$$
$$d_3^2 = 500\,000 \text{ hectolitres per year.}$$

Finally, the production plants have the following capacities:

$$p_1^1 = 1\,200\,000 \text{ hectolitres per year;}$$
$$p_2^1 = 1\,500\,000 \text{ hectolitres per year;}$$
$$p_1^2 = 500\,000 \text{ hectolitres per year;}$$
$$p_2^2 = 800\,000 \text{ hectolitres per year.}$$

To determine the optimal allocation of the two DCs, a TEMC model with $p = 2$ is built and solved. The decision variables s_{ijr}^k, $i = 1, 2$, $j = 1, \ldots, 4$, $r = 1, 2, 3$, $k = 1, 2$, represent the quantity (in hectolitres) of product k transported yearly from production plant i to demand point r through DC j, whereas transportation costs c^k_{ijr}, $i = 1, 2$, $j = 1, \ldots, 4$, $r = 1, 2, 3$, $k = 1, 2$, can be obtained from Tables 3.14 and 3.15.

In the optimal TEMC solution, the first and third DCs are opened ($z^*_1 = 1, z^*_2 = 0$, $z^*_3 = 1, z^*_4 = 0$) and the yearly cost amounts to 33.19 million euros. The first centre serves the first sales district, while the second one is used for the second and third sales district ($y^*_{11} = y^*_{32} = y^*_{33} = 1$). Moreover, the decision variables s^k_{ijr}, $i = 1, 2$, $j = 1, \ldots, 4$, $r = 1, 2, 3$, $k = 1, 2$, take the following values:

$$s^{1,*}_{111} = 800\,000 \text{ hectolitres per year};$$

$$s^{1,*}_{232} = 600\,000 \text{ hectolitres per year};$$

$$s^{1,*}_{233} = 700\,000 \text{ hectolitres per year};$$

$$s^{2,*}_{111} = 300\,000 \text{ hectolitres per year};$$

$$s^{2,*}_{133} = 100\,000 \text{ hectolitres per year};$$

$$s^{2,*}_{232} = 400\,000 \text{ hectolitres per year};$$

$$s^{2,*}_{233} = 300\,000 \text{ hectolitres per year}$$

(for the sake of brevity, the variables that take zero value are not reported).

In order to apply the Benders decomposition method, the following initial solution is used:

$$z^{(1)}_1 = 1;$$

$$z^{(1)}_2 = 1;$$

$$y^{(1)}_{11} = 1;$$

$$y^{(1)}_{22} = 1;$$

$$y^{(1)}_{23} = 1;$$

$$s^{1,(1)}_{111} = 800\,000 \text{ hectolitres per year};$$

$$s^{1,(1)}_{123} = 400\,000 \text{ hectolitres per year};$$

$$s^{1,(1)}_{222} = 600\,000 \text{ hectolitres per year};$$

$$s^{1,(1)}_{223} = 300\,000 \text{ hectolitres per year};$$

$$s^{2,(1)}_{111} = 300\,000 \text{ hectolitres per year};$$

$$s^{2,(1)}_{122} = 200\,000 \text{ hectolitres per year};$$

$$s^{2,(1)}_{222} = 200\,000 \text{ hectolitres per year};$$

$$s^{2,(1)}_{223} = 500\,000 \text{ hectolitres per year};$$

corresponding to an initial upper bound:

$$\text{UB} = U^{(1)} = 37.138 \text{ million euros per year.}$$

LB is set equal to $-\infty$. The solution $s_{ijr}^{k,(1)}$, $i = 1, 2$, $j = 1, \ldots, 4$, $r = 1, 2, 3$, $k = 1, 2$, corresponds to the following dual solution of the problem of demand allocation,

$$\mu_1^{1,(1)} = 1.34;$$
$$\mu_1^{2,(1)} = 1.64;$$
$$\pi_{11}^{1,(1)} = 4.02;$$
$$\pi_{22}^{1,(1)} = 4.02;$$
$$\pi_{23}^{1,(1)} = 7.37;$$
$$\pi_{11}^{2,(1)} = 4.92;$$
$$\pi_{22}^{2,(1)} = 4.92;$$
$$\pi_{23}^{2,(1)} = 9.02,$$

which allows us to initialize the set $T^{(1)}$ of constraints of the master problem. The master problem therefore provides the following optimal solution,

$$z_3^{(2)} = 1;$$
$$z_4^{(2)} = 1;$$
$$y_{32}^{(2)} = 1;$$
$$y_{41}^{(2)} = 1;$$
$$y_{43}^{(2)} = 1,$$

whose objective function value allows us to update LB:

$$\text{LB} = L^{(2)} = 16.757 \text{ million euros per year.}$$

The optimal solutions of the demand allocation problem and its dual are

$$s_{132}^{1,(2)} = 600\,000 \text{ hectolitres per year;}$$
$$s_{241}^{1,(2)} = 800\,000 \text{ hectolitres per year;}$$
$$s_{243}^{1,(2)} = 700\,000 \text{ hectolitres per year;}$$
$$s_{132}^{2,(2)} = 400\,000 \text{ hectolitres per year;}$$
$$s_{241}^{2,(2)} = 300\,000 \text{ hectolitres per year;}$$
$$s_{243}^{2,(2)} = 200\,000 \text{ hectolitres per year;}$$

Table 3.14 Transportation costs (in euros per hectolitre) c_{ijr}^k, $i = 1, 2$, $j = 1, \ldots, 4$, $r = 1, 2, 3$, for product $k = 1$ (Part I) in the K9 problem.

	$i = 1$			$i = 2$		
	$r = 1$	$r = 2$	$r = 3$	$r = 1$	$r = 2$	$r = 3$
1	2.68	4.02	7.37	5.36	6.7	10.05
2	4.02	2.68	6.03	5.36	4.02	7.37
3	6.7	4.02	6.03	5.36	2.68	4.69
4	10.72	8.04	6.03	8.04	5.36	3.35

Table 3.15 Transportation costs (in euros per hectolitre) c_{ijr}^k, $i = 1, 2$, $j = 1, \ldots, 4$, $r = 1, 2, 3$, for product $k = 2$ (Part II) in the K9 problem.

	$i = 1$			$i = 2$		
	$r = 1$	$r = 2$	$r = 3$	$r = 1$	$r = 2$	$r = 3$
1	3.28	4.92	9.02	6.56	8.2	12.3
2	4.92	3.28	7.38	6.56	4.92	9.02
3	8.2	4.92	7.38	6.56	3.28	5.74
4	13.12	9.84	7.38	9.84	6.56	4.1

$$\mu_2^{1,(2)} = 2.68; \qquad \pi_{43}^{1,(2)} = 6.03;$$

$$\mu_2^{2,(2)} = 3.28; \qquad \pi_{32}^{2,(2)} = 4.92;$$

$$\pi_{32}^{1,(2)} = 4.02; \qquad \pi_{41}^{2,(2)} = 13.12;$$

$$\pi_{41}^{1,(2)} = 10.72; \qquad \pi_{43}^{2,(2)} = 7.38.$$

Hence, $U^{(2)} = 38.984$ million euros per year. Since UB $< U^{(2)}$, the value of UB is not updated. The procedure continues until a ε-optimal solution is generated.

3.5 Logistics Facility Location in the Public Sector

As explained in Chapter 1, relevant logistics issues have to be tackled when planning a number of public services (fire fighting, transport of the disabled, ambulance dispatching, to name a few). In such contexts, it is often of primary importance to ensure not only a low logistics cost, but also an adequate service level to all users. As a result, specific models have to be applied when locating public facilities. For example, in the p-centre model described below, the service time of the most disadvantaged user is to be minimized, whereas in the *location-covering* model one has to determine the least-cost set of facilities such that each user can be reached within a given maximum travel time.

3.5.1 *p*-centre models

In the p-centre model the aim is to locate p facilities on a graph in such a way that the maximum travel time from a user to the closest facility is minimized. The p-centre model finds its application when it is necessary to ensure equity in servicing users spread on a wide geographical area.

The problem can be modelled on a directed, undirected or mixed graph $G(V, A, E)$, where V is a set of vertices representing both user sites and road intersections, while A and E (the set of arcs and edges, respectively) describe the road connections among the sites. Exactly p facilities have to be located either on a vertex or on an arc or edge. For $p \geqslant 2$, the p-centre model is NP-hard.

If G is a directed graph, there exists an optimal solution of the p-centre problem such that every facility location is a vertex (*vertex location property*). If G is undirected or mixed, the optimal location of a facility could be on an internal point of an edge. In what follows, a solution methodology is described for the 1-centre problem. The reader should consult specialized books for a discussion of the other cases. If G is directed, the 1-centre is simply the vertex associated with the minimum value of the maximum travel time to all the other vertices. In the case of an undirected or mixed graph, the 1-centre can correspond to a vertex or an internal edge point. To simplify the discussion, we will refer only to the case of the undirected graph ($A = \emptyset$), although the procedure can be easily applied in the case of mixed graph. For each $(i, j) \in E$, let a_{ij} be the traversal time of edge (i, j). Furthermore, for each pair of vertices $i, j \in V$, denote by t_i^j the shortest travel time between i and j, corresponding to the sum of the travel times of the edges of the shortest path between i and j. Note that, on the basis of the definition of travel time, the result is

$$t_i^j \leqslant a_{ij}, \quad (i, j) \in E.$$

Finally, denote by $\tau_h(p_{hk})$ the travel time along edge $(h, k) \in E$ between vertex $h \in V$ and a point p_{hk} of the edge. In this way, the travel time $\tau_h(p_{hk})$ along the edge (h, k) between the vertex $k \in V$ and p_{hk} results as (see Figure 3.9)

$$\tau_k(p_{hk}) = a_{hk} - \tau_h(p_{hk}).$$

The 1-centre problem can be solved by the following algorithm proposed by Hakimi.

Step 1. (*Computation of the travel time.*) For each edge $(h, k) \in E$ and for each vertex $\in V$, determine the travel time $T_i(p_{hk})$ from $i \in V$ to a point p_{hk} of the

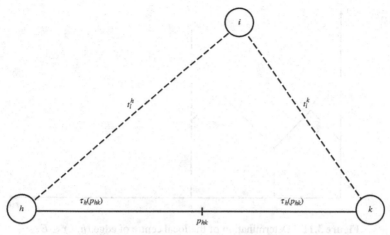

Figure 3.9 Computation of the travel time $T_i(p_{hk})$ from an user $i \in V$ to a facility in p_{hk}.

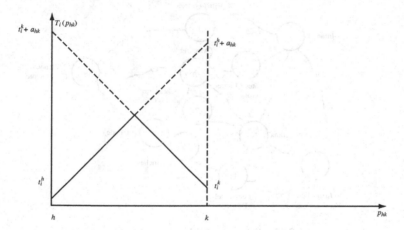

Figure 3.10 Travel time $T_i(p_{hk})$ versus the position of point p_{hk} along edge (h, k).

edge (h, k) (see Figure 3.10):

$$T_i(p_{hk}) = \min[t_i^h + \tau_h(p_{hk}), t_i^k + \tau_k(p_{hk})]. \tag{3.57}$$

Step 2. (*Finding the local centre.*) For each edge $(h, k) \in E$, determine the *local centre* p_{hk}^* as the point on (h, k) minimizing the travel time from the most disadvantaged vertex,

$$p_{hk}^* = \arg\min \max_{i \in V} \{T_i(p_{hk})\},$$

where $\max_{i \in V}\{T_i(p_{hk})\}$ corresponds to the superior envelope of the functions $T_i(p_{hk})$, $i \in V$ (see Figure 3.11).

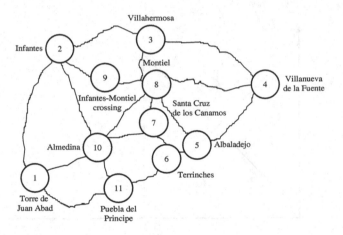

Figure 3.11 Determination of the local centre of edge $(h, k) \in E$.

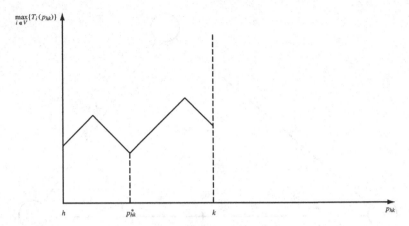

Figure 3.12 Location problem in La Mancha region.

Step 3. (*Determination of the 1-centre.*) The 1-centre p^* is the best local centre p^*_{hk}, $(h, k) \in E$, i.e.

$$p^* = \arg \min_{(h,k) \in E} \left\{ \min \max_{i \in V}\{T_i(p_{hk})\} \right\}.$$

In the La Mancha region of Spain (see Figure 3.12) a consortium of town councils, located in an underpopulated rural area, decided to locate a parking place for ambulances. A preliminary examination of the problem revealed that the probability of receiving a request for service during the completion of a previous call was extremely low because of the small number of the inhabitants of the zone. For this reason the

Table 3.16   Vertices of La Mancha 1-centre problem.

Vertex	Locality
1	Torre de Juan Abad
2	Infantes
3	Villahermosa
4	Villanueva de la Fuente
5	Albaladejo
6	Terrinches
7	Santa Cruz de los Canamos
8	Montiel
9	Infantes-Montiel crossing
10	Almedina
11	Puebla del Principe

team responsible for the service decided to use only one vehicle. In the light of this observation, the problem was modelled as a 1-centre problem on a road network G where all the connections are two-way streets (see Table 3.16). Travel times were calculated assuming a vehicle average speed of 90 km/h (see Table 3.17).

Travel times t_i^j, i, $j \in V$, are reported in Table 3.18. For each edge $(h, k) \in E$ and for each vertex $i \in V$ the travel time $T_i(p_{hk})$ from vertex i to a point p_{hk} of the edge (h, k) can be defined through Equation (3.57). This enables the construction, for each edge $(h, k) \in E$ of the function $\max_{i \in V}\{T_i(p_{hk})\}$, whose minimum corresponds to the local centre p_{hk}^*. For example, Figure 3.13 depicts the function $\max_{i \in V}\{T_i(p_{23})\}$, and Table 3.19 gives, for each edge $(h, k) \in E$, both the position of p_{hk}^* and the value $\max_{i \in V}\{T_i(p_{hk}^*)\}$.

Consequently the 1-centre corresponds to the point p^* on the edge $(8, 10)$. Therefore, the optimal positioning of the parking place for the ambulance should be located on the road between Montiel and Almedina, at 2.25 km from the centre of Montiel. The villages least advantaged by this location decision are Villanueva de la Fuente and Torre de Juan Abad, since the ambulance would take an average time of 11.5 min to reach them.

3.5.2   The location-covering model

In the location-covering model the aim is to locate a least-cost set of facilities in such a way that each user can be reached within a maximum travel time from the closest facility. The problem can modelled on a complete graph $G(V_1 \cup V_2, E)$, where vertices in V_1 and in V_2 represent potential facilities and customers, respectively, and each edge $(i, j) \in E = V_1 \times V_2$ corresponds to a least-cost path between i and j.

Table 3.17 Travel time (in minutes) on the road network edges in the La Mancha problem.

(i,j)	a_{ij}	(i,j)	a_{ij}
(1,2)	12	(1,11)	8
(2,3)	9	(2,9)	8
(3,4)	11	(2,10)	9
(4,5)	9	(3,9)	4
(5,6)	2	(4,8)	10
(6,7)	3	(5,8)	6
(7,8)	4	(6,11)	5
(8,9)	1	(7,10)	5
(8,10)	7	(1,10)	6
(10,11)	4		

Table 3.18 Travel times (in minutes) t_i^j, $i, j \in V$, in the La Mancha problem.

					i						
j	1	2	3	4	5	6	7	8	9	10	11
1	0	12	18	23	15	13	11	13	14	6	8
2	12	0	9	19	15	16	13	9	8	9	13
3	18	9	0	11	11	12	9	5	4	12	16
4	23	19	11	0	9	11	14	10	11	17	16
5	15	15	11	9	0	2	5	6	7	10	7
6	13	16	12	11	2	0	3	7	8	8	5
7	11	13	9	14	5	3	0	4	5	5	8
8	13	9	5	10	6	7	4	0	1	7	11
9	14	8	4	11	7	8	5	1	0	8	12
10	6	9	12	17	10	8	5	7	8	0	4
11	8	13	16	16	7	5	8	11	12	4	0

Let f_i, $i \in V_1$, be the fixed cost of potential facility i; p_j, $j \in V_2$, the penalty incurred if customer j is unserviced; t_{ij}, $i \in V_1$, $j \in V_2$, the least-cost travel time between potential facility i and customer j; a_{ij}, $i \in V_1$, $j \in V_2$, a binary constant equal to 1 if potential facility i is able to serve customer j, 0 otherwise (given a user-defined maximum time T, $a_{ij} = 1$, if $t_{ij} \leqslant T$, $i \in V_1$, $j \in V_2$, otherwise $a_{ij} = 0$). The decision variables are binary: y_i, $i \in V_1$, is equal to 1 if facility i is opened, 0 otherwise; z_j, $j \in V_2$, is equal to 1 if customer j is not served, otherwise it is 0.

The location-covering problem is modelled as follows:

Minimize

$$\sum_{i \in V_1} f_i y_i + \sum_{j \in V_2} p_j z_j \qquad (3.58)$$

Table 3.19 Distances of the local centres p^*_{hk} from vertices h (in kilometres) and $\max_{i \in V}\{T_i(p^*_{hk})\}$ (in minutes) in the La Mancha problem.

(h,k)	$\gamma_h(p^*_{hk})$	$\max_{i \in V}\{T_i(p^*_{hk})\}$	(h,k)	$\gamma_h(p^*_{hk})$	$\max_{i \in V}\{T_i(p^*_{hk})\}$
(1,2)	18.00	19.0	(1,11)	12.00	16.0
(2,3)	6.00	17.0	(2,9)	12.00	14.0
(3,4)	0.00	18.0	(2,10)	13.50	17.0
(4,5)	13.50	15.0	(3,9)	6.00	14.0
(5,6)	0.00	15.0	(4,8)	15.00	13.0
(6,7)	3.75	13.5	(5,8)	9.00	13.0
(7,8)	2.25	12.5	(6,11)	4.50	15.0
(8,9)	0.00	13.0	(7,10)	0.00	14.0
(8,10)	2.25	11.5	(1,10)	9.00	17.0
(10,11)	6.00	16.0			

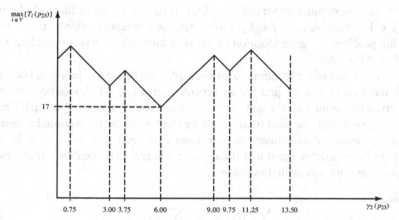

Figure 3.13 Time $\max_{i \in V}\{T_i(p_{23})\}$ versus position $\gamma_2(p_{23})$ of p_{23} in the La Mancha problem.

subject to

$$\sum_{i \in V_1} a_{ij} y_i + z_j \geqslant 1, \quad j \in V_2, \tag{3.59}$$

$$y_i \in \{0,1\}, \quad i \in V_1,$$

$$z_j \in \{0,1\}, \quad j \in V_2.$$

The objective function (3.58) is the sum of the fixed costs of the open facilities and the penalties corresponding to the unserviced customers. Constraints (3.59) impose that, for each $j \in V_2$, z_j is equal to 1 if the facilities opened do not cover customer j (i.e. if $\sum_{i \in V_1} a_{ij} y_i = 0$).

Table 3.20 Distances (in kilometres) between municipalities of
the consortium in Portugal (Part I).

	Almada	Azenha	Carregosa	Corroios	Lavradio
Almada	0.0	24.4	33.2	4.7	29.9
Azenha	24.4	0.0	13.2	18.2	14.8
Carregosa	33.2	13.2	0.0	27.1	11.1
Corroios	4.7	18.2	27.1	0.0	23.8
Lavradio	29.9	14.8	11.1	23.8	0.0
Macau	39.0	15.4	13.9	32.9	16.1
Moita	29.5	9.5	5.1	23.4	7.4
Montijo	38.8	18.7	7.2	32.7	16.6
Palmela	34.3	24.0	16.4	28.2	26.2
Pinhal Novo	39.1	16.9	10.9	33.0	14.2

If all customers must be served (i.e. if the penalties p_j are sufficiently high for
each $j \in V_2$), variables z_j, $j \in V_2$, can be assumed equal to 0. Hence, the location-
covering problem is a generalization of the well-known *set covering* problem and is
therefore NP-hard.

Several variants of the location-covering model can be used in practice. For example, if fixed costs f_i are equal for all potential sites $i \in V_1$, it can be convenient
to discriminate among all the solutions with the least number of open facilities the
one corresponding to the least total travelling time, or to the most equitable demand
distribution among the facilities. In the former case, let x_{ij}, $i \in V_1$, $j \in V_2$, be a
binary decision variable equal to 1 if customer j is served by facility i, 0 otherwise.
The problem can be modelled as follows:

Minimize

$$\sum_{i \in V_1} M y_i + \sum_{i \in V_1} \sum_{j \in V_2} t_{ij} x_{ij} \tag{3.60}$$

subject to

$$\sum_{i \in V_1} a_{ij} x_{ij} \geqslant 1, \quad j \in V_2, \tag{3.61}$$

$$\sum_{j \in V_2} x_{ij} \leqslant |V_2| y_i, \quad i \in V_1, \tag{3.62}$$

$$y_i \in \{0, 1\}, \quad i \in V_1, \tag{3.63}$$

$$x_{ij} \in \{0, 1\}, \quad i \in V_1, j \in V_2, \tag{3.64}$$

where M is a large positive constant. Constraints (3.61) guarantee that all the customers $j \in V_2$ are serviced, while constraints (3.62) ensure that if facility $i \in V_1$ is
not set up ($y_i = 0$), then no customer $j \in V_2$ can be served by it.

Table 3.21 Distances (in kilometres) between municipalities of
the consortium in Portugal (Part II).

	Macau	Moita	Montijo	Palmela	Pinhal Novo
Almada	39.0	29.5	38.8	34.3	39.1
Azenha	15.4	9.5	18.7	24.0	16.9
Carregosa	13.9	5.1	7.2	16.4	10.9
Corroios	32.9	23.4	32.7	28.2	33.0
Lavradio	16.1	7.4	16.6	26.2	14.2
Macau	0.0	9.1	12.0	6.9	1.7
Moita	9.1	0.0	10.6	11.7	7.2
Montijo	12.0	10.6	0.0	19.0	10.3
Palmela	6.9	11.7	19.0	0.0	8.8
Pinhal Novo	1.7	7.2	10.3	8.8	0.0

In Portugal, a consortium of 10 municipalities (Almada, Azenha, Carregosa, Corroios, Lavradio, Macau, Moita, Montijo, Palmela, Pinhal Novo), located in the neighbourhood of Lisbon, has decided to improve its fire-fighting service. The person responsible for the project has established that each centre of the community must be reached within 15 min from the closest fire station. Since the main aim is just to provide a first help in case of fire, the decision maker has decided to assign a single vehicle to each station. The annual cost of a station inclusive of the expenses of the personnel is €198 000. It is assumed that the average travelling speed is 60 km/h everywhere. In order to determine the optimal station location the location-covering model (3.60)–(3.64) is used, where $V_1 = V_2 = $ {Almada, Azenha, Carregosa, Corroios, Lavradio, Macau, Moita, Montijo, Palmela, Pinhal Novo}. Time coefficients t_{ij}, $i \in V_1$, $j \in V_2$, can be easily determined from the distances (in kilometres) reported in Tables 3.20 and 3.21. Coefficients a_{ij}, $i \in V_1$, $j \in V_2$, were obtained from Tables 3.20 and 3.21. The minimum number of fire stations turns out to be two. The facilities are located in Almada and Moita. The fire station located in Almada serves Corroios and Almada itself, the remaining ones are served by the fire station located in Moita.

Another interesting variant of the location-covering model arises when one must locate facilities to ensure double coverage of demand points. A classic case is ambulance location when users are better protected if two ambulances are located within their vicinity. If one of the two ambulances has to answer a call, there will remain one ambulance to provide coverage.

3.6 Data Aggregation

Modelling and solving a location problem often requires a considerable amount of data. For instance, the number of demand points of a large food producer often exceeds several thousands. Similarly, the items marketed by a distribution firm can be around tens of thousands. Therefore, in order to keep computation times and hardware requirements acceptable, data have to be aggregated. This can be done mainly in two ways.

Demand aggregation. Customers can be aggregated according to their position, service level or frequency of delivery. In the first case, demand points are clustered in such a way that customers close to one another (e.g. customers having the same zip code) are substituted by a *customer zone*. In the other cases, customers are aggregated into *classes*.

Item aggregation. Items are aggregated into a suitable number of *product groups*. This can be done according to their *distribution pattern* or their features (weight, volume, shape, cost, etc.). In the former case, products manufactured by the same plants and supplied to the same customers are treated as a single commodity. In the latter case, similar items (e.g. variants of the same basic product model) are aggregated.

Whatever the aggregation method, the reduced problem can be modelled with fewer variables and constraints. In what follows, a demand aggregation method for the CPL model (see Section 3.3.1) is analysed. The demand points of a subset $S \subset V_2$ are aggregated in the following way:

$$x_{ij} = x_{ik}, \quad i \in V_1, \ j \in S, \ k \in S.$$

Consequently, each point $j \in S$ receives the same fraction of demand from each facility $i \in V_1$. Let s be the customer zone resulting from the aggregation. Then the CPL model becomes:

Minimize

$$\sum_{i \in V_1} \sum_{j \in (V_2 \setminus S) \cup \{s\}} c_{ij} x_{ij} + \sum_{i \in V_1} f_i y_i$$

subject to

$$\sum_{i \in V_1} x_{ij} = 1, \quad j \in (V_2 \setminus S) \cup \{s\},$$

$$\sum_{j \in (V_2 \setminus S) \cup \{s\}} d_j x_{ij} \leqslant q_i y_i, \quad i \in V_1,$$

$$0 \leqslant x_{ij} \leqslant 1, \quad i \in V_1, \ j \in (V_2 \setminus S) \cup \{s\},$$

$$y_i \in \{0, 1\}, \quad i \in V_1,$$

where

$$c_{is} = \sum_{j \in S} c_{ij}, \quad i \in V_1,$$

and

$$d_s = \sum_{j \in S} d_j.$$

As a rule, the optimal solution value of the aggregated problem is worse than that of the original problem. It is useful to evaluate an upper bound on the error made resulting from the aggregation procedure. Let z^*_{CPL} and $z^{(a)}_{CPL}$ be the costs of the optimal solutions of the original problem and of the aggregate problem, respectively.

Property.   The following relation holds,

$$z^*_{CPL} \leqslant z^{(a)}_{CPL} \leqslant z^*_{CPL} + \varepsilon,$$

where

$$\varepsilon = \sum_{j \in S} \max_{i \in V_1} \left\{ \frac{d_j \sum_{r \in S} c_{ir}}{\sum_{r \in S} d_r} - c_{ij} \right\}. \tag{3.65}$$

Proof. In Equation (3.65), $\sum_{r \in S} c_{ir}$ represents the transportation cost when the whole demand of customers in S is served by facility $i \in V_1$, while $d_j / \sum_{r \in S} d_r$ represents the fraction of the total demand of customers in S. The difference

$$\left(d_j \sum_{r \in S} c_{ir} \Big/ \sum_{r \in S} d_r \right) - c_{ij}$$

represents, therefore, the variation (positive, negative or zero) of the cost of assigning the whole demand of customer $j \in V_2$ to facility $i \in V_1$ when the vertices of S are aggregated. Therefore, Equation (3.65) expresses the worst-case error made in the aggregation process as a sum of the maximum increases of the distribution costs of the various demand points in S.

It can be shown that bound (3.65) is *tight*, i.e. there are instances such that the aggregation error is nearly equal to ε.   □

If the demands of Brossard and Verdun are aggregated in the Goutte problem (see Section 3.3, where the maximum distance between plants and sales districts is assumed to be infinite), the worst-case error is, on the basis of Equation (3.65),

$$\varepsilon = \varepsilon_1 + \varepsilon_2 = 1578.92 \text{ Canadian dollars per year,}$$

where

$$\varepsilon_1 = \max_{i \in V_1} \left\{ \frac{q_1(c_{i1} + c_{i6})}{(q_1 + q_6)} - c_{i1} \right\} = 789.46 \text{ Canadian dollars per year,}$$

$$\varepsilon_2 = \max_{i \in V_1} \left\{ \frac{q_6(c_{i1} + c_{i6})}{(q_1 + q_6)} - c_{i6} \right\} = 789.46 \text{ Canadian dollars per year.}$$

The above aggregation technique can be obviously extended to the case of K disconnected subsets $S_1, \ldots, S_K$.

Property.   An *a priori* upper bound on the error is given by

$$z^*_{CPL} \leqslant z^{(a)}_{CPL} \leqslant z^*_{CPL} + \sum_{k=1}^{K} \varepsilon_k,$$

where

$$\varepsilon_k = \sum_{j \in S_k} \max_{i \in V_1} \left\{ \frac{d_j \sum_{r \in S_k} c_{ir}}{\sum_{r \in S_k} d_r} - c_{ij} \right\}.$$

3.7 Questions and Problems

3.1 Illustrate why crude oil refineries are customarily located near home heating and automotive fuel markets.

3.2 Describe how fixed costs f_i, $i \in V_1$, in the CPL model should be computed in order to take labour costs, property taxes and site-development costs into account.

3.3 Your company has to close 20 of its 125 warehouses. Suppose the CPL hypotheses hold. How would you define V_1? What is the value of p?

3.4 Modify the CPL model to take into account that a subset of already existing facilities $V_1' \subseteq V_1$ cannot be closed (but can be upgraded). Indicate the current fixed cost and capacity of facility $i \in V_1'$ as f_i' and q_i', respectively. Moreover, let f_i'' and q_i'' be the fixed cost and capacity if facility $i \in V_1'$ is upgraded, respectively.

3.5 Borachera is a major Spanish wine wholesaler currently operating two CDCs in Salamanca and Albacete, and a number of RDCs all over the Iberian peninsula. In order to reduce the overall logistics cost, the company wishes to redesign its distribution network by replacing its current RDCs with three (possibly new) RDCs. Based on a preliminary qualitative analysis, an RDC should be located in the Castilla-Leon region, either in Vallalolid, Burgos or Soria. A second RDC should be located in the Extremadura region, either in Badajoz, Plasencia or Caceres. Finally, the third RDC should be located in the Argon region, either in Barbastro, Saragossa or Teruel. Transportation costs from RDCs to retailers are charged to retailers. Formulate the Borachera problem as a modified CPL model.

3.6 As illustrated in Section 3.3, when designing a distribution network it is customary to remove excessively long transportation links from the model in order to allow for a timely delivery to customers. How should the CPL Lagrangian heuristic be modified in this case?

3.7 Modify the CPL Lagrangian heuristic for the case where demand is indivisible (i.e. a customer demand must be satisfied by a single warehouse). Is the modified heuristic still time-polynomial? How can it determine whether an instance of the modified problem is feasible?

3.8 Modify the TEMC model in Section 3.4 for the case where a customer can receive shipments from different DCs if these shipments are for different commodities. How should the proposed heuristics be changed in order to take this modification into account?

3.9 Modify the location-covering model in order to determine, among the least-cost solutions, the one associated with the most equitable demand distribution among the facilities.

3.10 Data aggregation is useful in facility location even if the available algorithms are able to solve the original problems. Why? (Hint: recall that demand forecasts at the account and product levels are generally mediocre.)

3.8 Annotated Bibliography

An extensive literature review of strategic production-distribution models is presented in the following article:

1. Vidal C and Goetschalckx M 1997 Strategic production-distribution models: a critical review with emphasis on global supply chain models. *European Journal of Operational Research* **98**, 1–18.

An in-depth examination of location theory in a broader context is reported in the following book:

2. Love RF, Morris JG and Wesolowsky GO 1988 *Facilities Location*. North-Holland, Amsterdam.

A review of the most important location-routing problems is presented in:

3. Laporte G 1988 Location-routing problems. In *Vehicle Routing: Methods and Studies* (ed. Golden BL and Assad AA). North-Holland, Amsterdam.

4

Solving Inventory Management Problems

4.1 Introduction

Inventories are stockpiles of items (raw materials, components, semi-finished and finished goods) waiting to be processed, transported or used at a point of the supply chain. As pointed out in Chapter 1, there are a number of reasons to have inventories, including improving service level, reducing overall logistics costs, coping with randomness in customer demand and lead times, making seasonal items available all year, speculating on price patterns, etc. At the same time, holding an inventory can be very expensive (the annual cost can be 30% of the value of the materials kept in stock, or even more). It is therefore crucial to manage inventories carefully.

This chapter deals with the solution of the most important inventory management problems. Inventory management amounts to deciding for each *stocking point* in the supply chain *when* to reorder and *how much* to order so that the expected annual cost is minimized while meeting a given service level. A useful formula that can be used to know whether a stocking point is well managed is the *inventory turnover ratio* (ITR), defined as the ratio between the annual sales priced at the value of the items in stock and the average inventory investment. Of course, high ITRs generally correspond to well-managed stocking points.

4.2 Relevant Costs

The costs relevant to the determination of an inventory policy can be classified into four broad categories.

Procurement costs. Procurement costs are those associated with the acquisition of goods. They typically include *fixed costs* (independent of the amount ordered) and *variable costs* (dependent of the amount acquired). They may include

Introduction to Logistics Systems Planning and Control   G. Ghiani, G. Laporte and R. Musmanno
© 2004 John Wiley & Sons, Ltd   ISBN: 0-470-84916-9 (HB) 0-470-84917-7 (PB)

- a (fixed) *reorder cost* (the cost of issuing and processing an order through the purchasing and accounting departments if the goods are bought, *or* the cost for setting up the production process if the goods are manufactured by the firm);

- a *purchasing cost* or a (variable) *manufacturing cost*, depending on whether the goods are bought from a supplier or manufactured by the firm;

- a transportation cost, if not included in the price of the purchased goods; for the sake of simplicity, we assume in the remainder of this chapter that fixed transportation costs are included in the reorder cost, while variable transportation costs are included in the purchasing cost;

- the cost of handling the goods at the receiving point.

Inventory holding costs.   Inventory *holding* costs are incurred when materials are stored for a period of time. They include the following.

- An *opportunity* (or *capital*) cost representing the *return on investment* the firm would have realized if money had been invested in a more profitable economic activity (e.g. on the stock market) instead of inventory. This cost is generally estimated on the basis of a standard banking interest rate.

- A *warehousing cost*. If the company runs its own warehouses, such costs include space and equipment costs, personnel wages, insurance on inventories, maintenance costs, energy costs and state taxes. Otherwise, warehousing costs amount to the fee paid for storing the goods in third-party warehouses.

Shortage costs.   *Shortage* costs are paid when customer orders are not met. Shortage costs depend heavily on customer behaviour and are difficult to evaluate accurately. They can be classified as follows.

- *Lost sales costs.* A lost sale is likely to occur if the unavailable items can be easily obtained from a competitor. Lost sales costs include the profit that would have made on the sale, and the negative effect that the shortage could have on future sales.

- *Back order costs.* When goods are difficult to replace, a shortage often results in a delayed sale. Apart from the negative effect on future sales, a back order could result in a penalty.

Obsolescence costs.   Obsolescence costs arise when stocked items lose some of their value over time. This happens, for example, when food inventories deteriorate, clothing items in stock go out of fashion, or newspapers are unsold. The value of an item at the end of its lifetime is usually referred to as its *salvage value*.

4.3   Classification of Inventory Management Models

Inventory management models can be classified according to a number of criteria.

Deterministic versus stochastic models.   In a *deterministic model*, where demands, prices and lead times are assumed to be known in advance, the focus is on balancing different categories of costs (e.g. the fixed costs of ordering and the costs of holding inventory). In a *stochastic model*, where some data are uncertain, it is impossible to always satisfy all the demand. As a result, a constraint on customer service level (stating, for example, that a customer is satisfied with a given probability) is usually imposed.

Fast-moving items versus slow-moving items.   *Fast-moving items* are those manufactured and sold regularly, and include most products on the market. The main issue when managing a fast-moving item inventory is to determine when stocking points have to be replenished and how much to order. On the other hand, *slow-moving items*, which are often spare parts of complex machineries, have a very low demand (e.g. a few units every 10 or 20 years). As a rule, manufacturing a spare part several years after the machinery is constructed is very expensive compared to producing it along with the machinery. Hence the main issue is to determine the number of items that have to be produced and stored at the beginning of the planning horizon.

Number of stocking points.   Optimal inventory policies can often be derived analytically for single stocking point models, while multiple stocking point situations are usually much harder to deal with.

Number of commodities.   When holding a multicommodity inventory, joint costs and constraints usually come into play. It is therefore not surprising that most multicommodity inventory management problems are NP-hard.

Instantaneous resupply versus noninstantaneous resupply.   A stocking point can be replenished almost instantaneously (e.g. as a consequence of a truck delivery) or noninstantaneously (e.g. at a constant rate through a manufacturing process).

4.4   Single Stocking Point: Single-Commodity Inventory Models under Constant Demand Rate

A single stocking point has to service a single-commodity constant rate demand, and places orders for the item from another facility. The planning horizon is assumed to be infinite (for the finite horizon case, see Problem 4.4). Since the demand rate is constant, a minimal cost policy is to replenish on a periodic basis. Let d be the constant demand rate, T the time lapse between two consecutive orders, q the order size (i.e. the amount of product ordered at each replenishment). These quantities are

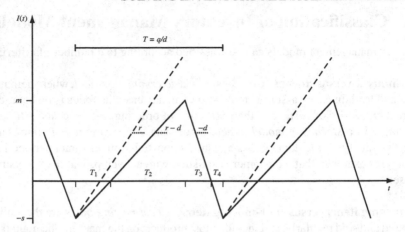

Figure 4.1 Inventory level as a function of time.

such that

$$q = dT. \tag{4.1}$$

Moreover, let c be the value of an item (assumed to be independent of the order size), k the fixed reorder cost, h the holding cost per item and per time unit, u the shortage cost per item, independent of the duration of the shortage, and v the shortage cost per item and per time unit. The holding cost h can be expressed as a fraction p of c:

$$h = pc. \tag{4.2}$$

The parameter p is a banking interest rate (measuring capital cost) increased to take into account warehousing costs. Moreover, let $I(t)$ be the inventory level at time t, m the maximum inventory level, s the maximum shortage, t_l the lead time, i.e. the time lapse between when an order is placed and when the items are received (see Chapter 1). The problem is to determine q (or, equivalently, T) and s in such a way that the overall average cost per time unit is minimum.

4.4.1   Noninstantaneous resupply

Let $T_r \geqslant 0$ be the *replenishment time*, i.e. the time to make a replenishment, and r the *replenishment rate*, i.e. the number of items per unit of time received during T_r. The following relation holds:

$$q = rT_r. \tag{4.3}$$

The inventory level $I(t)$ as a function of time t is shown in Figure 4.1. Dashed lines represent the cumulative number of items arriving at the stocking point during a replenishment (their slope is r). Since items are picked up at a rate d while a replenishment takes place, the net number of items stocked per unit of time during T_r is $r - d$. Finally, after a replenishment, the stock level decreases at a rate equal to

$-d$. Let T_1, T_2, T_3 and T_4 be the time the inventory level takes to go from $-s$ to 0, from 0 to m, from m to 0 and from 0 to $-s$, respectively.

The maximum inventory level m is given by

$$s + m = (r - d)T_r.$$

Therefore, from Equation (4.3),

$$m = (r - d)T_r - s = q(1 - d/r) - s.$$

The total average cost per time unit $\mu(q, s)$ is

$$\mu(q, s) = \frac{1}{T}(k + cq + h\bar{I}T + us + v\bar{S}T). \tag{4.4}$$

The quantity in parentheses in the right-hand side is the *average cost per period*, given by the fixed and variable costs of a resupply (k and cq, respectively), and the holding cost $h\bar{I}T$, plus the shortage costs (us and $v\bar{S}T$). The holding and shortage costs depend on the average inventory level $\bar{I}$, and on the average shortage level $\bar{S}$, respectively:

$$\bar{I} = \frac{1}{T}\int_0^T I_+(t)\, dt = \frac{1}{T}\left(\frac{m(T_2 + T_3)}{2}\right),$$

$$\bar{S} = \frac{1}{T}\int_0^T I_-(t)\, dt = \frac{1}{T}\left(\frac{s(T_1 + T_4)}{2}\right),$$

where

$$I_+(t) = \begin{cases} I(t) & \text{if } I(t) \geqslant 0, \\ 0 & \text{otherwise,} \end{cases} \quad \text{and} \quad I_-(t) = \begin{cases} -I(t) & \text{if } I(t) \leqslant 0, \\ 0 & \text{otherwise.} \end{cases}$$

Moreover, since

$$s = (r - d)T_1,$$
$$m = (r - d)T_2,$$
$$m = dT_3,$$
$$s = dT_4,$$

the time lapses T_1, T_2, T_3 and T_4 are given by

$$T_1 = \frac{s}{r - d},$$
$$T_2 = \frac{m}{r - d},$$
$$T_3 = \frac{m}{d},$$
$$T_4 = \frac{s}{d}.$$

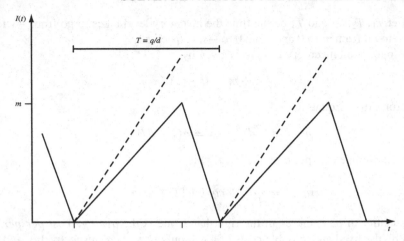

Figure 4.2   Inventory level as a function of time when shortage is not allowed.

Consequently,

$$\bar{I} = \frac{m^2}{2q(1 - d/r)} = \frac{[q(1 - d/r) - s]^2}{2q(1 - d/r)}, \tag{4.5}$$

$$\bar{S} = \frac{s^2}{2q(1 - d/r)}. \tag{4.6}$$

Finally, using Equations (4.1), (4.5) and (4.6), Equation (4.4) can be rewritten as

$$\mu(q, s) = \frac{kd}{q} + cd + \frac{h[q(1 - d/r) - s]^2}{2q(1 - d/r)} + \frac{usd}{q} + \frac{vs^2}{2q(1 - d/r)}. \tag{4.7}$$

If shortages are allowed, the minimum point (q^*, s^*) of the convex function $\mu(q, s)$ can be obtained by solving the equations:

$$\frac{\partial}{\partial q} \mu(q, s) \Big|_{q=q^*, s=s^*} = 0,$$

$$\frac{\partial}{\partial s} \mu(q, s) \Big|_{q=q^*, s=s^*} = 0.$$

As a result,

$$q^* = \sqrt{\frac{h + v}{v}} \sqrt{\frac{2kd}{h(1 - d/r)} - \frac{(ud)^2}{h(h + v)}} \tag{4.8}$$

and

$$s^* = \frac{(hq^* - ud)(1 - d/r)}{(h + v)}. \tag{4.9}$$

If shortages are not allowed (see Figure 4.2), Equation (4.7) can be simplified since $s = 0$:

$$\mu(q) = kd/q + cd + \tfrac{1}{2}hq(1 - d/r). \tag{4.10}$$

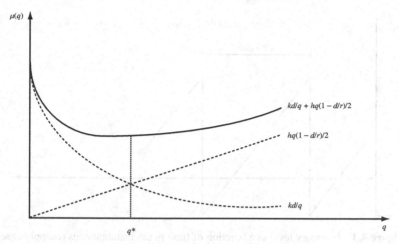

Figure 4.3 Average costs as a function of q.

Hence, a single equation has to be solved:

$$\frac{\mathrm{d}}{\mathrm{d}q}\mu(q)\bigg|_{q=q^*} = 0,$$

Finally, the optimal order size q^* is (see Figure 4.3)

$$q^* = \sqrt{\frac{2kd}{h(1 - d/r)}}. \tag{4.11}$$

Golden Food distributes tinned foodstuff in Great Britain. In a warehouse located in Birmingham, the demand rate d for tomato purée is 400 pallets a month. The value of a pallet is $c = £2500$ and the annual interest rate p is 14.5% (including warehousing costs). Issuing an order costs £30. The replenishment rate r is 40 pallets per day. Shortages are not allowed. The holding cost is given by

$$h = 0.145 \times 2500 = £362.5 \text{ per year per pallet}$$
$$= £30.2 \text{ per month per pallet}.$$

Therefore, from Equation (4.11),

$$q^* = \sqrt{\frac{2 \times 30 \times 400}{30.2[1 - 400/(40 \times 20)]}} = 39.9 \approx 40 \text{ pallets},$$

where it is assumed that the number of workdays in a month equals 20 (hence the

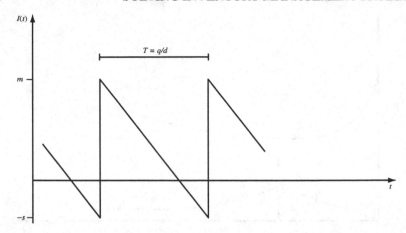

Figure 4.4   Inventory level as a function of time in the instantaneous resupply case.

demand rate d is 20 pallets per workday). Finally, from Equations (4.1) and (4.3),

$$T^* = 40/400 = 1/10 \text{ month} = 2 \text{ workdays,}$$
$$T_r^* = 40/40 = 1 \text{ workday.}$$

4.4.2   Instantaneous resupply

If resupply is instantaneous, the optimal inventory policy can be obtained by Equations (4.8), (4.9) and (4.11), taking into account that $r \to \infty$. If shortages are allowed (see Figure 4.4), then

$$q^* = \sqrt{\frac{h+v}{v}} \sqrt{\frac{2kd}{h} - \frac{(ud)^2}{h(h+v)}},$$
$$s^* = \frac{hq^* - ud}{(h+v)}.$$

If shortages are not allowed (see Figure 4.5), Equation (4.10) becomes

$$\mu(q) = kd/q + cd + \tfrac{1}{2}hq \tag{4.12}$$

and the optimal order size is given by

$$q^* = \sqrt{\frac{2kd}{h}}. \tag{4.13}$$

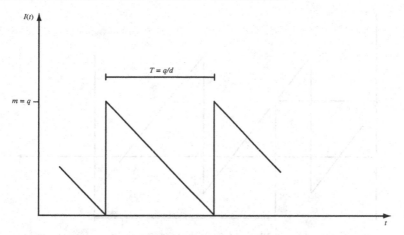

Figure 4.5   Inventory level as a function of time in the EOQ model.

This is the classical *economic order quantity* (EOQ) model introduced by F. W. Harris in 1913. The total cost per time unit of an EOQ policy is

$$\mu(q^*) = \sqrt{2kdh} + cd.$$

Optimal policies with no backlogging (in particular, the EOQ policy) satisfy the *zero inventory ordering* (ZIO) property, which states that an order is received exactly when the inventory level falls to zero.

Al-Bufeira Motors manufactures spare parts for aircraft engines in Saudi Arabia. Its component Y02PN, produced in a plant located in Jiddah, has a demand of 220 units per year and a unit production cost of $1200. Manufacturing this product requires a time-consuming set-up that costs $800. The current annual interest rate p is 18%, including warehousing costs. Shortages are not allowed. The holding cost is

$$h = 0.18 \times 1200 = 216 \text{ dollars/year per unit.}$$

Therefore, from Equation (4.13),

$$q^* = 40.37 \text{ units,}$$

and, from Equation (4.1),

$$T^* = 40.37/220 = 0.18 \text{ years} = 66.8 \text{ days.}$$

The total cost is given by Equation (4.12),

$$\frac{kd}{q^*} + \frac{hq^*}{2} = \frac{800 \times 220}{40.37} + \frac{216 \times 40.37}{2} = 8719.63 \text{ dollars per year,}$$

plus

$$cd = 220 \times 1200 = 264\,000 \text{ dollars per year.}$$

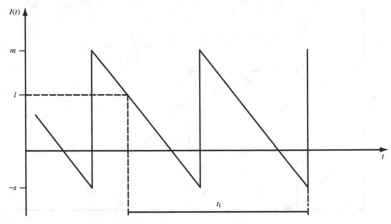

Figure 4.6   Reorder point.

In practice, the values of q^* and of T^* must be rounded as explained in Section 4.12.

4.4.3   Reorder point

An order has to be issued t_l time instants before the inventory level equals $-s$. In case resupplies are instantaneous, a replenishment is needed every time the inventory level equals the following *reorder point* l,

$$l = (t_l - \lfloor t_l/T \rfloor T)d - s,$$

where $\lfloor t_l/T \rfloor$ is the number of periods included in the lead time (see Figure 4.6).

In the Al-Bufeira Motors problem, a set-up has to be planned seven days in advance. Assuming $T = 67$ days, the reorder point l is

$$l = (7 - \lfloor 7/67 \rfloor \times 67)\, \frac{220}{365} - 0 \approx 4 \text{ units.}$$

4.5   Single Stocking Point: Single-Commodity Inventory Models under Deterministic Time-Varying Demand Rate

When orders are placed in advance, demands can be assumed to be deterministic, yet time-varying. Let $1, \ldots, T_H$ be a finite and discrete time horizon. In addition, let

d_t, $t = 1, \ldots, T_H$, be the demand at time period t, k the fixed reorder cost, and h the holding cost. The problem is to decide how much to order in each time period in such a way that the sum of reorder costs plus holding costs is minimized. No backlogging is allowed. In 1958, H. M. Wagner and T. M. Whithin formulated this problem as follows. The decision variables are the amount q_t, $t = 1, \ldots, T_H$, ordered at the beginning of time period t, the inventory level I_t, $t = 1, \ldots, T_H$, at the end of time period t; in addition let y_t, $t = 1, \ldots, T_H$, be a decision variable equal to 1 if an order is placed in time period t, 0 otherwise.

Minimize

$$\sum_{t=1,\ldots,T_H} (ky_t + hI_t) \tag{4.14}$$

subject to

$$I_t = I_{t-1} + q_t - d_t, \quad t = 1, \ldots, T_H, \tag{4.15}$$

$$q_t \leqslant y_t \sum_{r=t,\ldots,n} d_r, \quad t = 1, \ldots, T_H, \tag{4.16}$$

$$I_0 = 0, \tag{4.17}$$

$$I_t \geqslant 0, \quad t = 1, \ldots, T_H,$$

$$q_t \geqslant 0, \quad t = 1, \ldots, T_H,$$

$$y_t \in \{0, 1\}, \quad t = 1, \ldots, T_H,$$

where the objective function (4.14) is the total cost. Equations (4.15) are the *inventory-balance* constraints, inequalities (4.16) state that for each time period $t = 1, \ldots, T_H$, q_t is zero if y_t is zero, and equation (4.17) specifies the initial inventory.

An optimal solution of the Wagner–Within model can be obtained in $O(T_H^2)$ time through a dynamic programming procedure. This algorithm is based on the following theoretical results.

Proposition. Any optimal policy satisfies the ZIO property, i.e.

$$q_t I_{t-1} = 0, \quad t = 1, \ldots, T_H.$$

Proof. If $I_{t-1} = \delta > 0$ and $q_t > 0$, the total cost can be reduced by setting $I_{t-1} := 0$, $q_t := q_t + \delta$, and modifying the remaining variables accordingly. $\square$

Corollary. In an optimal policy, the amount ordered at each time period is the total demand of a set of consecutive subsequent periods.

The algorithm is as follows. Let $G = (V, A)$ be a directed acyclic graph, where $V = \{1, \ldots, T_H, T_H + 1\}$ is a vertex set and $A = \{(t, t') : t = 1, \ldots, T_H, t' = t+1, \ldots, T_H + 1\}$ is an arc set. With each arc (t, t') is associated the cost of ordering in time period t to satisfy the demands in time periods $t, t+1, \ldots, t' - 1$:

$$g_{tt'} = k + h \sum_{r=t,\ldots,t'-1} (r - t)d_r.$$

Then the shortest path between vertices 1 and $T_H + 1$ corresponds to a least-cost inventory policy.

Sao Vincente Chemical is a Portuguese company producing lubricants. In the following year, its product Serrado Oil is expected to have a demand of 720, 1410, 830 and 960 pallets in winter, spring, summer and autumn, respectively. Manufacturing this product requires a time-consuming set-up that costs €8900. The current annual interest rate p is 7.5%, including warehousing costs. The variable production cost amounts to \$350 per pallet while the initial inventory is zero. The holding cost is

$$h = 0.075 \times 350/4 = 6.56 \text{ euros/season per unit.}$$

Let $t = 1, 2, 3, 4$ represent the winter, spring, summer and autumn periods, respectively. By solving the Wagner–Within model, it follows that the optimal policy is to produce at the beginning of winter and summer of the next year. In particular, $y_1 = y_3 = 1$, $y_2 = y_4 = 0$, $q_1 = 2130$, $q_2 = 0$, $q_3 = 1790$, $q_4 = 0$, $I_1 = 1410$, $I_2 = 0$, $I_3 = 960$, $I_4 = 0$. Total holding and set-up costs amount to €33 353.

4.6   Models with Discounts

In the previous sections it has been assumed that the value of an item is always constant (equal to c). In practice, quantity discounts offered by suppliers, or economies of scale in the manufacturing processes, make the value of an item dependent on the order size q. In the following, *quantity-discounts-on-all-units* and *incremental quantity discounts* are examined under the EOQ hypothesis. Let $f(q)$ be the value of q items. If the value of an item is independent of q, then $f(q) = cq$. Otherwise, $f(q)$ is a concave nondecreasing function.

4.6.1   Quantity-discounts-on-all-units

The function $f(q)$ is assumed to be piecewise linear (see Figure 4.7),

$$f(q) = c_i q, \quad q_{i-1} \leqslant q < q_i, \quad i = 1, 2, \ldots,$$

where $q_0 = 0, q_1, \ldots$ are known *discount breaks* ($q_i < q_{i+1}$, $i = 1, 2, \ldots$), $f(q_0) = 0$, and $c_i > c_{i+1}$, $i = 1, 2 \ldots$. Hence, if the order size q is included between discount breaks q_{i-1} and q_i, the value of *every* item is c_i, $i = 1, 2, \ldots$. It is worth noting that, depending on c_i coefficients, $f(q)$ can be greater than $f(q')$ for $q < q'$ (see Figure 4.7). In practice, the *effective* cost function is

$$f(q) = \min\{c_i q, c_{i+1} q_i\}, \quad q_{i-1} \leqslant q < q_i, \quad i = 1, 2, \ldots.$$

The total average cost function $\mu(q)$ can be written as

$$\mu(q) = \mu_i(q), \quad q_{i-1} \leqslant q < q_i, \quad i = 1, 2, \ldots,$$

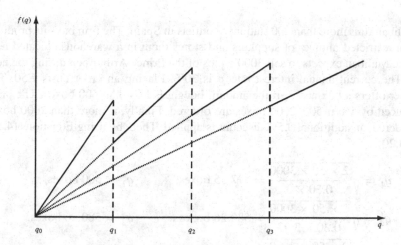

Figure 4.7 The value of q items: quantity-discounts-on-all-units.

where, as $h_i = pc_i$, $i = 1, 2, \ldots$,

$$\mu_i(q) = kd/q + c_i d + \tfrac{1}{2}h_i q, \quad i = 1, 2, \ldots. \tag{4.18}$$

Then, the optimal order size q^* can be obtained through the following procedure.

Step 1. By imposing

$$\left. \frac{d\mu_i(q_i)}{dq_i} \right|_{q_i = q_i'} = 0, \quad i = 1, 2, \ldots,$$

determine the order size q_i', $i = 1, 2, \ldots$, that minimizes $\mu_i(q)$,

$$q_i' = \sqrt{\frac{2kd}{h_i}}, \quad i = 1, 2, \ldots. \tag{4.19}$$

Let

$$q_i^* = \begin{cases} q_{i-1}, & \text{if } q_i' < q_{i-1}, \\ q_i', & \text{if } q_{i-1} \leqslant q_i' \leqslant q_i, \quad i = 1, 2, \ldots. \\ q_i, & \text{if } q_i' > q_i, \end{cases} \tag{4.20}$$

Step 2. Compute the optimal solution $q^* = q_{i^*}^*$, where

$$i^* = \arg \min_{i=1,2,\ldots} \{\mu(q_i^*)\}.$$

Maliban runs more than 200 stationery outlets in Spain. The firm buys its products from a restricted number of suppliers and stores them in a warehouse located near Sevilla. Maliban expects to sell 3000 boxes of the Prince Arthur pen during the next year. The current annual interest rate p is 30%. Placing an order costs €50. The supplier offers a box at €3, if the amount bought is less than 500 boxes. The price is reduced by 1% if 500–2000 boxes are ordered. Finally, if more than 2000 boxes are ordered, an additional 0.5% discount is applied. Then, by using Equations (4.19) and (4.20),

$$q_1' = \sqrt{\frac{2 \times 50 \times 3000}{0.30 \times 3}} = 577.35 \text{ boxes,} \qquad q_1^* = 500 \text{ boxes,}$$

$$q_2' = \sqrt{\frac{2 \times 50 \times 3000}{0.30 \times 2.97}} = 580.26 \text{ boxes,} \qquad q_2^* = 580.26 \text{ boxes,}$$

$$q_3' = \sqrt{\frac{2 \times 50 \times 3000}{0.30 \times 2.955}} = 581.73 \text{ boxes,} \qquad q_3^* = 2000 \text{ boxes.}$$

By comparing the corresponding annual average costs given by Equations (4.18), the optimal order size is $q^* = 580$ boxes ($\approx q_2^*$), corresponding to an annual cost of €9427.

4.6.2 Incremental quantity discounts

The function $f(q)$ is assumed to depend on q as follows (see Figure 4.8),

$$f(q) = f(q_{i-1}) + c_i(q - q_{i-1}), \quad q_{i-1} \leqslant q < q_i, \quad i = 1, 2, \ldots, \qquad (4.21)$$

where $q_0 = 0, q_1, \ldots$ are known discount breaks ($q_i < q_{i+1}, i = 1, 2, \ldots$), $f(q_0) = 0$, and $c_i > c_{i+1}, i = 1, 2, \ldots$. Consequently, if the order size q is included between discount breaks q_{i-1} and q_i, the value of $(q - q_{i-1})$ items is c_i, the value of $(q_{i-1} - q_{i-2})$ items is c_{i-1}, etc. The average total cost function $\mu(q)$ is

$$\mu(q) = \mu_i(q), \quad q_{i-1} \leqslant q < q_i, \quad i = 1, 2, \ldots,$$

where, on the basis of Equation (4.12),

$$\mu_i(q) = \frac{kd}{q} + \frac{f(q)d}{q} + p\frac{f(q)}{q}\frac{q}{2}, \quad i = 1, 2, \ldots.$$

Using Equation (4.21), $\mu_i(q), i = 1, 2, \ldots,$ can be rewritten as

$$\mu_i(q) = kd/q + [f(q_{i-1}) + c_i(q - q_{i-1})]d/q$$
$$+ \tfrac{1}{2}p[f(q_{i-1}) + c_i(q - q_{i-1})], \quad i = 1, 2, \ldots \qquad (4.22)$$

The optimal order size q^* can be computed through a procedure very similar to that used in the previous case.

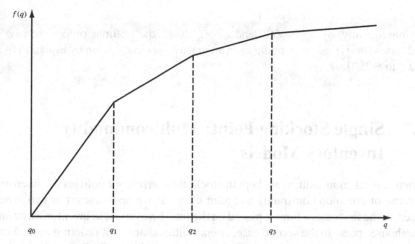

Figure 4.8 The value of q items: incremental quantity discounts.

Step 1. Determine the value q_i', $i = 1, 2, \ldots$, that minimizes $\mu_i(q)$ by imposing that

$$\left.\frac{d\mu_i(q_i)}{dq_i}\right|_{q_i=q_i'} = 0, \quad i = 1, 2, \ldots.$$

Hence

$$q_i' = \sqrt{\frac{2d[k + f(q_{i-1}) - c_i q_{i-1}]}{p c_i}}, \quad i = 1, 2, \ldots. \tag{4.23}$$

If $q_i' \notin [q_{i-1}, q_i]$, then let $\mu_i(q_i') = \infty$, $i = 1, 2, \ldots.$

Step 2. Compute the optimal solution $q^* = q_{i*}'$, where

$$i^* = \arg \min_{i=1,2,\ldots} \{\mu(q_i')\}.$$

If Maliban (see the previous problem) applies an incremental quantity discount policy, then, by using Equation (4.23),

$$q_1' = \sqrt{\frac{2 \times 3000 \times 50}{0.30 \times 3}} = 577.4 \text{ boxes,}$$

$$q_2' = \sqrt{\frac{2 \times 3000[50 + (3 \times 500) - (2.97 \times 500)]}{0.30 \times 2.97}} = 661.6 \text{ boxes,}$$

$$q_3' = \sqrt{\frac{2 \times 3000\{50 + [(3 \times 500) + (2.97 \times 1500)] - (2.955 \times 2000)\}}{0.30 \times 2.955}}$$

$$= 801.9 \text{ boxes.}$$

Consequently, as $q_1' > 500$ and $q_3' < 2000$, the optimal order size is $q^* = 662$ boxes ($\approx q_2'$), corresponding to an annual average cost, given by Equation (4.22), equal to €9501.67.

4.7   Single Stocking Point: Multicommodity Inventory Models

When several commodities are kept in stock, their inventory policies are intertwined because of common constraints and joint costs, as we now discuss in two separate cases. In the first case, a limit is placed on the total investment in inventories, or on the warehouse space. In the second case, commodities share joint ordering costs. For the sake of simplicity, both analysis will be performed under the EOQ model hypotheses.

4.7.1   Models with capacity constraints

Let n be the number of commodities in stock and q_j, $j = 1, \ldots, n$, the amount of commodity j ordered at each replenishment. The inventory management problem can be formulated as follows.

Minimize

$$\mu(q_1, \ldots, q_n) \tag{4.24}$$

subject to

$$g(q_1, \ldots, q_n) \leqslant b, \tag{4.25}$$
$$q_1, \ldots, q_n \geqslant 0, \tag{4.26}$$

where the objective function (4.24) is the total average cost per time unit. Under the EOQ hypothesis, the objective function $\mu(q_1, \ldots, q_n)$ can be written as

$$\mu(q_1, \ldots, q_n) = \sum_{j=1}^{n} \mu_j(q_j),$$

where, on the basis of Equation (4.12),

$$\mu_j(q_j) = k_j d_j / q_j + c_j d_j + \tfrac{1}{2} h_j q_j, \quad j = 1, \ldots, n,$$

and quantities k_j, d_j, c_j, h_j, $j = 1, \ldots, n$, are the fixed reorder cost, the demand rate, the value, and the holding cost of item j, respectively. As is customary, $h_j = p_j c_j$, $j = 1, \ldots, n$, where p_j is the interest rate of commodity j.

Equation (4.25) is a side constraint (referred to as a *capacity constraint*) representing both a budget constraint or a warehouse constraint. It can usually be considered as linear,

$$\sum_{j=1}^{n} a_j q_j \leqslant b, \tag{4.27}$$

where a_j, $j = 1, \ldots, n$, and b are constants. As a result, problem (4.24)–(4.26) has to be solved through iterative methods for NLP problems, such as the *conjugate gradient method*. Alternatively, the following simple heuristic can be used if the capacity constraint is linear and the interest rates are identical for all the commodities ($p_j = p$, $j = 1, \ldots, n$).

Step 1. Using Equation (4.13), compute the EOQ order sizes q'_j, $j = 1, \ldots, n$:

$$q'_j = \sqrt{\frac{2k_j d_j}{pc_j}}, \quad j = 1, \ldots, n. \tag{4.28}$$

If the capacity constraint (4.27) is satisfied, *STOP*, the optimal order size for each product j, $j = 1, \ldots, n$, has been determined.

Step 2. Increase the interest rate p of a δ quantity to be determined. Then, the order sizes become

$$q_j(\delta) = \sqrt{\frac{2k_j d_j}{(p + \delta)c_j}}, \quad j = 1, \ldots, n. \tag{4.29}$$

Determine the value δ^* satisfying the relation,

$$\sum_{j=1}^{n} a_j q_j(\delta^*) = b.$$

Hence,

$$\delta^* = \left(\frac{1}{b} \sum_{j=1}^{n} \left(a_j \sqrt{\frac{2k_j d_j}{c_j}} \right) \right)^2 - p. \tag{4.30}$$

Insert δ^* in Equations (4.29) in order to determine the order sizes $\bar{q}_j$, $j = 1, \ldots, n$.

New Frontier distributes knapsacks and suitcases in most US states. Its most successful models are the Preppie knapsack and the Yuppie suitcase. The Preppie knapsack has a yearly demand of 150 000 units, a value of $30 and a yearly holding cost equal to 20% of its value. The Yuppie suitcase has a yearly demand of 100 000 units, a value of $45 and a yearly holding cost equal to 20% of its value. In both cases, placing an order costs $250. The company's management requires the average capital invested in inventories does not exceed $75 000. This condition can be expressed by the following constraint,

$$30q_1/2 + 45q_2/2 \leqslant 75\,000,$$

where it is assumed, as a precaution, that the average inventory level is the sum of the average inventory levels of the two items. The EOQ order sizes, given by Equation (4.28),

$$q_1' = \sqrt{\frac{2 \times 250 \times 150\,000}{0.2 \times 30}} = 3535.53 \text{ units,}$$

$$q_2' = \sqrt{\frac{2 \times 250 \times 100\,000}{0.2 \times 45}} = 2357.02 \text{ units,}$$

do not satisfy the budget constraint. Applying the conjugated gradient method starting from the initial values $(q_1, q_2) = (1, 1)$, the following solution is obtained after 300 iterations,

$$\bar{q}_1 = 2500 \text{ units,}$$
$$\bar{q}_2 = 1666.66 \text{ units,}$$

whose total cost is \$9 045 000. Applying the heuristic procedure, by using Equation (4.30), the same solution is obtained. In effect,

$$\delta^* = \left[\frac{1}{75\,000} \left(\frac{30}{2} \sqrt{\frac{2 \times 250 \times 150\,000}{30}} + \frac{45}{2} \sqrt{\frac{2 \times 250 \times 100\,000}{45}} \right) \right]^2 - 0.2$$
$$= 0.2,$$

hence,

$$\bar{q}_1 = \sqrt{\frac{2 \times 250 \times 150\,000}{(0.2 + 0.2)\,30}} = 2500 \text{ units,}$$

$$\bar{q}_2 = \sqrt{\frac{2 \times 250 \times 100\,000}{(0.2 + 0.2)\,45}} = 1666.66 \text{ units.}$$

4.7.2 Models with joint costs

For the sake of simplicity, we assume in this section that only two commodities are kept in the inventory. Let k_1 and k_2 be the fixed costs for reordering the two commodities at different moments in time, and let k_{1-2} be the fixed cost for ordering both commodities at the same time ($k_{1-2} < k_1 + k_2$). In addition, let T_1 and T_2 be the time lapses between consecutive replenishments of commodities 1 and 2, respectively (see Figure 4.9). Then,

$$q_1 = d_1 T_1, \tag{4.31}$$
$$q_2 = d_2 T_2. \tag{4.32}$$

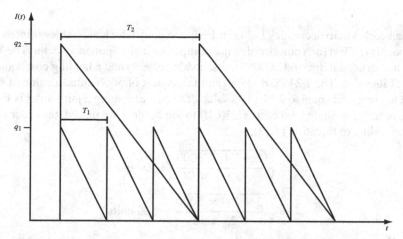

Figure 4.9   Inventory level as a function of time in case of synchronized orders.

The periodicity of a joint replenishment policy is

$$T = \max\{T_1, T_2\}.$$

In each period T, the orders issued for the two items are

$$N_1 = T/T_1,$$
$$N_2 = T/T_2.$$

N_1 and N_2 are positive integer numbers, one of them being equal to 1 (in the situation depicted in Figure 4.9, $N_1 = 3$ and $N_2 = 1$). During each period T, two items are ordered simultaneously exactly once. Moreover, $N_j - 1$ single orders are placed for each item j, $j = 1, 2$. Hence, the total average cost per time unit is

$$\mu(T, N_1, N_2)$$
$$= \frac{k_{1-2} + (N_1 - 1)k_1 + (N_2 - 1)k_2}{T} + c_1 d_1 + c_2 d_2 + \frac{h_1 d_1 T}{2N_1} + \frac{h_2 d_2 T}{2N_2}. \quad (4.33)$$

By solving the equation,

$$\frac{\partial}{\partial T}\mu(T, N_1, N_2)\bigg|_{T=T*} = 0,$$

the value $T^*(N_1, N_2)$ that minimizes $\mu(T, N_1, N_2)$ is obtained,

$$T^*(N_1, N_2) = \sqrt{\frac{2N_1 N_2[k_{1-2} + (N_1 - 1)k_1 + (N_2 - 1)k_2]}{h_1 d_1 N_2 + h_2 d_2 N_1}}, \quad (4.34)$$

as a function of N_1 and N_2.

Shamrock Microelectronics Ltd is an Irish company which assembles printed circuit boards (PCBs) for a number of major companies in the appliance sector. The $Y23$ PCB has an annual demand of 3000 units, a value of €30 and a holding cost equal to 20% of its value. The $Y24$ PCB has an annual request of 5000 units, a value of €40 and a holding cost equal to 25% of its value. The cost of issuing a joint order is €300 while ordering a single item costs €250. If no joint orders are placed, the order sizes are, according to Equation (4.13),

$$q_1^* = \sqrt{\frac{2 \times 250 \times 3000}{0.2 \times 30}} = 500 \text{ units},$$

$$q_2^* = \sqrt{\frac{2 \times 250 \times 5000}{0.25 \times 40}} = 500 \text{ units}.$$

From Equations (4.31) and (4.32):

$$T_1^* = 500/3000 = 1/6,$$

$$T_2^* = 500/5000 = 1/10.$$

This means that Shamrock would issue $1/T_1^* = 6$ orders per year of $Y23$ PCB and $1/T_2^* = 10$ orders per year of the $Y24$ PCB. Since

$$\mu_1(q_1^*) = \frac{250 \times 3000}{500} + 30 \times 3000 + \frac{0.2 \times 30 \times 500}{2}$$
$$= 93\,000 \text{ euros per year},$$

$$\mu_2(q_2^*) = \frac{250 \times 5000}{500} + 40 \times 5000 + \frac{0.25 \times 40 \times 500}{2}$$
$$= 205\,000 \text{ euros per year},$$

the average annual cost is €298 000 per year. If a joint order is placed and $N_1 = 1$, $N_2 = 2$, the periodicity of joint orders is, according to Equation (4.34),

$$T^* = \sqrt{\frac{2 \times 1 \times 2 \times (300 + 250)}{0.2 \times 30 \times 3000 \times 2 + 0.25 \times 40 \times 5000 \times 1}} = 0.16.$$

Shamrock would issue $1/T^* = 6.25$ joint orders per year. The annual average cost, computed through Equation (4.33), is equal to

$$\mu(T^*, 1, 2) = \frac{300 + 250}{0.16} + 30 \times 3000 + 40 \times 5000 + \frac{0.2 \times 30 \times 3000 \times 0.16}{2}$$
$$+ \frac{0.25 \times 40 \times 5000 \times 0.16}{2 \times 2}$$

$$= 296\,877.5 \text{ euros per year}.$$

4.8   Stochastic Models

Inventory problems with uncertain demand or lead times have quite a complex mathematical structure. In this section, a restricted number of stochastic models are illustrated. First, the classical *Newsboy Problem*, where a one-shot reorder decision has to be made, is examined. Then, (s, S) *policies* are introduced for a variant of the Newsboy Problem. Finally, the most common inventory policies used by practitioners (namely, the *reorder level* method, the *reorder cycle* method, the (s, S) method and the *two-bin* technique) are reviewed and compared. The first three policies make use of data forecasts, whereas the fourth policy does not require any data estimate.

4.8.1   The Newsboy Problem

In the Newsboy Problem, a resupply decision has to be made at the beginning of a period (e.g. a spring sales season) for a single commodity whose demand is not known in advance. The demand d is modelled as a random variable with a continuous cumulative distribution function $F_d(\delta)$. Let c be the purchasing cost or the variable manufacturing cost, depending on whether the goods are bought from an external supplier or produced by the company. Moreover, let r and u be the selling price and the salvage value per unit of commodity, respectively. Of course,

$$r > c > u.$$

There is no fixed reorder cost nor an initial inventory. In addition, shortage costs are assumed to be negligible. If the company orders q units of commodity, the *expected revenue* $\rho(q)$ is

$$\rho(q) = r \int_0^\infty \min(\delta, q) \, dF_d(\delta) + u \int_0^\infty \max(0, q - \delta) \, dF_d(\delta) - cq$$

$$= r \left(\int_0^q \delta \, dF_d(\delta) + q \int_q^\infty dF_d(\delta) \right) + u \int_0^q (q - \delta) \, dF_d(\delta) - cq.$$

By adding and subtracting $r \int_q^\infty \delta \, dF_d(\delta)$ to the right-hand side, $\rho(q)$ becomes

$$\rho(q) = r E[d] + r \int_q^\infty (q - \delta) \, dF_d(\delta) + u \int_0^q (q - \delta) \, dF_d(\delta) - cq, \qquad (4.35)$$

where $E[d]$ is the *expected demand*. It is easy to show that $\rho(q)$ is concave for $q \geqslant 0$, and $\rho(q) \to -\infty$ for $q \to \infty$. As a result, the maximum expected revenue is achieved when the derivative of $\rho(q)$ with respect to q is zero. Hence, by applying the Leibnitz rule, the optimality condition becomes

$$r(1 - F_d(q)) + u F_d(q) - c = 0,$$

where, by definition, $F_d(q)$ is the probability $\Pr(d \leqslant q)$ that the demand does not exceed q. As a result, the *optimal order quantity* S satisfies the following condition:

$$\Pr(d \leqslant S) = \frac{r - c}{r - u}. \tag{4.36}$$

Emilio Tadini & Sons is a hand-made shirt retailer, located in Rome (Italy), close to Piazza di Spagna. This year Mr Tadini faces the problem of ordering a new bright colour shirt made by a Florentine firm. He assumes that the demand is uniformly distributed between 200 and 350 units. The purchasing cost is $c = €18$ while the selling price is $r = €52$ and the salvage value is $u = €7$. According to Equation (4.36), $\Pr(d \leqslant S) = (S - 200)/(350 - 200)$ for $200 \leqslant S \leqslant 350$. Hence, Mr Tadini should order $S = 313$ units. According to Equation (4.35), the expected revenue is

$$\rho(q) = 52 \times 275 + 52 \int_q^{350} (q - \delta)\frac{1}{350 - 200}\, d\delta - 18q = 34q,$$

for $0 \leqslant q \leqslant 200$,

$$\rho(q) = 52 \times 275 + 52 \int_q^{350} (q - \delta)\frac{1}{350 - 200}\, d\delta$$
$$+ 7 \int_{200}^q (q - \delta)\frac{1}{350 - 200}\, d\delta - 18q$$
$$= -0.15q^2 + 94q - 6000,$$

for $200 < q \leqslant 350$, and

$$\rho(q) = 52 \times 275 + 7 \int_{200}^{350} (q - \delta)\frac{1}{350 - 200}\, d\delta - 18q = -11q + 12\,375,$$

for $q > 350$. Hence, the maximum expected revenue is equal to $\rho(313) = €8726.65$.

4.8.2   The (s, S) policy for single period problems

If there is an initial inventory q_0 and a fixed reorder cost k, the optimal replenishment policy can be obtained as follows. If $q_0 \geqslant S$, no reorder is needed. Otherwise, the best policy is to order $S - q_0$, provided that the expected revenue associated with this choice is greater than the expected revenue associated with not producing anything. Hence, two cases can occur:

   (i) if the expected revenue $\rho(S) - k - cq_0$ associated with reordering is greater than the expected revenue $\rho(q_0) - cq_0$ associated with not reordering, then $S - q_0$ units have to be reordered;

  (ii) otherwise, no order has to be placed.

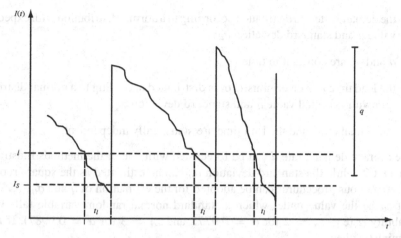

Figure 4.10 Reorder level inventory policy.

As a consequence, if $q_0 < S$, the optimal policy consists of ordering $S - q_0$ units if $\rho(q_0) \leqslant \rho(S) - k$. In other words, if s is the number such that

$$\rho(s) = \rho(S) - k,$$

the optimal policy is to order $S - q_0$ units if the initial inventory level q_0 is less than or equal to s, otherwise not to order. Policies like this are known as (s, S) policies. The parameter s acts as a *reorder point*, while S is called the *order-up-to-level*.

If $q_0 = 50$ and $k = €400$ in the Emilio Tadini & Sons problem, $\rho(s) = \rho(S) - k = €8526.65$ so that $s = 277$. As $q_0 < s$, the optimal policy is to order $S - q_0 = 253$ units.

4.8.3 The reorder point policy

In the *reorder point* policy (or *fixed order quantity policy*), the inventory level is kept under observation in an almost continuous way. As soon as its net value $I(t)$ (the amount in stock minus the unsatisfied demand plus the orders placed but not received yet) reaches a reorder point l, a constant quantity q is ordered (see Figure 4.10).

The reorder size q is computed through the procedures illustrated in the previous sections, by replacing d with $\bar{d}$. In particular, under the EOQ hypotheses:

$$q = \sqrt{\frac{2k\bar{d}}{h}}.$$

The reorder point l is obtained by requiring that the inventory level be nonnegative during t_l, with probability α. This is equivalent to assuming that demand should not exceed l during the interval t_l. In the following, it is assumed that

- the demand rate d is distributed according to a normal distribution with expected value $\bar{d}$ and standard deviation σ_d;

- $\bar{d}$ and σ_d are constant in time;

- the lead time t_l is deterministic or is distributed according to a normal distribution with expected value $\bar{t}_l$ and standard deviation σ_{t_l};

- the demand rate and the lead time are statistically independent.

The average demand rate $\bar{d}$ can be forecasted with one of the methods illustrated in Chapter 2, while the standard deviation σ_d can be estimated as the square root of MSE. Analogous procedures can be adopted for the estimation of $\bar{t}_l$ and σ_{t_l}.

Let z_α be the value under which a standard normal random variable falls with probability α (e.g. $z_\alpha = 2$ for $\alpha = 0.9772$ and $z_\alpha = 3$ for $\alpha = 0.9987$). If t_l is deterministic, then

$$l = \bar{d}t_l + z_\alpha\sigma_d\sqrt{t_l}, \tag{4.37}$$

where $\bar{d}t_l$ and $\sigma_d\sqrt{t_l}$ are the expected value and the standard deviation of the demand in an interval of duration t_l, respectively. If t_l is random, then,

$$l = \bar{d}\,\bar{t}_l + z_\alpha\sqrt{\sigma_d^2\,\bar{t}_l + \sigma_{t_l}^2\,\bar{d}^2},$$

where $\bar{d}\,\bar{t}_l$ and $\sqrt{\sigma_d^2\,\bar{t}_l + \sigma_{t_l}^2\,\bar{d}^2}$ are the expected value and the standard deviation of the demand in a time interval of random duration t_l, respectively.

The reorder point l minus the average demand in the reorder period constitutes a *safety stock* I_S. For example, in case t_l is constant, the safety stock is

$$I_S = l - \bar{d}\,t_l = z_\alpha\sigma_d\sqrt{t_l}. \tag{4.38}$$

Papier is a French retail chain. At the outlet located in downtown Lyon, the expected demand for mouse pads is 45 units per month. The value of an item in stock is €4, and the fixed reorder cost is equal to €30. The annual interest rate is 20%. The demand forecasting MSE is 25. Lead time is 1 month and a service level equal to 97.7% is required. On the basis of Equation (4.2), the holding cost is

$$h = 0.2 \times 4 = 0.8 \text{ euros/year per item} = 0.067 \text{ euros/month per item.}$$

Therefore, from Equation (4.13),

$$q^* = \sqrt{\frac{2 \times 30 \times 45}{0.067}} = 200.74 \approx 201 \text{ items.}$$

Moreover, σ_d can be estimated as follows:

$$\sigma_d = \sqrt{25} = 5.$$

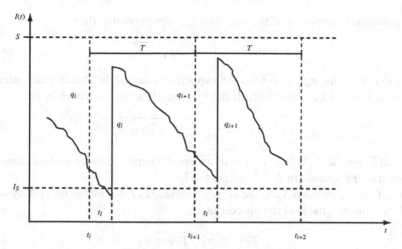

Figure 4.11 Reorder cycle inventory policy.

From Equation (4.37), the reorder point l is

$$l = 45 + 2 \times 5 = 55 \text{ units.}$$

Consequently, the safety stock I_S is

$$I_S = 55 - 45 = 10 \text{ units.}$$

4.8.4 The periodic review policy

In the reorder cycle policy (or *periodic review* policy) the stock level is kept under observation periodically at time instants t_i ($t_{i+1} = t_i + T$, $T \geqslant 0$). At time t_i, $q_i = S - I(t_i)$ units are ordered (see Figure 4.11). The parameter S (referred to as the *order-up-to-level*) represents the maximum inventory level in case lead time t_l is negligible.

The periodicity T of the sampling (*review period*) can be chosen using procedures analogous to those used for determining q^* in the deterministic models. For instance, under the EOQ hypotheses,

$$T = \sqrt{\frac{2k}{h\bar{d}}}. \tag{4.39}$$

The parameter S is determined in such a way that the probability that the inventory level becomes negative does not exceed a given value $(1 - \alpha)$. Since the *risk interval* is equal to T plus t_l, S is required to be greater than or equal to the demand in $T + \bar{t}_l$, with probability equal to α. If the lead time t_l is deterministic, then

$$S = \bar{d}(T + t_l) + z_\alpha \sigma_d \sqrt{T + t_l}, \tag{4.40}$$

where $\bar{d}(T + t_l)$ and $\sigma_d \sqrt{T + t_l}$ are the expected value and the standard deviation of the demand in $T + t_l$, respectively. If the lead time is a random variable, then

$$S = \bar{d}(T + \bar{t}_l) + z_\alpha \sqrt{\sigma_d^2(T + \bar{t}_l) + \sigma_{t_l}^2 \bar{d}^2},$$

where $\bar{d}(T + \bar{t}_l)$ and $\sqrt{\sigma_d^2(T + \bar{t}_l) + \sigma_{t_l}^2 \bar{d}^2}$ are the expected value and the standard deviation of the demand in $T + \bar{t}_l$, respectively.

The difference between S and the average demand in $T + \bar{t}_l$ makes up a safety stock I_S. For example, if the lead time is constant,

$$I_S = z_\alpha \sigma_d \sqrt{T + t_l}. \tag{4.41}$$

Comparing Equation (4.41) with Equation (4.38), it can be seen that the reorder cycle inventory policy involves a higher level of safety stock. However, such a policy does not require a continuous monitoring of the inventory level.

In the Papier problem, the parameters of the reorder cycle inventory policy, computed through Equations (4.39) and (4.40) are

$$T = \sqrt{\frac{2 \times 30}{0.067 \times 45}} = 4.47 \text{ months},$$

$$S = 45 \times (4.47 + 1) + 2 \times 5 \times \sqrt{4.47 + 1} = 269.54 \text{ units}.$$

The associated safety stock, given by Equation (4.41), is

$$I_S = 2 \times 5 \times \sqrt{4.47 + 1} = 23.39 \text{ units}.$$

4.8.5 The (s, S) policy

The (s, S) inventory policy is a natural extension of the (s, S) policy illustrated for the one-shot case. At time t_i, $S - I(t_i)$ items are ordered if $I(t_i) < s$ (see Figure 4.12). If s is large enough ($s \to S$), the (s, S) policy is similar to the reorder cycle inventory method. On the other hand, if s is small ($s \to 0$), the (s, S) policy is similar to a reorder level policy with a reorder point equal to s and a reorder quantity $q \cong S$. On the basis of these observations, the (s, S) policy can be seen as a good compromise between the reorder level and the reorder cycle policies. Unfortunately, parameters T, S and s are difficult to determine analytically. Therefore, simulation is often used in practice.

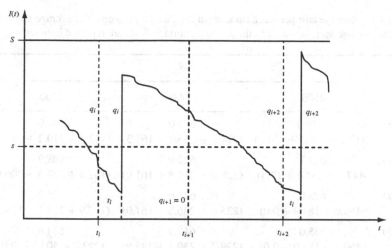

Figure 4.12 (s, S) policy.

Pansko, a Bulgarian chemical firm located in Plovdiv, supplies chemical agents to state clinical laboratories. Its product Merofosphine has a demand of 400 packages per week, a variable cost of 100 levs per unit, and a profit of 20 levs per unit. Every time the manufacturing process is set up, a fixed cost of 900 levs is incurred. The annual interest rate p is 20%. If the commodity is not available in stock, a sale is lost. In this case, a cost equal to the profit of the lost sale is incurred. The MSE forecast equals 2500. The lead time can be assumed to be constant and equal to a week. The inventory is managed by means of an (s, S) policy with a period T of two weeks. The values s and S are selected by simulating the system for all combinations of s (equal to 800, 900, 1000, 1100 and 1200, respectively) and S (equal to 1500, 2000 and 2500, respectively). According to the results reported in Table 4.1, $s = 1100$ and $S = 2000$ are the best choice. This would result in an average cost per week equal to 612.7 levs.

4.8.6   The two-bin policy

The *two-bin* policy can be seen as a variant of the reorder point inventory method where no demand forecast is needed, and the inventory level does not have to be monitored continuously. The items in stock are assumed to be stored in two identical bins. As soon as one of the two becomes empty, an order is issued for an amount equal to the bin capacity.

Browns supermarkets make use of the two-bin policy for tomato juice bottles. The capacity of each bin is 400 boxes, containing 12 bottles each. In a supermarket close

Table 4.1   Average cost per week (in levs) in the Pansko problem. The average fixed cost, the average variable cost and the average shortage costs are reported in brackets.

	S		
s	1500	2000	2500
800	1120.8	625.0	994.9
	$(337.8 + 168.5 + 614.5)$	$(224.8 + 236.8 + 163.3)$	$(152.3 + 330.2 + 512.3)$
900	644.7	622.9	908.9
	$(447.6 + 184.6 + 12.4)$	$(225.0 + 236.8 + 161.0)$	$(162.9 + 339.3 + 406.6)$
1000	625.0	623.0	724.3
	$(450.0 + 184.9 + 0.0)$	$(225.0 + 236.8 + 161.0)$	$(197.9 + 375.7 + 150.5)$
1100	635.0	612.7	634.6
	$(450.0 + 184.0 + 0.0)$	$(229.7 + 239.1 + 143.9)$	$(222.2 + 403.3 + 9.0)$
1200	635.0	622.7	631.8
	$(450.0 + 185.0 + 0.0)$	$(291.2 + 276.3 + 55.1)$	$(224.9 + 406.8 + 0.0)$

Table 4.2   Daily sales of tomato juice (in bottles) during the first week of December last in a Browns supermarket.

Day	Sales	Inventory level
1 Dec	850	8510
2 Dec	576	7934
3 Dec	932	7002
4 Dec	967	6035
5 Dec	945	5090
6 Dec	989	4101
7 Dec	848	3253

to Los Alamos (New Mexico, USA) the inventory level on 1 December last was 780 boxes of 12 bottles each. Last 6 December, the inventory level was less than 400 boxes and an order of 400 boxes was issued (see Table 4.2). The order was fulfilled the subsequent day.

4.9   Selecting an Inventory Policy

It is quite common for a warehouse to contain several hundreds (or even thousands) of items. In such a context, goods having a strong impact on the total cost have to be managed carefully while for less important goods it is wise to resort to simple and low-cost techniques.

The problem is generally tackled by clustering the goods into three categories (indicated with the symbols A, B and C) on the basis of the average value of the goods in stock. This method is often called the *ABC technique*. Category A is made up of products corresponding to a high percentage (e.g. 80%) of the total warehouse value. Category B is constituted by a set of items associated with an additional 15% of the warehouse value, while category C is formed by the remaining items. Goods are subdivided into these categories as follows: first, commodities are sorted by nonincreasing values with respect to the average value of the goods in stock; the items are then selected from the sorted list, to reach the pre-established cumulated value levels.

On the basis of the 80–20 principle (or *Pareto principle*), category A usually contains a small fraction (generally, 20–30%) of the goods whereas category C includes many products. This observation suggests that the goods of categories A and B should be managed with policies based on forecasts and a frequent monitoring (e.g. category A by means of the reorder level inventory method and category B through the reorder cycle inventory policy). Products in category C can be managed using the two-bin policy that does not require any forecast.

The Walloon Transportation Consortium (WTC) operates a Belgian public transportation service in the Walloon region. Buses are maintained in a facility located in Ans, close to a vehicle depot. The average inventory levels, the unit values and the total average value of the spare parts kept in stock are reported in Table 4.3. It was decided to allocate to category A the products corresponding approximately to the first 80% of the total value of the stock, to category B the items associated with the following 15%, and to category C the remaining commodities (see Table 4.4). It is worth noting that category A contains about 30% of the goods, while each of the categories B and C accounts for about 35% of the inventory. The cumulated percentage of the total value as a function of the cumulated percentage of the number of items (Pareto curve) is reported in Figure 4.13.

4.10 Multiple Stocking Point Models

Good inventory policies for multiple interdependent stocking points can be very difficult to devise. In this section a very simple model is described and analysed. In a *decentralized* logistics system, a market is divided into n identical sales districts, each of which is allocated to a warehouse, while in a *centralized* system every customer is serviced by a unique facility. Under the EOQ hypotheses, the average inventory levels of the two systems are linked by the following *square-root law*.

Property. If the EOQ hypotheses hold, and each warehouse in the decentralized system services the same demand, then the total average inventory level $\bar{I}^{(n)}$ in the

Table 4.3 Spare parts stocked by WTC.

Product code	Average stock	Average unit value (in euros)	Total average value (in euros)
AX24	137	50	6 850
BR24	70	2 000	140 000
BW02	195	250	48 750
CQ23	6	6 000	36 000
CR01	16	500	8 000
FE94	31	100	3 100
LQ01	70	2 500	175 000
MQ12	18	200	3 600
MW20	75	500	37 500
NL01	15	1 000	15 000
PE39	16	3 000	48 000
RP10	20	2 200	44 000
SP00	13	250	3 250
TA12	100	2 500	250 000
TQ23	10	5 000	50 000
WQ12	30	12 000	360 000
WZ34	30	15	450
ZA98	70	250	17 500

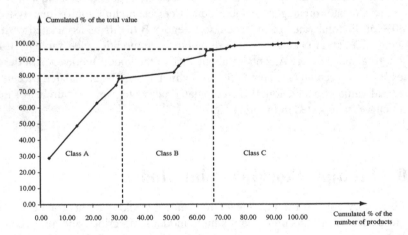

Figure 4.13 Pareto curve in the WTC problem.

decentralized system is

$$\bar{I}^{(n)} = \sqrt{n}\, \bar{I}^{(1)},$$

where $\bar{I}^{(1)}$ is the average inventory level in the centralized system.

Table 4.4 ABC classification of the spare parts in the WTC problem.

Product code	Fraction of total inventory	Cumulated % of the number of products	Total average value (in euros)	Total cumulated value (in euros)	Cumulated % of the total value	Class
WQ12	3.3	3.30	360 000	360 000	28.87	
TA12	10.8	14.10	250 000	610 000	48.92	
LQ01	7.6	21.70	175 000	785 000	62.95	A
BR24	7.6	29.30	140 000	925 000	74.18	
TQ23	1.1	30.40	50 000	975 000	78.19	
BW02	21.1	51.50	48 750	1 023 750	82.10	
PE39	1.8	53.30	48 000	1 071 750	85.95	
RP10	2.1	55.40	44 000	1 115 750	89.47	B
MW20	8.2	63.60	37 500	1 153 250	92.48	
CQ23	0.6	64.20	36 000	1 189 250	95.37	
ZA98	7.6	71.80	17 500	1 206 750	96.77	
NL01	1.6	73.40	15 000	1 221 750	97.98	
CR01	1.8	75.20	8 000	1 229 750	98.62	
AX24	14.8	90.00	6 850	1 236 600	99.17	C
MQ12	2.0	92.00	3 600	1 240 200	99.45	
SP00	1.4	93.40	3 250	1 243 450	99.72	
FE94	3.3	96.70	3 100	1 246 550	99.96	
WZ34	3.3	100.00	450	1 2470,00	100.00	

Proof. In the EOQ model the average inventory level is half the order size. Therefore, $\bar{I}^{(1)}$ is equal to $\frac{1}{2}\sqrt{2kd/h}$, where d is the demand of the whole market. In a decentralized system, $\bar{I}^{(n)}$ is the sum of the average inventory levels of the facilities, each of which services $1/n$ of demand d. Hence, $\bar{I}^{(n)} = \frac{1}{2}n\sqrt{2k(d/n)/h}$. $\qquad\Box$

Kurgantora distributes tyres in Russia and Kazakhstan. The distribution network currently includes 12 warehouses, each of which serves approximately the same demand. In an attempt to reduce the total inventory level by 30%, the company has decided to close some warehouses and allocate their demand to the remaining facilities. Applying the square-root law, we see that the number of stocking points should be reduced to 6, since

$$\bar{I}^{(12)} = \sqrt{12}\,\bar{I}^{(1)},$$

$$\bar{I}^{(n')} = \sqrt{n'}\,\bar{I}^{(1)},$$

$$\bar{I}^{(n')}/\bar{I}^{(12)} = \sqrt{n'}/\sqrt{12} = 0.7,$$

$$n' = 5.88.$$

4.11 Slow-Moving Item Models

As shown in the previous sections, a major issue for fast-moving product inventory management is determining how often reorders should take place. On the other hand, if demand is very low (e.g. a few units in 10–20 years), as in the case of spare parts of a complex machinery (*slow-moving products*), the main issue is determining the number of items to be purchased at the beginning of the machinery's life cycle.

In this section we examine a model in which item purchase cost and shortage penalties are taken into account while holding cost and salvage value (i.e. the value of unused spare parts at the end of the machinery lifetime) are negligible. Let c and u be the purchase cost of an item at the beginning of the planning horizon and during the planning horizon, respectively ($c < u$). If n units of product are purchased at the beginning of the planning period, and m units are demanded in the planning period, the total cost is

$$C(n, m) = cn, \quad \text{if } n \geqslant m;$$
$$C(n, m) = cn + u(m - n), \quad \text{if } n < m.$$

Let $P(m)$ be the probability that m items are demanded. Then, the expected cost $\bar{C}(n)$ in case n items are purchased is

$$\bar{C}(n) = \sum_{m=0}^{\infty} C(n, m) P(m) = cn + u \sum_{m=n+1}^{\infty} (m - n) P(m).$$

Hence,

$$\bar{C}(n - 1) = \bar{C}(n) - c + u[1 - F(n - 1)], \tag{4.42}$$
$$\bar{C}(n + 1) = \bar{C}(n) + c - u[1 - F(n)], \tag{4.43}$$

where $F(n)$ is the probability that n units (or less) are demanded. The minimum expected cost is achieved if n^* items are purchased at the beginning of the planning period:

$$\bar{C}(n^* - 1) \geqslant \bar{C}(n^*), \tag{4.44}$$
$$\bar{C}(n^* + 1) \geqslant \bar{C}(n^*). \tag{4.45}$$

Finally, combining Equation (4.44) and Equation (4.42), the following relation is obtained:

$$F(n^* - 1) \leqslant \frac{u - c}{u}.$$

Similarly, combining Equation (4.45) and Equation (4.43) for $n = n^*$ gives

$$F(n^*) \geqslant \frac{u - c}{u}.$$

Consequently, the optimal number of items n^* to be purchased is such that

$$F(n^* - 1) \leqslant \frac{u - c}{u} \leqslant F(n^*). \tag{4.46}$$

Hydro Ltd uses five LIF03 generators in its hydroelectric power plants located in Nigeria. Each piece of machinery has an average life of 20 years, during which the expected number of engine failures is equal to 1.4. The cost of a spare part, purchased when a generator is manufactured, is $60 000 while producing an additional unit costs around $300 000. The failure process is modelled as a Poisson probability distribution with expected value λ,

$$\lambda = 5 \times 1.4 = 7 \text{ faults per life cycle.}$$

Therefore, the probability $P(n)$ that the number of demanded spare parts equals n is given by

$$P(n) = \frac{e^{-\lambda}\lambda^n}{n!}, \quad n = 0, 1, \ldots,$$

while the cumulative probability $F(n)$ is

$$F(0) = P(0),$$

$$F(n) = \sum_{k=0}^{n} e^{-\lambda}\lambda^k/k! = F(n-1) + P(n), \quad n = 1, 2, \ldots.$$

The values of $P(n)$ and $F(n)$ for $n = 0, \ldots, 10$ are reported in Table 4.5.
Since

$$\frac{u - c}{c} = \frac{300\,000 - 60\,000}{300\,000} = 0.8,$$

on the basis of Equation (4.46), $n^* = 9$ spare parts should be purchased and stocked.

4.12 Policy Robustness

The inventory policies illustrated in the previous sections often have to be slightly modified in order to be used in practice. Fractional order sizes and shipment frequencies have to be suitably rounded up or down (e.g. $q^* = 14.43$ pallets should become 14 or 15 pallets). Fortunately, the total cost is not very sensitive to variations of the order size around the optimal value. For the EOQ model the following property holds.

Property. In the EOQ model, errors in excess of 100% on the optimal order size cause a maximum increase of the total cost equal to 25%.

Proof. Recall that the average cost in the EOQ model is given by

$$\mu(q) - cd = kd/q + hq/2.$$

If $q = q^* = \sqrt{2kd/h}$ (see Equation (4.13)), then

$$\mu(q^*) - cd = \sqrt{2kdh}.$$

Table 4.5 Probability distribution of spare part demand in the Hydro Ltd problem.

n	$P(n)$	$F(n)$
0	0.0009	0.0009
1	0.0064	0.0073
2	0.0223	0.0296
3	0.0521	0.0818
4	0.0912	0.1730
5	0.1277	0.3007
6	0.1490	0.4497
7	0.1490	0.5987
8	0.1304	0.7291
9	0.1014	0.8305
10	0.0710	0.9015

If $q = 2q^*$, then

$$\mu(2q^*) - cd = \tfrac{5}{4}\sqrt{2kdh}.$$

Therefore,

$$\frac{\mu(2q^*) - cd}{\mu(q*) - cd} = 1.25$$

and

$$\mu(q) - cd \leqslant 1.25(\mu(q^*) - cd), \quad q \in [q^*, 2q^*].$$

Similarly, it can be shown that rounding the reorder size to the closest power of 2 (*power-of-two policy*) induces a maximum cost increase of about 6%. ☐

In the Al-Bufeira Motors problem (see Section 4.4.2), the total cost associated with five shipments per year (i.e. $q = 44$ units),

$$\mu(q^*) = \frac{800 \times 220}{44} + \frac{216 \times 44}{2} = 8752 \text{ dollars per year,}$$

is higher than the optimal solution ($\mu(q^*) = 8719.63$ dollars per year) only by 0.37%.

4.13   Questions and Problems

4.1 In most industrialized countries the average ITR is around 20 for dairy products and around 5 for household electrical appliances. Discuss these figures.

4.2 Modify the EOQ formula for the case where the stocking point has a finite capacity Q.

4.3 Modify the EOQ formula for the case where the holding cost is a concave function of the number of items kept in inventory.

4.4 Devise an optimal inventory policy for the EOQ model with a finite time horizon T_H.

4.5 Modify the EOQ formula for the case where the order size q is delivered by a number of vehicles of capacity q_v each having a fixed cost k_v.

4.6 Draw the auxiliary graph used for solving the Sao Vincente Chemical problem as a shortest-path problem.

4.7 Modify the Wagner–Within model for the case where the stocking point is capacitated. Does the ZIO property still hold?

4.8 What is the optimal order quantity in the Newsboy Problem if the stocking point is capacitated?

4.9 If type A products are overstocked, the total cost increases dramatically, while if type C products are overstocked, the total cost does increase too much. Calculate the cost increase whenever the inventory level of A products is increased by 20%. Repeat the calculation for C products.

4.10 Show that the power-of-two policy induces a maximum cost increase of about 6%.

4.14 Annotated Bibliography

An in-depth treatment of inventory management can be found in:

1. Zipkin PH 2000 *Foundations of Inventory Management*. McGraw-Hill, New York.

A simplified approach is available in:

2. Lewis CD 1998 *Demand Forecasting and Inventory Control: A Computer Aided Learning Approach*. Wiley, New York.

An introduction to the use of simulation methods with some applications on the inventory management is:

3. Law L and Kelton WD 2000 *Simulation Modelling and Analysis*, 3rd edn. McGraw-Hill, New York.

5

Designing and Operating a Warehouse

5.1 Introduction

Warehouses are facilities where inventories are sheltered. They can be broadly classified into *production warehouses* and DCs. This chapter deals with warehouse design and operation, with an emphasis on DCs. In the following, a *product* is defined as a type of good, e.g. wine bottles of a specific brand. The individual units are called *items* (or *stock keeping units* (SKUs)). A *customer order* is made up of one or more items of one or more products.

Flow of items through the warehouse. Warehouses are often used not only to provide inventories a shelter, but also to sort or consolidate goods. In a typical DC, the products arriving by truck, rail, or internal transport are unloaded, checked and stocked. After a certain time, items are retrieved from their storage locations and transported to an order assembly area. In the simplest case (which occurs frequently in CDCs (see Chapter 1)), the main activity is the storage of the goods. Here, the merchandise is often received, stored and shipped in full pallets (all one product) and, as a result, material handling is relatively simple (see Figure 5.1). In the most complex case (which occurs frequently in RDCs (see Chapter 1)) large lots of products are received and shipments, containing small quantities of several items, have to be formed and dispatched to customers. Consequently, order picking is quite complex, and product sorting and consolidation play a mayor role in order assembly (see Figure 5.2).

Ownership of the warehouses. With respect to ownership, there are three main typologies of warehouses. *Company-owned warehouses* require a capital investment in the storage space and in the material handling equipment. They usually represent the least-expensive solution in the long run in the case of a substantial and constant demand. Moreover, they are preferable when a higher degree of control is required to ensure a high level of service, or when specialized personnel and equipment are

Introduction to Logistics Systems Planning and Control   G. Ghiani, G. Laporte and R. Musmanno
© 2004 John Wiley & Sons, Ltd   ISBN: 0-470-84916-9 (HB) 0-470-84917-7 (PB)

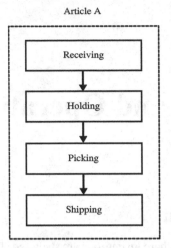

Figure 5.1   The flow of items through the warehouse; the goods are received and shipped in full pallets or in full cartons (all one product).

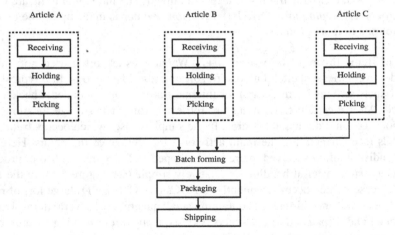

Figure 5.2   The flow of items through the warehouse; the goods are received in full pallets or in full cartons (all one product) and shipped in less than full pallets or cartons.

needed. Finally, they can be employed as a depot for the company's vehicles or as a base for a sales office. *Public warehouses* are operated by firms providing services to other companies on a short-term basis. As a rule, public warehouses have standardized equipment capable of handling and storing specific types of merchandise (e.g. bulk materials, temperature-controlled goods, etc.). Here, all warehousing costs are variable, in direct proportion to the storage space and the services required. As a result, it is easy and inexpensive to change warehouse locations as demand varies. For these reasons, public warehouses can suitably accommodate seasonal inventories.

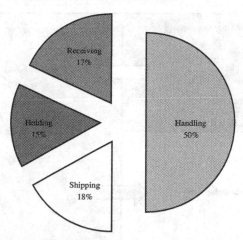

Figure 5.3   Common warehouse costs.

Finally, *leased warehouse space* is an intermediate choice between short-term space rental and the long-term commitment of a company-owned warehouse.

Warehouse costs.   The total annual cost associated with the operation of a warehouse is the result of four main activities: receiving the products, holding inventories in storage locations, retrieving items from the storage locations, assembling customer orders and shipping. These costs depend mainly on the storage medium, the storage/retrieval transport technology and its policies. As a rule, receiving the incoming goods and, even more so, forming the outgoing lots, are operations that are difficult to automate and often turn out to be labour-intensive tasks. Holding inventories depends mostly on the storage medium, as explained in the following. Finally, picking costs depend on the storage/retrieval transport system which can range from a fully manual system (where goods are moved by human pickers travelling on foot or by motorized trolleys) to fully automated systems (where goods are moved by devices under the control of a centralized computer). Common warehouse costs are reported in Figure 5.3.

5.1.1   Internal warehouse structure and operations

The structure of a warehouse and its operations are related to a number of issues:

- the physical characteristics of the products (on which depends whether the products have to be stored at room temperature, in a refrigerated or ventilated place, in a tank, etc.);

- the number of products (which can vary between few units to tens of thousands);

- the volumes handled in and out of the warehouse (which can range between a few items per month to hundreds of pallets per day).

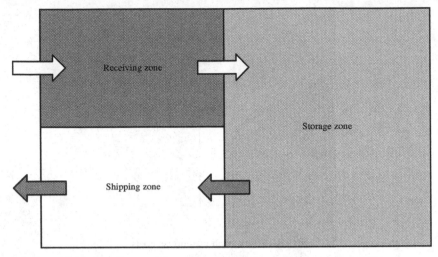

Figure 5.4 Warehouse with a single receiving zone and a single shipping zone.

Typically, in each DC there are (see Figure 5.4)

- one or more *receiving zones* (each having one or more rail or truck *docks*), where incoming goods are unloaded and checked;

- a *storage zone*, where SKUs are stored;

- one or more *shipping zones* (each having one or more rail or truck *docks*), where customer orders are assembled and outgoing vehicles are loaded.

The storage zone is sometimes divided into a large *reserve zone* where products are stored in the most economical way (e.g. as a stack of pallets), and into a small *forward zone*, where goods are stored in smaller amounts for easy retrieval by order pickers (see Figure 5.5). The transfer of SKUs from the reserve zone to the forward zone is referred to as a *replenishment*. If the reserve/forward storage is well-designed, the reduction in picking time is greater than replenishment time.

5.1.2   Storage media

The choice of a storage medium is strongly affected by the physical characteristics of the goods in stock and by the average number of items of each product in a customer order. Briefly, when storing solid goods three main alternatives are available: stacks, racks and drawers. In the first case, goods are stored as cartons or as pallets, and aisles are typically 3.5–4 m wide (see Figure 5.6). Stacks do not require any capital investment and are suitable for storing low-demand goods, especially in reserve zones. In the second case, goods are stored as boxes or pallets on metallic shelves separated by aisles. Here quick picking of single load units is possible. When SKUs are moved by forklifts, the racks (see Figure 5.7) are usually 5–6 m tall and aisles are around

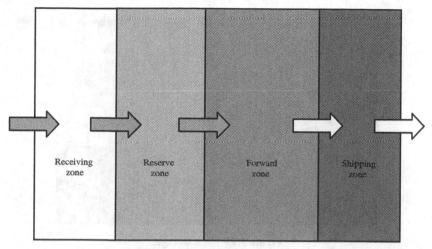

Figure 5.5   Warehouse with reserve and forward storage zones.

Figure 5.6   Block stacking system.

3.5 m wide. Instead, as explained in the following, in *automated storage and retrieval systems* (AS/RSs), racks are typically 10–12 m tall and aisles are usually 1.5 m wide (see Figure 5.8). Finally, in the third case, items are generally of small size (e.g. metallic small parts), and are kept in fixed or rotating drawers.

5.1.3   Storage/retrieval transport mechanisms and policies

A common way of classifying warehouses is the method by which items are retrieved from storage.

Figure 5.7 Rack storage.

Figure 5.8 An AS/RS.

Picker-to-product versus product-to-picker systems. The picking operations can be made

- by a team of human order pickers, travelling to storage locations (*picker-to-product* system);

- by an automated device, delivering items to stationery order pickers (*order-to-picker* system).

Clearly, mixed solutions are possible. For instance, in *picker-to-belt* systems, the items are retrieved by a team of human order pickers and then transported to the order assemblers by a belt conveyor. Picker-to-product systems can be further classified according to the mode of travel inside the warehouse. In *person-aboard AS/RS*, pickers

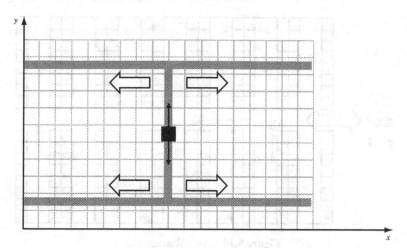

Figure 5.9   Item retrieval by trolley.

are delivered to storage locations by automated devices which are usually restricted to a single aisle. In *walk/ride and pick systems* (W/RPSs), pickers travel on foot or by motorized trolleys and may visit multiple aisles (see Figure 5.9).

The most popular order-to-picker systems are the AS/RSs. An AS/RS consists of a series of storage aisles, each of which is served by a single storage and retrieval (S/R) machine or crane. Each aisle is supported by a *pick-up and delivery station* customarily located at the end of the aisle and accessed by both the S/R machine and the external handling system. Therefore, assuming that the speeds v_x and v_y along the axes x and y (see Figure 5.10) are constant, travel times t satisfy the Chebychev metric,

$$t = \max\left\{ \frac{\Delta x}{v_x}, \frac{\Delta y}{v_y} \right\},$$

where Δx and Δy are the distances travelled along the x- and y-axes.

AS/RSs were introduced in the 1950s to eliminate the walking that accounted for nearly 70% of manual retrieval time. They are often used along with high racks and narrow aisles (see Figure 5.11). Hence, their advantages include savings in labour costs, improved throughput and high floor utilization.

Unit load retrieval systems.   In some warehouses it is possible to move a single load at a time (*unit load retrieval system*), because of the size of the loads, or of the technological restrictions of the machinery (as in AS/RSs). In AS/RSs, an S/R machine usually operates in two modes:

- *single cycle*: storage and retrieval operations are performed one at a time;

- *dual cycle*: pairs of storage and retrieval operations are made in sequence in an attempt to reduce the overall travel time.

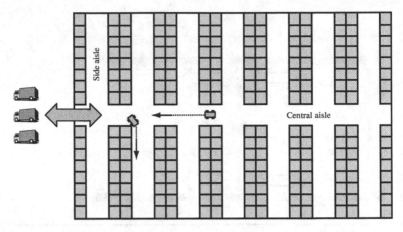

Figure 5.10 An S/R machine.

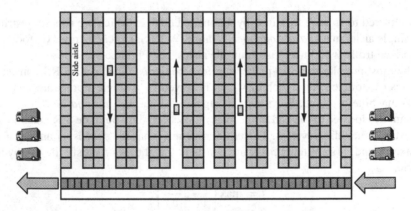

Figure 5.11 Item storage and retrieval by an AS/RS and a belt conveyor.

There exist systems in which it is possible to store or pick up several loads at the same time (*multi-command cycle*).

Strict order picking versus batch picking. In *multiple load retrieval systems*, customer orders can be assigned to pickers in two ways:

- each order is retrieved individually (*strict order picking*);

- orders are combined into batches.

In the latter case, each batch may be retrieved by a single picker (*batch picking*). Otherwise, the warehouse is divided into a given number of zones and each picker is in charge of retrieving items from a specific zone (*zone picking*).

5.1.4 Decisions support methodologies

Warehouses are highly dynamic environments where resources have to be allocated in real-time to satisfy customer orders. Because orders are not fully known in advance, design and operational decisions are affected by uncertainty. To overcome the inherent difficulty of dealing with complex queueing models or stochastic programs, the following approach is often used.

Step 1. A limited number of alternative solutions are selected on the basis of experience or by means of simple relations linking the decision variables and simple statistics of customer orders (the average number of orders per day, the average number of items per order, etc.).

Step 2. Each alternative solution generated in Step 1 is evaluated through a detailed simulation model and the best solution (e.g. with respect to throughput) is selected.

5.2 Warehouse Design

Designing a warehouse amounts to choosing its building shell, as well as its layout and equipment. In particular, the main design decisions are

- determining the length, width and height of the building shell;

- locating and sizing the receiving, shipping and storage zones (e.g. evaluating the number of I/O ports, determining the number, the length and the width of the aisles of the storage zone and the orientation of stacks/racks/drawers);

- selecting the storage medium;

- selecting the storage/retrieval transport mechanism.

The objective pursued is the minimization of the expected annual operating cost for a given throughput, usually subject to an upper bound on capital investment.

In principle, the decision maker may choose from a large number of alternatives. However, in practice, several solutions can be discarded on the basis of a qualitative analysis of the physical characteristics of the products, the number of items in stock and the rate of storage and retrieval requests. In addition, some design decisions are intertwined. For instance, when choosing an AS/RS as a storage/retrieval transport mechanism, rack height can be as high as 12 m, but when traditional forklifts are used racks must be much lower. As a result, each design problem must be analysed as a unique situation.

Following the general framework introduced at the end of Section 5.1, we will make in the remainder of this section a number of qualitative remarks and we will illustrate two simple analytical procedures.

5.2.1 Selecting the storage medium and the storage/retrieval transport mechanism

The choice of storage and retrieval systems is influenced by the physical characteristics of the goods, their packaging at the arrival and the composition of the outgoing lots. For example, in a single storage zone warehouse, palletized goods are usually stocked on racks if their demand is high enough, otherwise, stacks are used. Automated systems are feasible if the goods can be automatically identified through bar codes or other techniques. They have low space and labour costs, but require a large capital investment. Hence, they are economically convenient provided that the volume of goods is large enough.

5.2.2 Sizing the receiving and shipment subsystems

The receiving zone is usually wider than the shipping area. This is because the incoming vehicles are not under the control of the warehouse manager, while the formation of the outgoing shipments can be planned in order to avoid congesting the output stations.

Determining the number of truck docks

Goods are usually received and shipped by rail or by truck. In the latter case, the number of docks n_D can be estimated through the following formula,

$$n_D = \left\lceil \frac{dt}{qT} \right\rceil,$$

where d is the daily demand from all orders, t is the average time required to load/unload a truck, q is the truck capacity, and T is the daily time available to load/unload trucks.

Sintang is a third-party Malaysian firm specialized in manufacturing electronic devices. A new warehouse has been recently opened in Kuching. It is used for storing digital satellite receivers, whose average daily demand is $d = 27\,000$ units. Outgoing shipments are performed by trucks, with a capacity equal to 850 boxes. Since the average time to load a truck is $t = 280$ min and 15 working hours are available every day, the warehouse has been designed with the following number n_D of docks:

$$n_D = \left\lceil \frac{27\,000 \times 280}{850 \times 900} \right\rceil = 10.$$

5.2.3 Sizing the storage subsystems

The area of the storage zone must be large enough to accommodate goods in peak periods. On the other hand, if the storage zone exceeds the real needs of the firm, storage and retrieval times become uselessly high. This could decrease throughput or increase material handling costs.

Determining the capacity of a storage area

The size of a storage area depends on the storage policy. In a *dedicated storage policy*, each product is assigned a pre-established set of positions. This approach is easy to implement but causes an underutilization of the storing space. In fact, the space required is equal to the sum of the maximum inventory of each product in time. Let n be the number of products and let $I_j(t)$, $j = 1, \ldots, n$, be the inventory level of item j at time t. The number of required storage locations m_d in a dedicated storage policy is

$$m_d = \sum_{j=1}^{n} \max_t I_j(t). \tag{5.1}$$

In a *random storage policy*, item allocation is decided dynamically on the basis of the current warehouse occupation and on future arrival and request forecast. Therefore, the positions assigned to a product are variable in time. In this case the number of storage locations m_r is

$$m_r = \max_t \sum_{j=1}^{n} I_j(t) \leqslant m_d. \tag{5.2}$$

The random storage policy allows a higher utilization of the storage space, but requires that each item be automatically identified through a bar code (or a similar technique) and a database of the current position of all items kept at stock is updated at every storage and every retrieval.

In a *class-based storage policy*, the goods are divided into a number of categories according to their demand, and each category is associated with a set of zones where the goods are stored according to a random storage policy. The class-based storage policy reduces to the dedicated storage policy if the number of categories is equal to the number of items, and to the random storage policy if there is a single category.

Potan Up bottles two types of mineral water. In the warehouse located in Hangzhou (China), inventories are managed according to a reorder level policy (see Chapter 4). The sizes of the lots and of the safety stocks are reported in Table 5.1. Inventory levels as a function of time are illustrated in Figures 5.12 and 5.13. The company is currently using a dedicated storage policy. Therefore, the number of storage locations is given by Equation (5.1):

$$m_d = 600 + 360 = 960.$$

The firm is now considering the opportunity of using a random storage policy. The number of storage locations required by this policy would be (see Equation (5.2))

$$m_r = 600 + 210 = 810.$$

Table 5.1   Lots and safety stocks (both in pallets) in the Potan Up problem.

Product	Lot	Safety stock
Natural water	500	100
Sparkling water	300	60

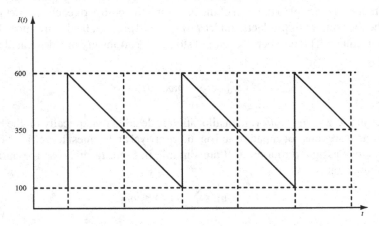

Figure 5.12   Inventory level of natural mineral water in the Potan Up problem.

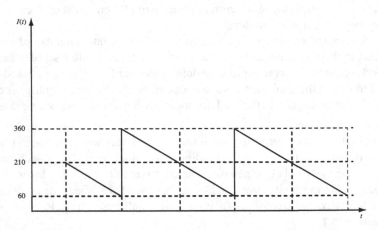

Figure 5.13   Inventory level of sparkling mineral water in the Potan Up problem.

Determining length, width and height of a storage zone

In this section a methodology for determining length, width and height of a storage zone (see Figure 5.14) is described. The same methodology can be easily extended to other types of storage zone. As explained in the introductory section, the maximum

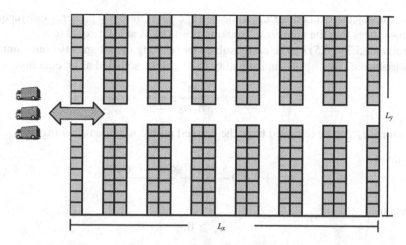

Figure 5.14 A traditional storage zone.

height of the racks/stacks/drawers is determined by the storage technology. Therefore, the sizing decision amounts to calculating the length and the width. Let m be the required number of stocking positions; α_x and α_y the occupation of a unit load (e.g. a pallet or a cartoon) along the directions x and y, respectively; w_x and w_y, the width of the side aisles and of the central aisle, respectively; n_z the number of stocking zones along the z-direction allowed by the storage technology; v the average speed of a picker. The decision variables are n_x, the number of storage locations along the x-direction, and n_y, the number of storage locations along the y-direction.

The extension L_x of the stocking zone along the direction x is given by the following relation,

$$L_x = (\alpha_x + \tfrac{1}{2}w_x)n_x,$$

where, for the sake of simplicity, n_x is assumed to be an even number. Similarly, the extension L_y is

$$L_y = \alpha_y n_y + w_y.$$

Therefore, under the hypothesis that a handling operation consists of storing or the retrieving a single load, and all stocking points have the same probability of being accessed, the average distance covered by a picker is: $2(L_x/2 + L_y/4) = L_x + L_y/2$. Hence, the problem of sizing the storage zone can be formulated as follows.

Minimize

$$(\alpha_x + \tfrac{1}{2}w_x)\frac{n_x}{v} + \frac{\alpha_y n_y + w_y}{2v} \tag{5.3}$$

subject to

$$n_x n_y n_z \geqslant m \tag{5.4}$$

$$n_x, \ n_y \geqslant 0, \ \text{integer}, \tag{5.5}$$

where the objective function (5.3) is the average travel time of a picker, while inequality (5.4) states that the number of stocking positions is at least equal to m.

Problem (5.3)–(5.5) can be easily solved by relaxing the integrality constraints on the variables n_x and n_y. Then, inequality (5.4) will be satisfied as an equality:

$$n_x = \frac{m}{n_y n_z}. \tag{5.6}$$

Therefore, n_x can be removed from the relaxed problem in the following way.

Minimize

$$(\alpha_x + \tfrac{1}{2} w_x)\frac{m}{n_y n_z v} + \frac{\alpha_y n_y + w_y}{2v} \tag{5.7}$$

subject to

$$n_y \geqslant 0.$$

Since the objective function (5.7) is convex, the minimizer n'_y can be found through the following relation:

$$\frac{\mathrm{d}}{\mathrm{d}(n_y)}\left((\alpha_x + \tfrac{1}{2}w_x)\frac{m}{n_y n_z v} + \frac{\alpha_y n_y + w_y}{2v}\right)\Bigg|_{n_y=n'_y} = 0.$$

Hence,

$$n'_y = \sqrt{\frac{2m(\alpha_x + \tfrac{1}{2}w_x)}{\alpha_y n_z}}. \tag{5.8}$$

Finally, replacing n_y in Equation (5.6) by the n'_y value given by Equation 5.8, n'_x is determined:

$$n'_x = \sqrt{\frac{m\alpha_y}{2n_z(\alpha_x + \tfrac{1}{2}w_x)}}. \tag{5.9}$$

Consequently, a feasible solution $(\bar{n}_x, \bar{n}_y)$ is

$$\bar{n}_x = \lceil n'_x \rceil \quad \text{and} \quad \bar{n}_y = \lceil n'_y \rceil.$$

Alternatively, a better solution could be found by setting $\bar{n}_x = \lfloor n'_x \rfloor$ (or $\bar{n}_y = \lfloor n'_y \rfloor$), provided that Equation (5.4) is satisfied.

Wagner Bros is going to build a new warehouse near Sidney (Australia) in order to supply its sales points in New South Wales. On the basis of a preliminary analysis of the problem, it has been decided that the facility will accommodate at least 780 90×90 cm^2 pallets. The goods will be stored onto racks and transported by means of traditional trolleys. Each rack has four shelves, each of which can store a single pallet. Each pallet occupies a 1.05×1.05 m^2 area. Racks are arranged as in Figure 5.14, where side aisles are 3.5 m wide, while the central aisle is 4 m wide. The average

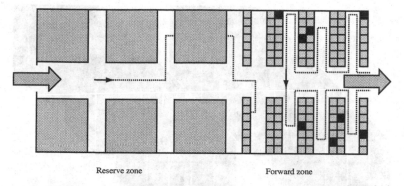

Reserve zone   Forward zone

Figure 5.15 Warehouse with a reserve/forward storage system.

speed of a trolley is 5 km/h. Using Equations (5.8) and (5.9), variables n_x and n_y are determined:

$$n'_x = \sqrt{\frac{780 \times 1.05}{2 \times 4 \times (1.05 + \frac{3.5}{2})}} = 6.05,$$

$$n'_y = \sqrt{\frac{2 \times 780 \times (1.05 + \frac{3.5}{2})}{1.05 \times 4}} = 32.25.$$

Assuming $\bar{n}_x = 6$ and $\bar{n}_y = 33$, the total number of storage locations turns out to be 792, while $L_x = [1.05 + (3.5/2)] \times 6 = 16.8$ m and $L_y = 1.05 \times 33 + 4 = 38.65$ m.

Sizing a forward area

In a reserve/forward storage system (see Figure 5.15), the main decision is to determine how much space must be assigned to each product in the forward area. In principle, once this decision has been made, the problem of determining the length, width and height of the pick-up zone should be solved. However, since each picking route in the forward area usually collects small quantities of several items, at every trip a large portion of the total length of the aisles is usually covered (see Figure 5.15). Hence, the dimensions of the pick-up zone are not critical and can be selected quite arbitrarily.

If the number of items stored in the forward area increases, replenishments are less frequent. However, at the same time the extension of the forward area increases and, consequently, the average picking time also goes up.

Let (see Figure 5.16) n be the number of products, o the average number of orders per time period; d the average number of orders in a batch; o_j, $j = 1, \ldots, n$, the average number of orders containing product j; u_j, $j = 1, \ldots, n$, the average number of items of product j in an order; v the average speed of a picker in the forward area; h the cost of a picker per time period; k the area and equipment cost per unit of lane

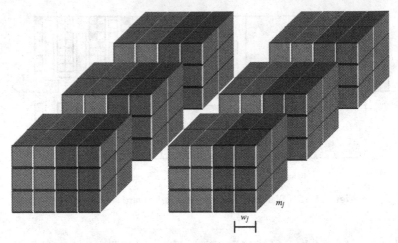

Figure 5.16 Storage locations assigned to an item j, $j = 1, \ldots, n$.

length and per time period; f_j, $j = 1, \ldots, n$, the fixed cost of a replenishment of product j; g_j, $j = 1, \ldots, n$, the variable cost for replenishing a unit of product j; w_j, $j = 1, \ldots, n$, the length of a portion of aisle occupied by an item of product j; m_j, $j = 1, \ldots, n$, the number of items of product j that can be stored in an aisle position. The decision variables are the number of aisle positions s_j, $j = 1, \ldots, n$, assigned to each product j.

The contribution of product j, $j = 1, \ldots, n$, to the cost of picking items from the reserve area is

$$c_{1j}(s_j) = h\frac{(o/d)w_j s_j}{v}, \tag{5.10}$$

where o/d represents the average number of batches picked up per time period (i.e. the average number of times a picker passes in front of a position per time period); $w_j s_j$ is the total length of the portion of aisle assigned to the product j, $[(o/d)w_j s_j]/v$ represents the average time spent during a time period by the picker because of product j.

The contribution of product j, $j = 1, \ldots, n$, to the average cost per time period of replenishing the forward area is

$$c_{2j}(s_j) = f_j\frac{u_j o_j}{m_j s_j} + g_j u_j o_j,$$

where $u_j o_j$ represents the average demand per time period of product j, while $m_j s_j$ is the number of storage locations assigned to product j, and $(u_j o_j)/(m_j s_j)$ is the average number of resupplies of product j per time period.

The portion of the space and equipment costs due to product j, $j = 1, \ldots, n$, is

$$c_{3j}(s_j) = kw_j s_j.$$

Moreover, let $s_j^{\min}$ and $s_j^{\max}$, $j = 1, \ldots, n$, be lower and upper bounds on the number of positions of product j, respectively. The problem of sizing the forward area can be modelled as follows.

Minimize

$$\sum_{j=1}^{n} [c_{1j}(s_j) + c_{2j}(s_j) + c_{3j}(s_j)] \tag{5.11}$$

subject to

$$s_j^{\min} \leqslant s_j \leqslant s_j^{\max}, \quad j = 1, \ldots, n, \tag{5.12}$$

$$s_j \geqslant 0, \text{ integer}, \quad j = 1, \ldots, n, \tag{5.13}$$

where the objective function (5.11) is the sum of the costs due to the various products, while constraints (5.12) impose lower and upper bounds on the number of aisle positions assigned to each item j, $j = 1, \ldots, n$.

Problem (5.11)–(5.13) can be decomposed into n subproblems, one for each product j, $j = 1, \ldots, n$, and solved by exploiting the convexity of the objective function.

Step 1. Determine the value s'_j, $j = 1, \ldots, n$, that minimizes the total cost $c_j(s_j) = c_{1j}(s_j) + c_{2j}(s_j) + c_{3j}(s_j)$ due to product j:

$$\left. \frac{dc_j(s_j)}{ds_j} \right|_{s_j = s'_j} = 0, \quad j = 1, \ldots, n,$$

$$s'_j = \sqrt{\frac{f_j u_j o_j}{m_j [how_j / dv + k w_j]}}, \quad j = 1, \ldots, n.$$

Set $\bar{s}_j = \lfloor s'_j \rfloor$, if $c_j(\lfloor s'_j \rfloor) < c_j(\lceil s'_j \rceil)$, otherwise $\bar{s} = \lceil s'_j \rceil$, $j = 1, \ldots, n$.

Step 2. Compute the optimal solution s_j^*, $j = 1, \ldots, n$, as follows:

$$s_j^* = \begin{cases} s_j^{\min}, & \text{if } \bar{s}_j < s_j^{\min}, \\ \bar{s}_j, & \text{if } s_j^{\min} \leqslant \bar{s}_j \leqslant s_j^{\max}, \\ s_j^{\max}, & \text{if } \bar{s}_j > s_j^{\max}, \end{cases} \quad j = 1, \ldots, n.$$

The total length of the aisles w^{tot} can then be obtained by the following relation:

$$w^{\text{tot}} = \sum_{j=1}^{n} w_j s_j^*. \tag{5.14}$$

An estimate of the number of pickers can be computed by dividing the total workload per time period $[(o/d)w^{\text{tot}}]/v$ by the duration of a work shift. This approach underestimates the number of pickers since it assumes that the orders are uniformly distributed in time in such a way that pickers are never idle. A more realistic estimate can be heuristically obtained by reducing v by an appropriate 'utilization coefficient' empirically estimated, or by using an appropriate simulation model.

Table 5.2 Characteristics of the products at stock in the Wellen warehouse.

Item (j)	o_j	u_j	f_j (euros per supply)	g_j (euros per unit of items)	w_j (m)	m_j
1	40	4	5	0.2	0.75	12
2	60	2	5	0.2	1.00	8
3	35	5	5	0.2	0.75	12
4	45	3	5	0.2	1.00	8
5	95	2	5	0.2	1.00	8
6	65	4	5	0.2	0.75	12
7	45	2	5	0.2	0.75	12
8	50	4	5	0.2	1.00	8
9	65	3	5	0.2	1.00	8
10	45	5	5	0.2	0.75	12

Wellen is a Belgian firm manufacturing and distributing mechanical parts for numerical control machines. Its warehouse located in Herstal consists of a wide reserve zone (where the goods are stocked as stacks), and of a forward zone. At present 10 products are stored (see Table 5.2). Then, $o = 400$ orders per day, $d = 3$ orders per lot, $v = 12\,000$ m per day, $h = €75$ per day, while the area and equipment cost per unit of aisle length k is assumed to be negligible. The minimum and the maximum number of positions for the various products are reported in Table 5.3. The number of positions s_j^*, $j = 1, \ldots, 10$, to be assigned to the different products in the forward zone can be obtained through the two-stage procedure previously described. The results are reported in Table 5.4. Consequently, the total length w^{tot} of the aisles of the forward zone is equal to 98.5 m (see Equation (5.14)), while the workload is around 1.09 working days, so that at least two pickers are required.

5.3 Tactical Decisions

The main tactical decision consists of allocating products to space. In this section, the problem is modelled as a structured LP problem, namely, the classical *transportation problem*.

5.3.1 Product allocation

The allocation of products within a warehouse is based on the principle that fast-moving products must be placed closer to the I/O ports in order to minimize the overall handling time. In the sequel, the case of a dedicated storage policy is examined. The

Table 5.3 Minimum and maximum number of aisle positions in the Wellen warehouse.

Item (j)	$s_j^{\min}$	$s_j^{\max}$
1	9	14
2	5	12
3	13	15
4	8	10
5	7	20
6	9	11
7	10	13
8	8	15
9	11	16
10	14	19

Table 5.4 Number of aisle positions in the forward zone of the Wellen warehouse.

Item (j)	s_j'	$\lfloor s_j' \rfloor$	$c_j(\lfloor s_j' \rfloor)$	$\lceil s_j' \rceil$	$c_j(\lceil s_j' \rceil)$	$\bar{s}_j$	s_j^*
1	10.33	10	44.92	11	44.94	10	10
2	9.49	9	39.83	10	39.83	9	9
3	10.80	10	48.54	11	48.50	11	13
4	10.06	10	43.77	11	43.84	10	10
5	11.94	11	57.96	12	57.90	12	12
6	13.17	13	68.46	14	68.49	13	11
7	7.75	7	27.73	8	27.69	8	10
8	12.25	12	60.42	13	60.45	12	12
9	12.09	12	59.16	13	59.21	12	12
10	12.25	12	60.31	13	60.34	12	14

allocation problem amounts to assigning each of the m_d storage locations available to a product. Let n be the number of products; m_j, $j = 1, \ldots, n$, the number of storage locations required for product j (in a dedicated storage policy, relation $\sum_{j=1}^{n} m_j \leq m_d$ holds); R the number of I/O ports of the warehouse; p_{jr}, $j = 1, \ldots, n$, $r = 1, \ldots, R$, the average number of handling operations on product j through I/O port r per time period; t_{rk}, $r = 1, \ldots, R$, $k = 1, \ldots, m_d$, the travel time from I/O port r and storage location k.

Under the hypothesis that all storage locations have an identical utilization rate, it is possible to compute the cost c_{jk}, $j = 1, \ldots, n$, $k = 1, \ldots, m_d$, of assigning storage location k to product j,

$$c_{jk} = \sum_{r=1}^{R} \frac{p_{jr}}{m_j} t_{rk}, \qquad (5.15)$$

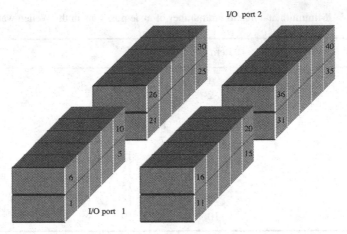

Figure 5.17   Warehouse of the Malabar company.

where p_{jr}/m_j represents the average number of handling operations per time period on product j between I/O port r and anyone of the storage locations assigned to the product. Consequently, $(p_{jr}/m_j)t_{rk}$ is the average travel time due to storage location k if it is assigned to product j.

Let x_{jk}, $j = 1, \ldots, n$, $k = 1, \ldots, m_d$, be a binary decision variable, equal to 1 if storage location k is assigned to product j, 0 otherwise. The problem of seeking the optimal product allocation to the storage locations can then be modelled as follows.

Minimize

$$\sum_{j=1}^{n}\sum_{k=1}^{m_d} c_{jk}x_{jk} \tag{5.16}$$

subject to

$$\sum_{k=1}^{m_d} x_{jk} = m_j, \quad j = 1, \ldots, n, \tag{5.17}$$

$$\sum_{j=1}^{n} x_{jk} \leqslant 1, \quad k = 1, \ldots, m_d, \tag{5.18}$$

$$x_{jk} \in \{0, 1\}, \quad j = 1, \ldots, n, \ k = 1, \ldots, m_d, \tag{5.19}$$

where constraints (5.17) state that all the items at stock must be allocated, while constraints (5.18) impose that each storage location k, $k = 1, \ldots, m_d$, can be assigned to at most one product.

It is worth noting that because of the particular structure of constraints (5.17) and (5.18), relations (5.19) can be replaced with the simpler nonnegativity conditions,

$$x_{jk} \geqslant 0, \quad j = 1, \ldots, n, \ k = 1, \ldots, m_d, \tag{5.20}$$

Table 5.5 Features of the products of the Malabar company.

Item	Storage locations	Number of storages and retrievals per day in the Malabar warehouse	
		I/O port 1	I/O port 2
1	12	25	18
2	6	16	26
3	8	14	30
4	4	24	22
5	8	22	22

Table 5.6 Distance (in metres) between storage locations and I/O port 1 in the Malabar warehouse.

Storage location	Distance	Storage location	Distance	Storage location	Distance	Storage location	Distance
1	2	11	2	21	14	31	14
2	4	12	4	22	16	32	16
3	6	13	6	23	18	33	18
4	8	14	8	24	20	34	20
5	10	15	10	25	22	35	22
6	3	16	3	26	15	36	15
7	5	17	5	27	17	37	17
8	7	18	7	28	19	38	19
9	9	19	9	29	21	39	21
10	11	20	11	30	23	40	23

Table 5.7 Distance (in metres) between storage locations and I/O port 2 in the Malabar warehouse.

Storage location	Distance	Storage location	Distance	Storage location	Distance	Storage location	Distance
1	22	11	22	21	10	31	10
2	20	12	20	22	8	32	8
3	18	13	18	23	6	33	6
4	16	14	16	24	4	34	4
5	14	15	14	25	2	35	2
6	23	16	23	26	11	36	11
7	21	17	21	27	9	37	9
8	19	18	19	28	7	38	7
9	17	19	17	29	5	39	5
10	15	20	15	30	3	40	3

Table 5.8 Cost coefficients c_{jk}, $j = 1, \ldots 5$, $k = 1, \ldots, 20$, in the Malabar problem.

	Assignment cost				
Storage location k	Product $j = 1$	Product $j = 2$	Product $j = 3$	Product $j = 4$	Product $j = 5$
1	37.17	100.67	86.00	133.00	66.00
2	38.33	97.33	82.00	134.00	66.00
3	39.50	94.00	78.00	135.00	66.00
4	40.67	90.67	74.00	136.00	66.00
5	41.83	87.33	70.00	137.00	66.00
6	40.75	107.67	91.50	144.50	71.50
7	41.92	104.33	87.50	145.50	71.50
8	43.08	101.00	83.50	146.50	71.50
9	44.25	97.67	79.50	147.50	71.50
10	45.42	94.33	75.50	148.50	71.50
11	37.17	100.67	86.00	133.00	66.00
12	38.33	97.33	82.00	134.00	66.00
13	39.50	94.00	78.00	135.00	66.00
14	40.67	90.67	74.00	136.00	66.00
15	41.83	87.33	70.00	137.00	66.00
16	40.75	107.67	91.50	144.50	71.50
17	41.92	104.33	87.50	145.50	71.50
18	43.08	101.00	83.50	146.50	71.50
19	44.25	97.67	79.50	147.50	71.50
20	45.42	94.33	75.50	148.50	71.50

since it is known *a priori* that there exists an optimal solution of problem (5.16)–(5.18), (5.20) in which the variables take 0/1 values.

Malabar Ltd is an Indian company having a warehouse with two I/O ports and 40 storage locations, arranged in four racks (see Figure 5.17). The characteristics of the products at stock are reported in Table 5.5, while the distances between the two I/O ports and the storage locations are indicated in Tables 5.6 and 5.7. The optimal product allocation can found through model (5.16)–(5.18), (5.20), in which $n = 5$, $m_d = 40$, while m_j, $j = 1, \ldots, 5$ are calculated on the basis of the second column of Table 5.5. Cost coefficients c_{jk}, $j = 1, \ldots, 5$, $k = 1, \ldots, 40$, are indicated in Tables 5.8 and 5.9 and calculated using Equation (5.15), where it is assumed that travel time t_{rk} from I/O port $r = 1, 2$, to storage location k, $k = 1, \ldots, 40$, is directly proportional to the corresponding distance. The optimal solution is reported in Table 5.10. It is worth noting that two storage locations (locations 26 and 27) are not used since the positions available are 40, while $\sum_{j=1}^{5} m_j = 38$.

Table 5.9 Cost coefficients c_{jk}, $j = 1, \ldots 5$, $k = 21, \ldots, 40$, in the Malabar problem.

	Assignment cost				
Storage location k	Product $j = 1$	Product $j = 2$	Product $j = 3$	Product $j = 4$	Product $j = 5$
21	44.17	80.67	62.00	139.00	66.00
22	45.33	77.33	58.00	140.00	66.00
23	46.50	74.00	54.00	141.00	66.00
24	47.67	70.67	50.00	142.00	66.00
25	48.83	67.33	46.00	143.00	66.00
26	47.75	87.67	67.50	150.50	71.50
27	48.92	84.33	63.50	151.50	71.50
28	50.08	81.00	59.50	152.50	71.50
29	51.25	77.67	55.50	153.50	71.50
30	52.42	74.33	51.50	154.50	71.50
31	44.17	80.67	62.00	139.00	66.00
32	45.33	77.33	58.00	140.00	66.00
33	46.50	74.00	54.00	141.00	66.00
34	47.67	70.67	50.00	142.00	66.00
35	48.83	67.33	46.00	143.00	66.00
36	47.75	87.67	67.50	150.50	71.50
37	48.92	84.33	63.50	151.50	71.50
38	50.08	81.00	59.50	152.50	71.50
39	51.25	77.67	55.50	153.50	71.50
40	52.42	74.33	51.50	154.50	71.50

Table 5.10 Optimal allocation of products in the Malabar warehouse.

Storage location	Product	Storage location	Product	Storage location	Product	Storage location	Product
1	1	11	1	21	5	31	5
2	4	12	4	22	2	32	2
3	4	13	4	23	2	33	2
4	5	14	5	24	2	34	2
5	5	15	5	25	3	35	3
6	1	16	1	26	—	36	5
7	1	17	1	27	—	37	5
8	1	18	1	28	3	38	3
9	1	19	1	29	3	39	3
10	1	20	1	30	3	40	3

If the warehouse has a single I/O port ($R = 1$), the problem solving methodology can be simplified. In fact, under this hypothesis, cost coefficients c_{jk}, $j = 1, \ldots, n$, $k = 1, \ldots, m_d$, take the following form,

$$c_{jk} = \frac{p_{j1}}{m_j} t_{1k} = a_j b_k,$$

where $a_j = p_{j1}/m_j$ and $b_k = t_{1k}$ depend only on product j and on storage location k, respectively. Then, the optimal product allocation can be determined by using the following procedure.

Step 1. Construct a vector α of $\sum_{j=1}^{n} m_j$ components, in which there are m_j copies of each a_j, $j = 1, \ldots, n$. Sort the vector α by nonincreasing values of its components. Define $\sigma_\alpha(i)$ in such a way that $\sigma_\alpha(i) = j$ if $\alpha_i = a_j$, $i = 1, \ldots, \sum_{r=1}^{n} m_r$.

Step 2. Let b be the vector of m_d components corresponding to values b_k, $k = 1, \ldots, m_d$. Sort the vector b by nondecreasing values of its components. Let β be the vector of $\sum_{j=1}^{n} m_j$ components, corresponding to the first $\sum_{j=1}^{n} m_j$ components of the sorted vector b. Define $\sigma_\beta(i)$ in such a way that $\sigma_\beta(i) = k$ if $\beta_i = b_k$, $i = 1, \ldots, \sum_{r=1}^{n} m_r$.

Step 3. Determine the optimal solution of problem (5.16)–(5.18), (5.20) as

$$x^*_{\sigma_\alpha(i),\sigma_\beta(i)} = 1, \quad i = 1, \ldots, \sum_{j=1}^{n} m_j$$

and $x^*_{jk} = 0$, for all the remaining components.

This procedure is based on the fact that the minimization of the scalar product of two vectors α and β is achieved by ordering α by nonincreasing values and β by nondecreasing values.

If the warehouse of Malabar company (see the previous problem) has a single I/O port (corresponding to port 1 in Figure 5.17), the coefficients a_j, $j = 1, \ldots, 5$, are those reported in Table 5.11. For the sake of simplicity, travel times are assumed to be equal to distances (see Table 5.6). Values of α_i, $\sigma_\alpha(i)$, β_i, $\sigma_\beta(i)$, $i = 1, \ldots, 38$, are reported in Table 5.12. The optimal solution is reported in Table 5.13. It is worth noting that no item is allocated to the storage locations farthest from the I/O port (locations 30 and 40).

5.4 Operational Decisions

Operational decisions comprise storage and retrieval planning as well as order assembly. Because of the randomness of the orders, these decisions have to be made in real-time. They may include one or more of the following activities depending on which

Table 5.11 Characteristics of products in the Malabar problem.

Product (j)	Storage locations	Number of storages and retrievals per day	a_j
1	12	43	3.58
2	6	42	7.00
3	8	44	5.50
4	4	46	11.50
5	8	44	5.50

Table 5.12 Values of α_i, $\sigma_\alpha(i)$, β_i, $\sigma_\beta(i)$, for $i = 1, \ldots, 38$, in the Malabar problem.

i	α_i	$\sigma_\alpha(i)$	β_i	$\sigma_\beta(i)$	i	α_i	$\sigma_\alpha(i)$	β_i	$\sigma_\beta(i)$
1	11.50	4	2	1	20	5.50	5	11	20
2	11.50	4	2	11	21	5.50	5	14	21
3	11.50	4	3	6	22	5.50	5	14	31
4	11.50	4	3	16	23	5.50	5	15	26
5	7.00	2	4	2	24	5.50	5	15	36
6	7.00	2	4	12	25	5.50	5	16	22
7	7.00	2	5	7	26	5.50	5	16	32
8	7.00	2	5	17	27	3.58	1	17	27
9	7.00	2	6	3	28	3.58	1	17	37
10	7.00	2	6	13	29	3.58	1	18	23
11	5.50	3	7	8	30	3.58	1	18	33
12	5.50	3	7	18	31	3.58	1	19	28
13	5.50	3	8	4	32	3.58	1	19	38
14	5.50	3	8	14	33	3.58	1	20	24
15	5.50	3	9	9	34	3.58	1	20	34
16	5.50	3	9	19	35	3.58	1	21	29
17	5.50	3	10	5	36	3.58	1	21	39
18	5.50	3	10	15	37	3.58	1	22	25
19	5.50	5	11	10	38	3.58	1	22	35

technologies and policies are used: the formation of batches (see Section 5.4.1), picker routing (see Section 5.4.2), S/R machine scheduling (see Problem 5.4) and vehicle loading (see Section 5.4.3).

5.4.1 Batch formation

If batch picking is used, customer orders are combined into batches. Batches can be formed in a number of ways depending on whether the warehouse is zoned or not. In what follows a simple two-stage procedure for the single-zone case is presented.

Table 5.13   Optimal product allocation to storage locations in the Malabar problem.

Storage location	Product	Storage location	Product	Storage location	Product	Storage location	Product
1	4	11	4	21	5	31	5
2	2	12	2	22	5	32	5
3	2	13	2	23	1	33	1
4	3	14	3	24	1	34	1
5	3	15	3	25	1	35	1
6	4	16	4	26	5	36	5
7	2	17	2	27	1	37	1
8	3	18	3	28	1	38	1
9	3	19	3	29	1	39	1
10	5	20	5	30	—	40	—

In the first stage, the optimal batch size d^* is estimated in an attempt to balance the picking and sorting efforts. Then in Step 2 batches are created according to a '*first come first served*' (FCFS) policy by aggregating d^* consecutive orders.

Batch sizing.   Batches are sized in an attempt to minimize the total workload, which is the sum of the total picking and sorting times. In what follows, this approach is illustrated for the configuration in Figure 5.18, in which goods are retrieved by one or more pickers and then transported to the shipping zone by a belt conveyor.

Let o be the average number of orders per time period, u the average number of items in an order, t_1 the time needed to make a path including all storage locations, t_2 the traversal time on foot of the shipping zone, where orders are assembled. The decision variable is the average number of orders in a lot d.

Under the hypothesis that the items are uniformly distributed and that each batch is made up of many items, the time spent for picking a batch is approximately t_1. As o/d is the average number of pickings per time period, the time devoted to picking operations is ot_1/d per time period. On the other hand, the time spent for sorting the items in the shipping zone is $\alpha uot_2 d$, where $\alpha \in (0, 1)$ is a parameter to be defined either empirically or through a simulation model. In order to determine the optimal d value, the following model has to be solved.

Minimize

$$\frac{ot_1}{d} + \alpha uot_2 d \tag{5.21}$$

subject to

$$d \geqslant 0, \text{ integer,} \tag{5.22}$$

where the objective function (5.21) is the workload per time period. The optimal solution of problem (5.21)–(5.22) can be obtained by means of the following procedure.

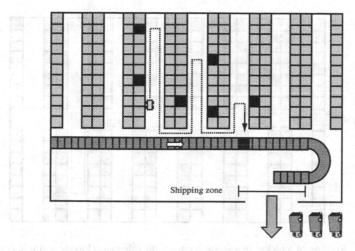

Figure 5.18 Batch picking in a warehouse equipped with a belt conveyor.

Step 1. Determine the minimum point d' of function $c(d) = ot_1/d + \alpha uot_2 d$, by imposing that the first derivative of $c(d)$ becomes zero:

$$d' = \sqrt{\frac{t_1}{\alpha u t_2}}.$$

Step 2. Let $d^* = \lfloor d' \rfloor$ if $c(\lfloor d' \rfloor) < c(\lceil d' \rceil)$, otherwise $d^* = \lceil d' \rceil$.

Hence, d^* increases as the size of the storage zone increases, and decreases as the average number of items in an order increases.

Clavier distributes French ties in Brazil. Its warehouse located in Manaus has a layout similar to the one represented in Figure 5.18, with 15 aisles, each one of which is 25 m long and 3.5 m wide. The area occupied by a pallet is 1.05×1.05 m^2. The vehicles used for picking goods move at about 3.8 km/h, while the time to traverse the shipping zone on foot is about 1.5 min. The average number of orders handled in a day is 300, and the average number of items in an order is 10. The parameter α has been empirically set equal to 0.1. Hence,

$$t_1 = \frac{1.05 + 25 \times 30 + 3.5 \times 15 + 1.05 \times 28}{3800} \times 60 = 13.15 \text{ min.}$$

and

$$d' = \sqrt{\frac{13.15}{0.1 \times 10 \times 1.5}} = 2.96.$$

Finally, since $c(\lfloor d' \rfloor) > c(\lceil d' \rceil)$, the batch size is $d^* = 3$.

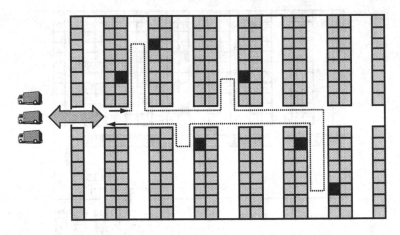

Figure 5.19   Routing of a picker in a storage zone with side aisles having a single entrance (the dark-coloured storage locations are the ones where a retrieval has to be performed).

5.4.2   Order picker routing

In W/RPSs, where pickers travel on foot or by motorized trolley and may visit multiple aisles, order picker routing is a major issue. Picker routing is part of a large class of combinatorial optimization problems known as *vehicle routing problems* (VRPs), which will be examined extensively in Chapter 7 in the context of distribution management. In this section a single picker problem, known as *road travelling salesman problem* (RTSP), is illustrated. The RTSP is a slight variant of the classical *travelling salesman problem* (TSP) (see Section 7.3) and consists of determining a least-cost tour including a subset of vertices of a graph. The RTSP is NP-hard, but, in the case of warehouses, it is often solvable in polynomial time due to the particular characteristics of the travel network. If each aisle has a single entrance, the least duration route is obtained by first visiting all the required storage locations placed in the upper side aisles and then the required storage locations situated in the lower side aisles (see Figure 5.19). On the other hand, if the side aisles have some interruptions (i.e. if there is more than one cross aisle), the problem can be solved to optimality by using the Ratliff and Rosenthal dynamic programming algorithm, whose worst-case computational complexity is a linear function of the number of side aisles. However, if there are several cross aisles, the number of states and transitions increases rapidly and the use of the dynamic programming procedure becomes impractical. Therefore, in what follows, two simple heuristics are illustrated. The reader interested in the Ratliff and Rosenthal algorithm is referred to the list at the end of the chapter.

S-shape heuristic.   Any aisle containing at least one item to be retrieved is traversed entirely. Aisles where there are no items to be picked are skipped.

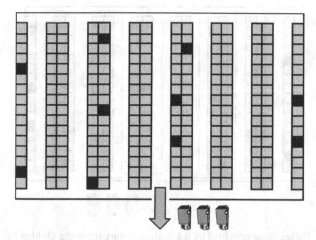

Figure 5.20   Routing of a picker in a storage zone with side aisles having two entrances (the dark-coloured storage locations are the ones where a retrieval has to be performed).

Largest gap heuristic.   In this context, a *gap* is the distance between any two adjacent items to be retrieved in an aisle, or the distance between the last item to be retrieved in an aisle and the closest cross aisle. The picker goes to the front of the side aisle, closest to the I/O port, including at least one item to be retrieved. Then the picker traverses the aisle entirely, while the remaining side aisles are entered and left once or twice, both times from the same side, depending on the largest gap of the aisle.

Golden Fruit is an Honduran company manufacturing fruit juice. Last 17 September, a picker routing problem had to be solved at 10:30 a.m. in the warehouse located in Puerto Lempira (see Figure 5.20). Travel times were assumed to be proportional to distances. The routes provided by the S-shape and largest gap heuristics are shown in Figures 5.21 and 5.22, respectively, and the least-cost route is illustrated in Figure 5.23.

5.4.3   Packing problems

Packing problems arise in warehouses when preparing the outgoing shipments. Depending on the characteristics of the products and on the transportation mode, items or cartoons have to be mounted onto a pallet or inserted in a container, pallets have to be loaded onto trucks, or containers have to be put on a ship or on a plane. All these problems share a common mathematical structure as in both cases some 'objects' (named items in the following) have to packed into a set of bins. The objective is to minimize the cost associated with using the bins, or simply to minimize the number

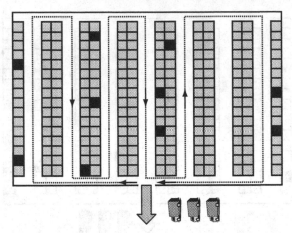

Figure 5.21   Picker route provided by the S-shape heuristic in the Golden Fruit warehouse.

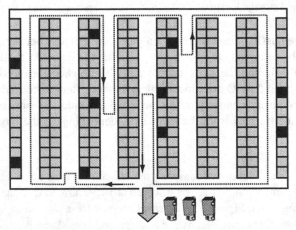

Figure 5.22   Picker route provided by the largest gap heuristic in
the Golden Fruit warehouse.

of required bins. Constraints on the stability of the load are sometimes imposed. From a mathematical point of view, packing problems are mostly NP-hard, so that in most decision support systems heuristics are used.

Classification.   In some packing problems, not all physical characteristics of the items have to be considered when packing. For instance, when loading high-density goods onto a truck, items can be characterized just by their weight, without any concern for their length, width and height. As a result, packing problems can be classified according to the number of parameters needed to characterize an item.

- *One-dimensional packing problems* often arise when dealing with high-density items, in which case weight is binding.

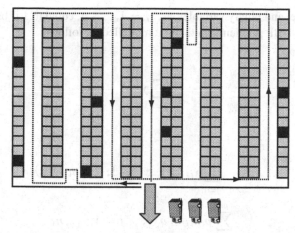

Figure 5.23 Least cost picker route in the Golden Fruit warehouse.

- *Two-dimensional packing problems* usually arise when loading a pallet with items having the same height.

- *Three-dimensional packing problems* occur when dealing with low-density items, in which case volume is binding.

In the following, it is assumed, for the sake of simplicity, that items are rectangles in two-dimensional problems, and that their sides must be parallel or perpendicular to the sides of the bins in which they are loaded. Similarly, in three-dimensional problems, the items are assumed to be parallelepipeds and their surfaces are parallel or perpendicular to the surfaces of the bins in which they are loaded. These assumptions are satisfied in most settings.

Packing problems are usually classified as *off-line* and *on-line* problems, depending on whether the items to be loaded are all available or not when packing starts. In the first case, item characteristics *can* be preprocessed (e.g. items can be sorted by nondecreasing weights) in order to improve heuristic performance. A heuristic using such preprocessing is said to be *off-line*, otherwise it is called *on-line*. Clearly, an on-line heuristic can be used for solving an off-line problem, but an off-line heuristic cannot be used for solving an on-line problem.

One-dimensional packing problems

The simplest one-dimensional packing problem is known as *bin packing* (1-BP) problem. It amounts to determining the least number of identical capacitated bins in which a given set of weighted items can be accommodated. Let m be the number of items to be loaded; n the number of available bins (or an upper bound on the number of bins in an optimal solution); p_i, $i = 1, \ldots, m$, the weight of item i; q_j the capacity of bin j, $j = 1, \ldots, n$. The problem can be modelled by means of binary variables x_{ij}, $i = 1, \ldots, m$, $j = 1, \ldots, n$, each of them equal to 1 if item i is assigned to bin

j or 0 otherwise, and binary variables y_j, $j = 1, \ldots, n$, equal to 1 if bin j is used, 0 otherwise. The 1-BP problem can then be modelled as follows.

Minimize

$$\sum_{j=1}^{n} y_j \qquad (5.23)$$

subject to

$$\sum_{j=1}^{n} x_{ij} = 1, \quad i = 1, \ldots, m, \qquad (5.24)$$

$$\sum_{i=1}^{m} p_i x_{ij} \leqslant q y_j, \quad j = 1, \ldots, n, \qquad (5.25)$$

$$x_{ij} \in \{0, 1\}, \quad i = 1, \ldots, m, \ j = 1, \ldots, n,$$

$$y_j \in \{0, 1\}, \quad j = 1, \ldots, n.$$

The objective function (5.23) is the number of bins used. Constraints (5.24) state that each item is allocated to exactly one bin. Constraints (5.25) guarantee that bin capacities are not exceeded.

A lower bound $\underline{z}(I)$ on the number of bins in any 1-BP feasible solution can be easily obtained as

$$\underline{z}(I) = \lceil (p_1 + p_2 + \cdots + p_m)/q \rceil. \qquad (5.26)$$

The lower bound (5.26) can be very poor if the average number of items per bin is low (see Problem 5.8 for an improved lower bound). Such lower bounds can be used in a branch-and-bound framework or to evaluate the performance of heuristic methods. In the remainder of this section, four heuristics are illustrated. The first two procedures (the *first fit* (FF) and the *best fit* (BF) algorithms) are on-line heuristics while the others are off-line heuristics.

FF algorithm.

Step 0. Let S be the list of items, V the list of available bins and T the list of bins already used. Initially, T is empty.

Step 1. Extract an item i from the top of list S and insert it into the first bin $j \in T$ having a residual capacity greater than or equal to p_i. If no such bin exists, extract from the top of list V a new bin k and put it at the bottom of T; insert item i into bin k.

Step 2. If $S = \emptyset$, STOP, all bins have been loaded. Then T is the list of bins used, while V provides the list of bins unused. If $S \neq \emptyset$, go back to step 1.

BF algorithm.

Step 0. Let S be the list of items to be packed, V the list of available bins and T the list of bins already used. Initially, T is empty.

Step 1. Extract an item i from the top of list S and insert it into the bin $j \in T$ whose residual capacity is greater than or equal to p_i, and closer to p_i. If no such bin exists, extract a new bin k from the top of V and put it at the bottom of T; insert item i into bin k.

Step 2. If $S = \emptyset$, *STOP*, all the items have been loaded, T represents the list of the bins used, while V is the list of bins unused. If $S \neq \emptyset$, go back to step 1.

The two procedures can both be implemented so that the computational complexity is equal to $O(m \log m)$.

It is useful to characterize the performance ratios of such heuristics. Recall that the performance ratio R^H of a heuristic H is defined as

$$R^H = \sup_I \left\{ \frac{z^H(I)}{z^*(I)} \right\},$$

where I is a generic instance of the problem, $z^H(I)$ is objective function value of the solution provided by heuristic H for instance I, while $z^*(I)$ represents the optimal solution value for the same instance.

This means that

- $z^H(I)/z^*(I) \leqslant R^H, \ \forall I$;

- there are some instances I such that $z^H/z^*(I)$ is arbitrarily close to R^H.

Unfortunately, the worst-case performance ratios of the FF and BF heuristics are not known, but it has been proved that

$$R^{FF} \leqslant \tfrac{7}{4} \quad \text{and} \quad R^{BF} \leqslant \tfrac{7}{4}.$$

The FF and BF algorithms can be easily transformed into off-line heuristics, by preliminary sorting the items by nonincreasing weights, yielding the *first fit decreasing* (FFD) and the *best fit decreasing* (BFD) algorithms. Their complexity is still equal to $O(m \log m)$, while their performance ratios are

$$R^{FFD} = R^{FBFD} = \tfrac{3}{2}.$$

It can be proved that this is the minimum worst-case performance ratio that a polynomial 1-BP heuristic can have.

Al Bahar is an Egyptian trucking company located in Alexandria which must plan the shipment of 17 parcels, whose characteristics are reported in Table 5.14. For these shipments the company can use a single van whose capacity is 600 kg. Applying the

Table 5.14   Weight of the parcels (in kilograms) in the Al Bahar problem.

Number of parcels	Weight
4	252
3	228
3	180
3	140
4	120

Table 5.15   Sorted list of the parcels in the Al Bahar problem
(parcel weights are expressed in kilograms).

Parcel	Weight	Parcel	Weight
1	252	10	180
2	252	11	140
3	252	12	140
4	252	13	140
5	228	14	120
6	228	15	120
7	228	16	120
8	180	17	120
9	180		

BFD heuristic, the parcels are sorted by nonincreasing weights (see Table 5.15) and the solution reported in Table 5.16 is obtained. The number of trips is six. The lower bound on the number of trips given by Equation (5.26) is $\lceil 3132/600 \rceil = 6$. Hence, the BFD heuristic solution is optimal.

Two-dimensional packing problems

The simplest two-dimensional packing problem (referred to as the 2-BP problem in the following) consists of determining the least number of identical rectangular bins in which a given set of rectangular items can be accommodated. It is also assumed that no item rotation is allowed. Let L and W be the length and the width of a bin, respectively, and let l_i and w_i, $i = 1, \ldots, m$, be the length and the width of item i.

A lower bound $\underline{z}(I)$ on the number of bins in any feasible solution is

$$\underline{z}(I) = \lceil (l_1 w_1 + l_2 w_2 + \cdots + l_m w_m)/LW \rceil.$$

Most heuristics for the 2-BP problem are based on the idea of forming layers of items inside the bins. Each layer has a width W, and a length equal to that of its

Table 5.16 Parcel-to-trip allocation in the optimal solution of the Al Bahar problem (parcel weights are expressed in kilograms).

Parcel	Weight	Trip	Parcel	Weight	Trip
1	252	1	10	180	5
2	252	1	11	140	3
3	252	2	12	140	5
4	252	2	13	140	5
5	228	3	14	120	5
6	228	3	15	120	6
7	228	4	16	120	6
8	180	4	17	120	6
9	180	4			

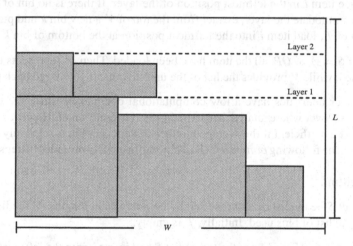

Figure 5.24 Layers of items inside a bin.

longest item. All the items of a layer are located on its bottom, which corresponds to the level of the longest item of the previous layer (see Figure 5.24).

Here we illustrate two off-line heuristics, named *finite first fit* (FFF) and *finite best fit* (FBF) heuristics.

FFF algorithm.

Step 0. Let S be the list of items, sorted by nonincreasing lengths, V the list of bins and T the list of bins used. Initially, T is empty.

Step 1. Extract an item i from the top of S and insert it into the leftmost position of the first layer (which can accommodate it) of the first bin $j \in T$. If no such layer exists, create a new one in the first bin of T (which can accommodate it) and introduce item i in the leftmost position of the layer. If there is no bin of T which can accommodate the layer, extract from the top of V a new bin k and put it at the bottom of T, load item i into the leftmost position at the bottom of bin k.

Step 2. If $S = \emptyset$, *STOP*, all the item have been loaded. Then, T represents the list of bins used, while V provides the list of the unused bins. If $S \neq \emptyset$, go back to step 1.

FBF algorithm.

Step 0. Let S be the list of items, sorted by nonincreasing lengths, V the list of bins and T the list of bins used. Initially, T is empty.

Step 1. Extract an item i from the top of S and insert it into the leftmost position of the layer of a bin $j \in T$ whose residual width is greater than or equal to, and closer to, item width. If no such layer exists, create a new one in the bin of T whose residual length is greater than or equal to, and closer to, the length of item i. Then, introduce item i in the leftmost position of the layer. If there is no bin of T which can accommodate the layer, extract from the top of V a new bin k and put it at the bottom of T, load item i into the leftmost position at the bottom of bin k.

Step 2. If $S = \emptyset$, *STOP*, all the item have been loaded. Then, T represents the list of bins used, while V provides the list of the unused bins. If $S \neq \emptyset$, go back to step 1.

Such layer heuristics have a low computational complexity, since the effort for selecting the layer where an item has to be inserted is quite small. However, they can turn out to be inefficient if the average number of items per bin is relatively small. In such a case, the following *bottom left* (BL) algorithm usually provides better solutions.

BL algorithm.

Step 0. Let S be the list of items, sorted by nonincreasing lengths, V the list of bins and T the list of bins used. Initially, T is empty.

Step 1. Extract an item i from the top of list S and insert it into the leftmost position at the bottom of the first bin $j \in T$ able to accommodate it. If no such bin exists, extract a new bin k from the top of V, and put it at the bottom of T; load item i into the leftmost position at the bottom of bin k.

Step 2. If $S = \emptyset$, *STOP*, all the items have been loaded, T represents the list of bins used, while V provides the list of bins unused. If $S \neq \emptyset$, go back to step 1.

Kumi is a South Korean company manufacturing customized office furniture in Pusan. Outgoing products for overseas customers are usually loaded into containers ISO 40, whose characteristics are reported in Table 5.17. Once packaged, parcels are 2 or 1 m high. They are mounted on wooden pallets so that they cannot be rotated at loading time. The list of parcels shipped last 14 May is reported in Table 5.18. Parcels that are 1 m high are coupled in order to form six pairs of (1×1) m^2 parcels and five

Table 5.17 Characteristics of ISO 40 container.

Length (m)	Width (m)	Height (m)	Capacity (m³)	Capacity (kg)
12.069	2.373	2.405	68.800	26.630

Table 5.18 Parcels shipped by Kumi company.

Quantity	Length (m)	Width (m)	Height (m)
6	1.50	1.50	2.00
5	1.20	1.70	2.00
13	1.00	1.00	1.00
11	0.80	0.50	1.00

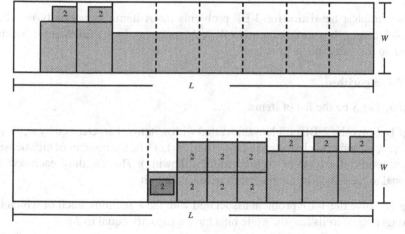

Figure 5.25 Parcels allocated to the two containers shipped by Kumi company (2 indicates two overlapped parcels).

pairs of (0.8×0.5) m² parcels. Then, each such a pair is considered as a single item. Applying the FBF algorithm, the solution reported in Figure 5.25 is obtained.

Three-dimensional packing problems

The simplest three-dimensional packing problem (referred to as the 3-BP problem in the following) consists of determining the least number of identical parallelepipedic

Table 5.19 Parcels loaded at McMillan company.

Type	Quantity	Length (m)	Width (m)	Height (m)	Volume (m^3)	Weight (kg)
1	2	2.50	0.75	1.30	2.4375	75.00
2	4	2.10	1.00	0.95	1.9950	68.00
3	7	2.00	0.65	1.40	1.8200	65.50
4	4	2.70	0.70	0.95	1.7900	63.00
5	3	1.40	1.50	0.80	1.6800	61.50

bins in which a given set of parallelepipedic items can be accommodated. It is also assumed that no item rotation is allowed. Let L, W and H be the length, width and height of a bin, respectively, and let l_i, w_i and h_i, $i = 1, \ldots, m$, be the length, width and height of item i.

A lower bound $\underline{z}(I)$ on the number of bins is

$$\underline{z}(I) = \lceil l_1 w_1 h_1 + l_2 w_2 h_2 + \cdots + l_m w_m h_m)/LWH \rceil. \qquad (5.27)$$

The simplest heuristics for 3-BP problems insert items sequentially into layers parallel to some bin surfaces (e.g. to $W \times H$ surfaces). In the following a heuristics based on this principle is illustrated.

3-BP-L algorithm.

Step 0. Let S be the list of items.

Step 1. Solve the 2-BP problem associated with m items characterized by w_i, h_i, $i = 1, \ldots, m$, and bins characterized by W and H. Let k be the number of bidimensional bins used (referred to as *sections* in the following). The length of each section is equal to the length of the largest item loaded into it.

Step 2. Solve the 1-BP problem associated with the k sections, each of which has a weight equal to its length, while bins have a capacity equal to L.

If the items are all available when bin loading starts, it can be useful to sort list S by nonincreasing values of the volume. However, unlike one-dimensional problems, more complex procedures are usually needed to improve solution quality.

McMillan company is a motor carrier headquartered in Bristol (Great Britain). The firm has recently semi-automatized the procedure for allocating outgoing parcels to vehicles, using a decision support system. This software tool uses the 3-BP-L algorithm as a basic heuristic, and then applies a local search procedure. Last 26 January the parcels to be loaded were those reported in Table 5.19. The characteristics of the vehicles are indicated in Table 5.20. The parcels are mounted on pallets and

Table 5.20 Characteristics of the vehicles in the McMillan problem.

Length (m)	Width (m)	Height (m)	Capacity (m^3)	Capacity (kg)
6.50	2.40	1.80	28.08	12.30

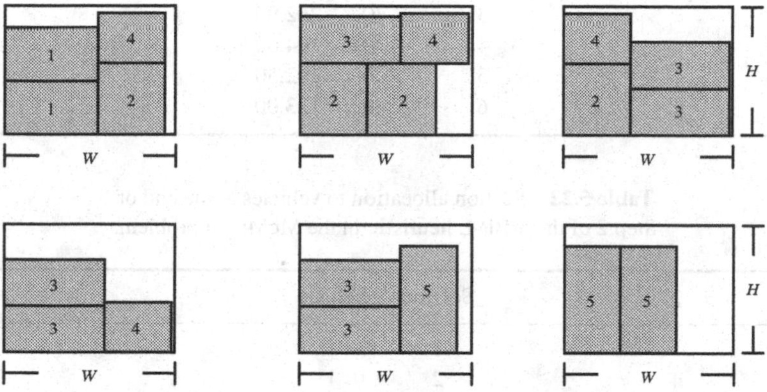

Figure 5.26 Sections generated at the end of Step 1 of the 3-BP-L heuristic in the McMillan problem.

cannot be rotated. First, the parcels are sorted by nonincreasing volumes. Then, the 3-BP-L algorithm (in which 2-BP problems are solved through the BL heuristic) is used. The solution is made up of six (2.4×1.8) m^2 sections, loaded as reported in Figure 5.26. Finally, a 1-BP problem is solved by means of the BFD heuristic. In the solution (see Table 5.22) three vehicles are used, the most loaded of which carries a weight of 545.5 kg, much less than the weight capacity. It is worth noting that the lower bound provided by Equation (5.27) is $\lceil 37.795/28.08 \rceil = 2$.

5.5 Questions and Problems

5.1 Show that a warehouse can be modelled as a queueing system.

5.2 A warehouse stores nearly 20 000 pallets. Goods turn about five times a year. How much is the required labour force? Assume two eight-hour shifts per day and about 250 working days per year. (Hint: apply Little's Law stating that for a queueing system in steady state the average length L_Q of the queue equals the average arrival rate λ times the average waiting time T_W, $L_Q = \lambda T_W$.)

Table 5.21   Width of the sections generated at the end of
Step 1 of the 3-BP-L heuristic in the McMillan problem.

Section	Width (m)	Weight (kg)
1	2.70	281.00
2	2.70	264.50
3	2.70	262.00
4	2.70	194.00
5	2.00	192.50
6	1.40	123.00

Table 5.22   Section allocation to vehicles at the end of
Step 2 of the 3-BP-L heuristic in the McMillan problem.

Section	Vehicle
1	1
2	1
3	2
4	2
5	3

5.3 Let d be the daily demand from all orders, l_C the average length of a rail car, q the capacity of a rail car and n_C the number of car changes per day. Estimate the length of rail dock l_D needed by a warehouse.

5.4 Show that scheduling an S/R machine can be modelled as a rural postman problem on a directed graph. The rural postman problem consists of determining a least-cost route traversing a subset of required arcs of a graph at least once (see Section 7.6.2 for further details).

5.5 Show that an optimal picker route cannot traverse an aisle (or a portion of an aisle) more than twice. Illustrate how this property can be used to devise a dynamic programming algorithm.

5.6 Demonstrate that both the FF and BF heuristics for the 1-BP problem take $O(m \log m)$ steps.

5.7 Devise a branch-and-bound algorithm based on formula (5.26).

5.8 Devise an improved 1-BP lower bound.

5.9 Modify the heuristics for the 1-BP problem for the case where each bin j, $j = 1, \ldots, n$, has a capacity q_j and a cost f_j. Apply the modified version of the BFD algorithm to the following problem. Brocard is a road carrier operating

Table 5.23 List of the parcels to load and corresponding weight
(in kilograms) in the Brocard problem.

Parcel	Weight	Parcel	Weight
1	228	18	170
2	228	19	170
3	228	20	170
4	217	21	170
5	217	22	95
6	217	23	95
7	217	24	95
8	210	25	95
9	210	26	75
10	210	27	75
11	210	28	75
12	195	29	75
13	195	30	75
14	195	31	75
15	170	32	55
16	170	33	55
17	170	34	55

mainly in France and in the Benelux. The vehicle fleet comprises 14 vans of
capacity equal to 800 kg and 22 vans of capacity equal to 500 kg. The company
has to deliver on behalf of the EU 34 parcels of different sizes from Paris to
Frankfurt (the distance between these cities is 592 km). The characteristics
of the parcels are reported in Table 5.13. As only five company-owned vans
(all having capacity of 800 kg) will be available on the day of the delivery,
Brocard has decided to hire additional vehicles from a third-party company.
The following additional vehicles will be available:

- two trucks with a capacity of 3 tons each, whose hiring total cost (inclusive
  of drivers) is €1.4 per kilometre;

- one truck with trailer, with a capacity of 3.5 tons, whose hiring total cost
  (inclusive of drivers) is €1.6 per kilometre.

Which trucks should Brocard hire?

5.10 Determine a lower bound on the optimal solution cost in the Brocard problem
by suitably modifying Equation (5.26). Also determine the optimal shipment
decision by solving a suitable modification of the 1-BP problem.

5.6   Annotated Bibliography

A recent survey on warehouse design and control is:

1. Jeroen P and Van den Berg L 1999 A literature survey on planning and control of warehousing systems. *IIE Transactions* **31**, 751–762.

An in-depth treatment of warehouse management is:

2. Bartholdi JJ and Hackman ST 2002 *Warehouse and Distribution Science.* (Available at http://www.warehouse-science.com.)

A dynamic programming algorithm for the picker routing problem in Section 5.4.2 is presented in:

3. Ratliff HD and Rosenthal AS 1983 Order picking in a rectangular warehouse: a solvable case of the travelling salesman problem. *Operations Research* **31**, 507–521.

A survey of packing problems is:

4. Dowsland KA and Dowsland WB 1992 Packing problems. *European Journal of Operational Research* **56**, 2–14.

An annotated bibliography on packing problems is:

5. Dyckhoff H, Scheithauer G and Terno J 1997 Cutting and packing. In *Annotated Bibliographies in Combinatorial Optimization* (ed. Dell'Amico M, Maffioli F and Martello S). Wiley, Chichester.

For a detailed treatment of 1-BP, Chapter 8 of the following book is recommended:

6. Martello S and Toth P 1990 *Knapsack Problems: Algorithms and Computer Implementations.* Wiley, Chichester.

Two recent papers on exact algorithms for two and three-dimensional packing problems are:

7. Martello S and Vigo D 1998 Exact solution of the two-dimensional finite bin packing problem. *Management Science* **44**, 388–399.

8. Martello S, Pisinger D and Vigo D 2000 The three-dimensional bin packing problem. *Operations Research* **48**, 256–267.

6

Planning and Managing Long-Haul Freight Transportation

6.1 Introduction

Freight transportation plays a fundamental role in every modern supply chain. It is essential to move raw materials from sources to plants, semi-finished products between factories, and final goods to customers and retail outlets. As pointed out in Chapter 1, transportation systems are rather complex organizations which require considerable human, financial and material resources. Transportation cost accounts for a significant part (often between one-third and two-thirds) of the logistics cost in several industries.

Players. Several players are involved in freight transportation. *Shippers*, which include both producers and *brokers*, originate the demand for transportation. *Carriers*, such as railways, motor carriers and shipping lines, supply transportation services. Some shippers operate their own transportation fleet so that they act as a dedicated carrier. *Governments* construct and operate several transportation infrastructures, such as rail facilities, roads, ports and airports, and regulate several aspects of the industry. This chapter and the next deal with freight transportation planning and management from a shipper's or a carrier's point of view. Strategic planning activities performed on a regional, national or even international scale, by governments and international organizations, are beyond the scope of this book.

Long-haul versus short-haul transportation problems. In *long-haul* freight transportation, goods are moved over relatively long distances, between terminals or other facilities (plants, warehouses, etc.). Commodities may be transported by truck, rail, ship or any combination of modes. On the other hand, in *short-haul* freight transportation, goods are transported, usually by truck, between pick-up and delivery points

Introduction to Logistics Systems Planning and Control   G. Ghiani, G. Laporte and R. Musmanno
© 2004 John Wiley & Sons, Ltd   ISBN: 0-470-84916-9 (HB) 0-470-84917-7 (PB)

situated in the same area (e.g. between a warehouse, or a terminal, and a set of customers). Such tasks are of short duration (much shorter than a work shift) and vehicle tours are to be built through a sequence of tasks. Long-haul transportation is examined in this chapter, while short-haul problems are dealt with in the next chapter.

Classification of long-haul transportation services. As explained in Chapter 1, long-haul transportation services can be broadly classified into two groups, depending on whether the movement of goods is performed by the shipper or by a *public carrier*. In *privately operated transportation*, freight has to be moved from a restricted number of origins (e.g. plants and warehouses) to several destinations (e.g. retail locations and customers). This is the case of private truck fleet operations and of *industrial cargo shipping*. These *few-to-many* transportation systems are relatively simple to manage compared to *many-to-many* systems operated by *public carriers*. In such systems, transportation demand is usually made up of several *traffic classes*, each characterized by an origin, a destination, a commodity class and a freight tonnage. Public carrier services can be customized or consolidation-based. In *customized transportation*, such as TL *trucking* and *tramp cargo shipping*, a vehicle serves a single shipper request at a time. When a shipper calls, a vehicle and a crew are sent to a pick-up point. Goods are then loaded and the vehicle moves to a delivery point. Finally, the vehicle is unloaded and the driving team is asked to move empty to a new location (either a new pick-up point or a staging point). Hence, vehicle routes are built in an ongoing fashion as customer requests arrive. In *consolidation-type* transportation, such as LTL *trucking* and *liner cargo shipping*, service is not individualized. A vehicle may therefore move freight of different shippers with possibly different origin–destination pairs. Carriers establish regular routes, each of which is characterized by a given frequency. For instance, a route may be a container ship line operated once a month from Rotterdam (The Netherlands) to Mumbay (India) with an intermediate stop at Cape Town (South Africa). A consolidation-type transportation system is made up of a network of terminals connected by physical (e.g. roads or rail tracks) or virtual (e.g. air or sea lines) links. *End-of-line terminals* are places where small shipments are brought (usually by a fleet of trucks) and consolidated into larger shipments. Moreover, in end-of-line terminals, incoming consolidated shipments are broken and delivered to their final destinations by a fleet of vans. In *breakbulk terminals*, freight traffic from several end-of-line terminals is sorted (or classified) and consolidated.

6.2   Relevant Costs

Before illustrating the main decision problems arising in freight transportation, it is useful to describe the various *motion* costs. Motion costs can be classified as *transportation costs* and *handling costs*.

The cost of operating a fleet. The main costs in operating a fleet of vehicles are related to crews' wages, fuel consumption, vehicle depreciation, maintenance, insurance, administration and occupancy. Wages and insurance are time dependent, fuel consumption and maintenance are distance dependent, depreciation depends on both time and distance, while administration and occupancy costs are customarily allocated as a fixed annual charge.

The cost paid by a carrier for transporting a shipment. The cost paid by a carrier for transporting a shipment is somewhat arbitrary because different shipments usually share some common costs. For example, in LTL trucking where several shipments are moved jointly by the same vehicle, it is unclear how much of a trip cost has to be assigned to each shipment. Similarly, in TL trucking, a shipment cost is not well defined because of empty trips necessary to move trucks from each delivery point to the subsequent pick-up point.

The cost paid for hiring a vehicle. Hire charges include, in addition to the costs paid by the carrier, an undisclosed mark-up.

The cost of a shipment when a public carrier is used. When a shipper uses a public carrier, the cost for transporting a shipment can be calculated on the basis of the rates published by the carrier. For customized transportation, the cost of a full load depends on both the origin and destination of the movement, as well as on the size and equipment of the vehicle required. For consolidation-based transportation, each shipment is given a rating (called a *class*) which depends on the physical character-istics (weight, density, etc.) of the goods. For example, in North America the railway classification includes 31 classes, while the National Motor Freight Classification (NMFC) comprises 23 classes. *Rates* (i.e. the transportation cost per unit of weight) depend on the origin and the destination of the movement, as well as on the shipment weight and its class (see Figures 6.1–6.3). Rates are usually reported in tables, or can be calculated through *rating engines* available on the Internet. In Table 6.1, the LTL rates published by a USA carrier for two NMFC classes are shown. Costs often present discontinuities, as illustrated in Figure 6.4 (cost may decrease by adding extra weight).

Handling costs. Handling costs are incurred when inserting individual items into a bin (e.g. a pallet or a container), loading the bin onto an outbound vehicle, and reversing these operations at destination.

6.3 Classification of Transportation Problems

In principle, managing a transportation system gives rise to several decision problems. However, the way these issues are addressed is greatly influenced by the nature of the operational constraints. A typical example is the *vehicle and crew scheduling* problem, which amounts to finding a least-cost allocation over time of vehicles and crews to transportation tasks in such a way that rules and regulations on vehicle

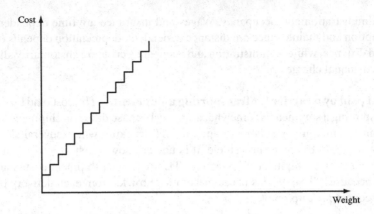

Figure 6.1 Transportation rates for parcels.

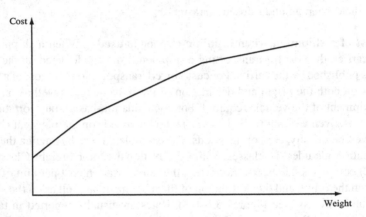

Figure 6.2 Transportation rates for LTL trucking.

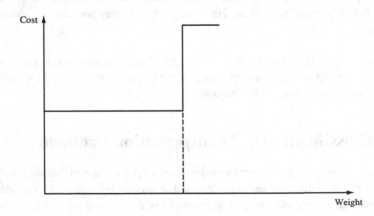

Figure 6.3 Transportation rates for TL trucking.

Table 6.1 LTL rates (dollars per 100 pounds) from New York to Los Angeles published by the USA National Classification Committee (classes 55 and 70 correspond to products having densities higher than 15 and 35 pounds per cubic feet, respectively).

Weight (W)	Class 55	Class 70
$0 \leqslant W < 500$	129.57	153.82
$500 \leqslant W < 1000$	104.90	124.60
$1000 \leqslant W < 2000$	89.43	106.10
$2000 \leqslant W < 5000$	75.17	89.24
$5000 \leqslant W < 10\,000$	64.82	76.95
$10\,000 \leqslant W < 20\,000$	53.13	63.05
$20\,000 \leqslant W < 30\,000$	46.65	55.37
$30\,000 \leqslant W < 40\,000$	40.15	47.67
$W \geqslant 40\,000$	37.58	44.64

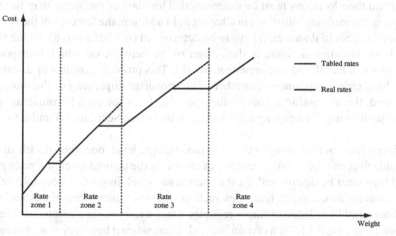

Figure 6.4 Transportation rates for LTL trucking.

maintenance and crew rests are satisfied. In air and rail transportation, where regular lines are usually operated, vehicle and crew scheduling is a tactical problem solved a few times a year. On the other hand, in TL trucking both vehicles and crews are allocated to tasks in an on-going fashion as customer requests arrive. As a result, freight transportation planning and management problems come in a large number of variants. Some of them are common to all transportation systems, while others are specific to a transportation mode or to a way of operating the system.

Common decision problems. Common decision problems include, at the strategic level, a broad definition of the operating strategy of the system, the design of the physical network (if any exists) and the acquisition of expensive resources, such as airplanes. The tactical level covers the allocation of existing resources (vehicles, crews, etc.) as well as the purchase of additional capacity to cope with variations in demand. At the operational level, the focus is on adjusting vehicle and crew schedules in order to take into account last-minute events, such as order modifications, equipment failures, strikes and unfavourable weather conditions.

Privately operated transportation systems.   The decisions to be made are relatively simple. If demand varies over the year, one must determine the optimal mix between owned and hired vehicles (see Section 6.4). Moreover, on a short-term basis, decisions have to be made on order consolidation and on shipment scheduling (see Section 6.7). The objective is usually to minimize total cost while meeting a pre-established service level.

Consolidation-based transportation systems.   At the strategic level, a carrier has to decide what kind of commodities to transport and which origin–destination pairs to serve. Moreover, the number of terminals (e.g. crossdocks or railway terminals) to be used and their locations must be determined. This class of problems may be tackled by using the methods illustrated in Chapter 3. In addition, the features of the terminals (shape, number of doors, etc.) have to be determined (see Section 6.8). At the tactical level, an important decision is the design of the network on which transportation services will be offered (the *service network*). This problem consists of determining the characteristics (frequency, number of intermediate stops, etc.) of the routes to be operated, the way traffic is routed, the operating rules for each terminal, as well as the repositioning of empty vehicles and containers (see Sections 6.5 and 6.6).

Tailored transportation systems.   At the strategic level, one must decide the most suitable fleet and the required number of crews. At the tactical level, the price of full-load trips must be determined. At the operational level, important decisions relate to the dynamic allocation of resources such as tractors, trailers, containers and crews, without a full knowledge of future requests. Once a resource is allocated to a request, it is no longer available for a certain interval. Then, when it becomes available again, it is usually located in a different place. Therefore, in such settings one must also decide which requests have to be accepted and which have to be rejected, as well as how the accepted requests have to be serviced and how idle resources (e.g. idle tractors, empty trailers and empty containers) have to be repositioned (see Sections 6.9 and 6.10). A further operational problem is *spot pricing*, i.e. the pricing of unallocated vehicle capacity. In both tailored and consolidation-based transportation, a carrier aims at maximizing the expected profit over a pre-established planning horizon.

6.4   Fleet Composition

When demand varies over the year, carriers usually cover the baseload of demand through an owned fleet, while using hired vehicles to cover peak periods. In what follows, the least-cost mix of owned and hired vehicles is determined under the

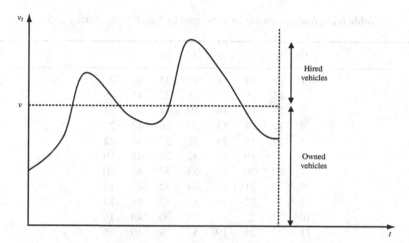

Figure 6.5   Fleet composition when demand varies over the year.

assumption that all vehicles are identical. Let n be the number of time periods into which the time horizon of a year is decomposed (for example, $n = 52$ if the time period corresponds to a week); v be the decision variable corresponding to the number of owned vehicles; v_t, $t = 1, \ldots, n$, be the required number of vehicles at time period t; m be the number of time periods per year in which $v_t > v$. Moreover, let c_F and c_V be the fixed and variable cost per time period of an owned vehicle, respectively, and let c_H be the cost per time period of hiring a vehicle (clearly, $c_F + c_V < c_H$). Then, the annual transportation cost as a function of the number of owned vehicles is

$$C(v) = n c_F v + c_V \sum_{t=1}^{n} \min(v_t, v) + c_H \sum_{t:v_t>v} (v_t - v), \qquad (6.1)$$

where the right-hand side is the sum of the annual fixed cost, the annual variable cost of the owned vehicles and the annual cost of hiring vehicles to cover peak demand. The minimum annual transportation cost is achieved when the derivative of $C(v)$ with respect to v is zero. As the two summations in Equation (6.1) are equal to the areas below and above the line $v_t = v$, respectively (see Figure 6.5), then their derivatives are equal to m and $-m$, respectively. Consequently, $C(v)$ is minimal when

$$n c_F + c_V m - c_H m = 0.$$

Hence, the optimal fleet size can be determined by requiring that

$$m = n \frac{c_F}{c_H - c_V}. \qquad (6.2)$$

Table 6.2   Weekly number of vans used by Fast Courier during 2002.

t	v_t	t	v_t	t	v_t	t	v_t
1	12	14	18	27	23	40	25
2	15	15	17	28	22	41	25
3	16	16	16	29	24	42	24
4	17	17	14	30	26	43	22
5	17	18	13	31	27	44	22
6	18	19	13	32	28	45	19
7	20	20	14	33	30	46	20
8	20	21	15	34	32	47	18
9	21	22	16	35	32	48	17
10	22	23	17	36	30	49	16
11	24	24	19	37	29	50	16
12	22	25	21	38	28	51	14
13	20	26	22	39	26	52	13

Fast Courier is a USA transportation company located in Wichita (Kansas), specializing in door-to-door deliveries. The company owns a fleet of 14 vans and turns to third parties for hiring vans when service demand exceeds fleet capacity. In 2002, the number of vans used weekly for meeting all the transportation demand is reported in Table 6.2. For 2003, the company has decided to redesign the fleet composition, with the aim of reducing the annual transportation cost. Assuming that $c_F = \$350$, $c_V = \$150$ and $c_H = \$800$, Equation (6.2) gives $m = 28$. Hence, the number of owned vehicles is $v^* = 19$, which corresponds to an annual transportation cost $C(v^*) = \$606\,600$. The saving with respect to the previously adopted solution is $\$31\,850$.

6.5   Freight Traffic Assignment Problems

Freight *traffic assignment problems* (TAPs) consist of determining a least-cost routing of goods over a network of transportation services from their origins (e.g. manufacturing plants) to their destinations (e.g. retail outlets). In a sense, the demand allocation problems illustrated in Chapter 3 are particular freight TAPs. From a mathematical point of view, TAPs can be cast as NF problems. NF problems include, as special cases, several remarkable network optimization problems, such as the *shortest-path problem* and the *transportation problem*.

Classification.   TAPs can be classified as *static* or *dynamic*. Static models are suitable when the decisions to be made are not affected explicitly by time. They are formulated on a directed graph (or multigraph) $G = (V, A)$, where the vertex set V often corresponds to a set of facilities (terminals, plants, warehouses) and the arcs in the set A represent possible transportation services linking the facilities. Some vertices represent *origins* of transportation demand for one or several products, while others are *destinations*, or act as a *transshipment points*. Let K be the set of traffic classes (or simply, *commodities*). With each arc is associated a cost (possibly dependent on the amount of freight flow on the arc) and a capacity. Cost functions may represent both monetary costs (see, for example, Figures 6.1–6.3) and congestion effects arising at terminals.

In dynamic models, a time dimension is explicitly taken into account by modelling the transportation services over a given planning horizon through a *time-expanded* directed graph. In a time-expanded directed graph, the horizon is divided into a number of time periods $t_1, t_2, \ldots$, and the physical network (containing terminals and other material resources) is replicated in each time period. Then, temporal links are added. A temporal link connecting two representations of the same terminal at two different time periods may represent freight waiting to be loaded onto an incoming vehicle, or the time required for freight classification at the terminal. On the other hand, a temporal link connecting two representations of different terminals may describe a transportation service. Further vertices and arcs may be added to model the arrival of commodities at destinations and impose penalties in case of delays. With each link may be associated a capacity and a cost, similar to those used in static formulations. An example of static transportation service network is shown in Figure 6.6, while an associated time-expanded network is reported in Figure 6.7. In the static network there are three terminals (A, B, C) and four transportation services operating from A to B, from B to A, from B to C and from C to A. The travel durations are 2, 2, 1 and 1 days, respectively. If the planning horizon includes four days, a dynamic representation has four vertices for each terminal (A_i, $i = 1, \ldots, 4$, describes terminal A at the ith day). Some arcs (such as (A_1, B_3)) represent transportation services, while others (such as (B_1, B_2)) describe commodities standing idle at terminals. In addition, there may be supersinks for some terminals (such as terminal C in Figure 6.7), in which case the costs on the arcs entering the supersinks reflect service penalties.

6.5.1 Minimum-cost flow formulation

Let $O(k)$, $k \in K$, be the set of origins of commodity k; $D(k)$, $k \in K$, the set of destinations of commodity k; $T(k)$, $k \in K$, the set of *transshipment* points with respect to commodity k; o_i^k, $i \in O(k)$, $k \in K$, the supply of commodity k of vertex i; d_i^k, $i \in D(k)$, $k \in K$, the demand of commodity k of vertex i; u_{ij}, $(i, j) \in A$, the capacity of arc (i, j) (i.e. the maximum flow that arc (i, j) can carry); u_{ij}^k, $(i, j) \in A$, $k \in K$, the maximum flow of commodity k on arc (i, j). The variables x_{ij}^k, $(i, j) \in A$, $k \in K$, represent the flow of commodity k on arc (i, j). Moreover, let $C_{ij}^k(x_{ij}^k)$, $(i, j) \in A$, $k \in K$, be the cost for transporting x_{ij}^k flow units of commodity k on arc (i, j).

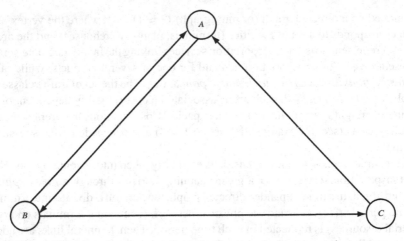

Figure 6.6   A static representation of a three-terminal transportation system.

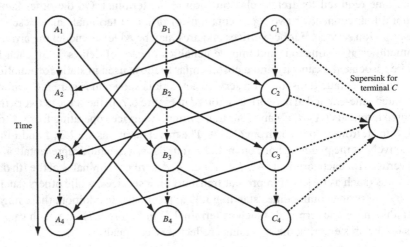

Figure 6.7   Dynamic network representation of the transportation system
illustrated in Figure 6.6.

In the following, G is assumed to be a strongly connected directed graph. The
extension to the case where G is a multigraph, or a collection of strongly connected
directed subgraphs, is straightforward. A quite general multicommodity minimum-
cost flow (MMCF) formulation is as follows:

Minimize

$$\sum_{k \in K} \sum_{(i,j) \in A} C_{ij}^k(x_{ij}^k) \qquad (6.3)$$

subject to

$$\sum_{\{j \in V : (i,j) \in A\}} x_{ij}^k - \sum_{\{j \in V : (j,i) \in A\}} x_{ji}^k = \begin{cases} o_i^k, & \text{if } i \in O(k), \\ -d_i^k, & \text{if } i \in D(k), \\ 0, & \text{if } i \in T(k), \end{cases} \quad i \in V, \ k \in K,$$

(6.4)

$$x_{ij}^k \leqslant u_{ij}^k, \quad (i,j) \in A, \ k \in K, \tag{6.5}$$

$$\sum_{k \in K} x_{ij}^k \leqslant u_{ij}, \quad (i,j) \in A, \tag{6.6}$$

$$x_{ij}^k \geqslant 0, \quad (i,j) \in A, \ k \in K.$$

The objective function (6.3) is the total cost, constraints (6.4) correspond to the flow conservation constraints holding at each vertex $i \in V$ and for each commodity $k \in K$. Constraints (6.5) impose that the flow of each commodity $k \in K$ does not exceed capacity u_{ij}^k on each arc $(i,j) \in A$. Constraints (6.6) (*bundle constraints*) require that, for each $(i,j) \in A$, the total flow on arc (i,j) is not greater than the capacity u_{ij}.

It is worth noting that o_i^k, $k \in K$, $i \in O(k)$ and d_i^k, $k \in K$, $i \in D(k)$, must satisfy the following conditions:

$$\sum_{i \in O(k)} o_i^k = \sum_{i \in D(k)} d_i^k, \quad k \in K,$$

otherwise the problem is infeasible.

In the remainder of this section some of the most relevant solution methods for some special cases of the MMCF problem are illustrated.

6.5.2   Linear single-commodity minimum-cost flow problems

The linear single-commodity minimum-cost flow (LMCF) model can be formulated as follows.

Minimize

$$\sum_{(i,j) \in A} c_{ij} x_{ij} \tag{6.7}$$

subject to

$$\sum_{\{j \in V : (i,j) \in A\}} x_{ij} - \sum_{\{j \in V : (j,i) \in A\}} x_{ji} = \begin{cases} o_i, & \text{if } i \in O, \\ -d_i, & \text{if } i \in D, \\ 0, & \text{if } i \in T, \end{cases} \quad i \in V, \tag{6.8}$$

$$x_{ij} \leqslant u_{ij}, \quad (i,j) \in A, \tag{6.9}$$

$$x_{ij} \geqslant 0, \quad (i,j) \in A. \tag{6.10}$$

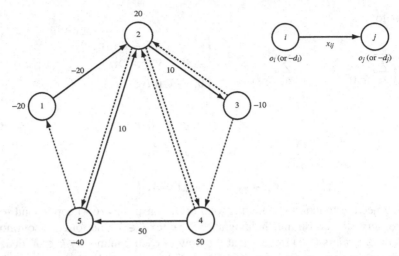

Figure 6.8 A spanning tree (full line arcs) and the associated (infeasible) basic solution $(x_{12} = -20, \ x_{23} = 10, \ x_{45} = 50, \ x_{52} = 10,$ all other variables are equal to 0).

The LMCF model is a structured LP problem and, as such, can be solved through the simplex algorithm or any other LP procedure. Instead of using a general-purpose algorithm, it is common to employ a tailored procedure, the (primal) *network simplex* algorithm, a specialized version of the classical simplex method, which takes advantage of the particular structure of the coefficient matrix associated with constraints (6.8) (corresponding to the vertex-arc incidence matrix of the directed graph G).

We first examine the case where there are no capacity constraints (6.9). In such a case, it is useful to exploit the following characterization of the basic solutions of the system of equations (6.8), which is stated without proof.

Property. The basic solutions of the system of equations (6.8) have $|V| - 1$ basic variables. Moreover, each basic solution corresponds to a tree spanning G and vice versa.

In order to find a basic solution of problem (6.7), (6.8), (6.10) it is therefore sufficient to select a tree spanning G, set to zero the variables associated with the arcs which are not part of the tree, and then solve the system of linear equations (6.8). The latter step can be easily accomplished through a substitution method. Of course, the basic solution associated with a spanning tree is not always feasible, since the nonnegativity constraints (6.10) may be violated (see Figure 6.8).

The network simplex algorithm has the same structure as the standard simplex procedure. However, the optimality test and the pivot operations are performed in a simplified way.

Step 1. Find an initial basic feasible solution $x^{(0)}$. Set $h = 0$.

Step 2. Determine the reduced costs $c'^{(h)}$ associated with $x^{(h)}$.

Step 3. If $c_{ij}^{\prime(h)} \geqslant 0$, $(i, j) \in A$, STOP, $x^{(h)}$ is an optimal solution; otherwise choose a variable x_{vw} such that $c_{vw}^{\prime(h)} < 0$.

Step 4. Select a variable x_{pq} coming out of the basis, make a pivot in order to substitute x_{pq} for x_{vw} in the basis; set $h = h + 1$ and go back to Step 2.

The particular structure of problem (6.7), (6.8), (6.10) and of its dual,

Maximize

$$\sum_{i \in O} o_i \pi_i - \sum_{i \in D} d_i \pi_i$$

subject to

$$\pi_i - \pi_j \leqslant c_{ij}, \quad (i, j) \in A,$$

enables the execution of Steps 2–4 as follows. At Step 2, the reduced costs can be computed through the formula:

$$c_{ij}^{\prime(h)} = c_{ij} - \pi_i^{(h)} + \pi_j^{(h)}, \quad (i, j) \in A, \tag{6.11}$$

where $\pi^{(h)} \in \mathfrak{R}^{|V|}$ can be determined by requiring that the reduced costs of the basic variables be zero:

$$c_{ij}^{\prime(h)} = c_{ij} - \pi_i^{(h)} + \pi_j^{(h)} = 0, \quad (i, j) \in A : x_{ij}^{(h)} \text{ is a basic variable.}$$

At Step 3, if $c_{ij}^{\prime(h)} \geqslant 0$, $(i, j) \in A$, then $\pi_i^{(h)} - \pi_j^{(h)} \leqslant c_{ij}$, $(i, j) \in A$, i.e. solution $\pi^{(h)} \in \mathfrak{R}^{|V|}$, is feasible for the dual problem. Then, $x^{(h)}$ and $\pi^{(h)}$ are optimal for the primal and the dual problems, respectively.

On the other hand, if there is a variable x_{vw} whose reduced cost is negative at iteration h, then arc (v, w) does not belong to the spanning tree associated with iteration h. It follows that, by adding (v, w) to the tree, a single cycle Ψ is created. In order to decrease the objective function value as much as possible, the flow on arc (v, w) has to be increased as much as possible while satisfying constraints (6.8) and (6.10).

Let Ψ^+ be the set of arcs in Ψ oriented as (v, w), and let Ψ^- be the set of the arcs in Ψ oriented in the opposite direction (obviously, $\Psi = \Psi^+ \cup \Psi^-$). If the flow on arc (v, w) is increased by t units, then constraints (6.8) require that the flow on all arcs $(i, j) \in \Psi^+$ be increased by t units, and the flow on all arcs $(i, j) \in \Psi^-$ be decreased by the same amount.

The maximum increase of flow on (v, w) is therefore equal to the minimum flow on the arcs oriented in the opposite direction as (v, w), i.e.

$$t = \min_{(i,j) \in \Psi^-} \{x_{ij}^{(h)}\}.$$

The arc $(p, q) \in \Psi^-$ for which such a condition holds determines which variable x_{pq} will come out from the basis.

The previous description shows that an iteration of the network simplex algorithm requires only a few additions and subtractions. As a result, this procedure is much faster than the standard simplex method and, in addition, does not make rounding errors.

In order to find a feasible solution (if any exists), the *big M method* can be used. A new vertex $i_0 \in T$ and $|V|$ dummy arcs between vertex i_0 and all the other vertices $i \in V$ are introduced. If $i \in O$, then a dummy arc (i, i_0) is inserted. Otherwise, an arc (i_0, i) is added. Let $A^{(a)}$ be the set of dummy arcs. With each dummy arc is associated an arbitrarily large cost M.

The dummy problem is as follows.

Minimize

$$\sum_{(i,j)\in A} c_{ij} x_{ij} + M \sum_{(i,i_0)\in A^{(a)}} x_{ii_0} + M \sum_{(i_0,i)\in A^{(a)}} x_{i_0 i} \qquad (6.12)$$

subject to

$$\sum_{\{j\in V:(i,j)\in A\cup A^{(a)}\}} x_{ij} - \sum_{\{j\in V:(j,i)\in A\cup A^{(a)}\}} x_{ji} = \begin{cases} o_i, & \text{if } i \in O, \\ -d_i, & \text{if } i \in D, \\ 0, & \text{if } i \in T, \end{cases} \quad i \in V \cup \{i_0\},$$

$$(6.13)$$

$$x_{ij} \geqslant 0, \quad (i, j) \in A \cup A^{(a)}. \qquad (6.14)$$

Of course, the $|V|$ dummy arcs make up a spanning tree of the modified directed graph, corresponding to the following basic feasible solution to problem (6.12)–(6.14) (see Figure 6.9):

$$x_{ii_0}^{(0)} = o_i, \qquad i \in O;$$

$$x_{i_0 i}^{(0)} = d_i, \qquad i \in D;$$

$$x_{i_0 i}^{(0)} = 0, \qquad i \in T;$$

$$x_{ij}^{(0)} = 0, \qquad (i, j) \in A.$$

By solving the dummy problem (6.12)–(6.14), a basic feasible solution to the original problem (6.7), (6.8), (6.10) is then obtained.

NTN is a Swiss intermodal carrier located in Lausanne. When a customer has to transport goods between an origin and a destination, NTN supplies it with one or more empty containers in which the goods can be loaded. Once arrived at destination, the goods are unloaded and the empty containers have to be transported to the pick-up points of new customers. As a result, NTN management needs to reallocate the empty containers periodically (in practice, on a weekly basis). Empty container transportation is very expensive (its cost is nearly 35% of the total operating cost). Last 13 May, several empty ISO 20 containers had to be reallocated among the terminals in Amsterdam (The Netherlands), Berlin (Germany), Munich (Germany), Paris

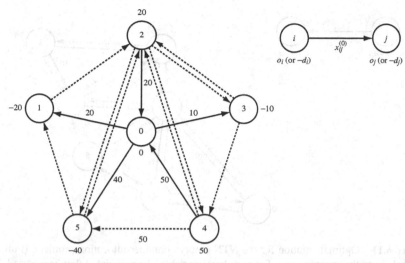

Figure 6.9 Dummy directed graph for the original directed graph in Figure 6.8 (0 is the dummy vertex. Full line arcs belong to the spanning tree. For each basic variable, the associated flow is reported).

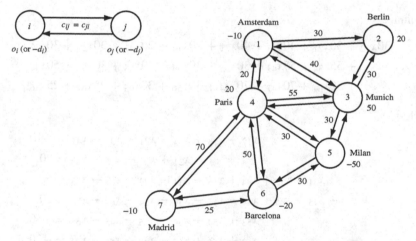

Figure 6.10 Graph representation of the empty container allocation problem.

(France), Milan (Italy), Barcelona (Spain) and Madrid (Spain). The number of empty containers available or demanded at the various terminals is reported, along with transportation costs (in euros/container), in Figure 6.10.

The problem can be formulated as follows.

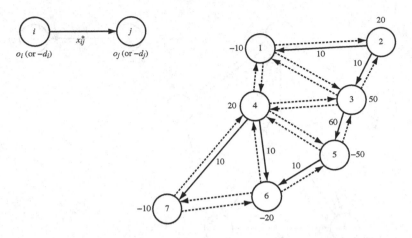

Figure 6.11 Optimal solution for the NTN empty container allocation problem. (Full line arcs belong to the spanning tree. For each basic variable, the associated flow is reported. The optimal cost is equal to €3900.)

Minimize

$$30x_{12} + 30x_{21} + 40x_{13} + 40x_{31} + 20x_{14} + 20x_{41} + 30x_{23} + 30x_{32}$$
$$+ 55x_{34} + 55x_{43} + 30x_{35} + 30x_{53} + 30x_{45} + 30x_{54} + 50x_{46}$$
$$+ 50x_{64} + 70x_{47} + 70x_{74} + 30x_{56} + 30x_{65} + 25x_{67} + 25x_{76}$$

subject to

$$x_{12} + x_{13} + x_{14} - x_{21} - x_{31} - x_{41} = -10,$$
$$x_{21} + x_{23} - x_{12} - x_{32} = 20,$$
$$x_{31} + x_{32} + x_{34} + x_{35} - x_{13} - x_{23} - x_{43} - x_{53} = 50,$$
$$x_{41} + x_{43} + x_{45} + x_{46} + x_{47} - x_{14} - x_{34} - x_{54} - x_{64} - x_{74} = 20,$$
$$x_{53} + x_{54} + x_{56} - x_{35} - x_{45} - x_{65} = -50,$$
$$x_{64} + x_{65} + x_{67} - x_{46} - x_{56} - x_{76} = -20,$$
$$x_{74} + x_{76} - x_{47} - x_{67} = -10,$$

$$x_{12}, x_{21}, x_{13}, x_{31}, x_{14}, x_{41}, x_{23}, x_{32}, x_{34}, x_{43}, x_{35},$$
$$x_{53}, x_{45}, x_{54}, x_{46}, x_{64}, x_{47}, x_{74}, x_{56}, x_{65}, x_{67}, x_{76} \geqslant 0.$$

Using the network simplex method, the optimal solution illustrated in Figure 6.11 is obtained.

The above procedure can be easily adapted to the case of capacitated arcs. To this

Table 6.3 Transportation costs (in euros) per item from the warehouses to the sales districts in the Boscheim problem.

	London	Birmingham	Leeds	Edinburgh
Bristol	9.6	7.0	15.2	28.5
Middlesborough	19.5	13.3	5.0	11.3

purpose, constraints (6.9) are rewritten by introducing auxiliary variables $\gamma_{ij} \geq 0$:

$$x_{ij} + \gamma_{ij} = u_{ij}, \quad (i, j) \in A.$$

If the variable x_{ij} is equal to u_{ij}, then the associated auxiliary variable γ_{ij} takes the value zero and is therefore out of the basis (if the solution is not degenerate). Based on this observation, the following optimality conditions can be derived (the proof is omitted for the sake of brevity).

Theorem 6.1. *A basic feasible solution* $x^{(h)}$ *is optimal for problem (6.7)–(6.10) if, for each nonbasic variable* $x_{ij}^{(h)}$, $(i, j) \in A$, *the following conditions hold,*

$$x_{ij}^{(h)} = 0, \qquad if\, c_{ij}'^{(h)} \geq 0,$$
$$x_{ij}^{(h)} = u_{ij}, \qquad if\, c_{ij}'^{(h)} \leq 0,$$

where $c_{ij}'^{(h)}$ *are the reduced costs defined by (6.11).*

Let $x^{(h)}$ be the basic feasible solution at iteration h of the network simplex method (for simplicity, $x^{(h)}$ is assumed to be nondegenerate). If the value of a nonbasic variable $x_{ij}^{(h)}$, $(i, j) \in A$ is increased, the objective function value increases if the reduced cost $c_{ij}'^{(h)}$ is negative. On the other hand, if $x_{ij}^{(h)} = u_{ij}$, then a decrease in the objective function value is obtained if the reduced cost $c_{ij}'^{(h)}$ is positive.

Let x_{vw} be the variable entering the basis at iteration h (Step 4). If $x_{vw}^{(h)} = 0$, then $c_{vw}'^{(h)} < 0$ and arc $(v, w) \in A$ is not part of the spanning tree associated with $x^{(h)}$. By adding the arc (v, w) to the tree, a single cycle Ψ is formed. In the new basic feasible solution, the variable x_{vw} will take a value t equal to

$$t = \min \left\{ \min_{(i,j) \in \Psi^+} \{u_{ij} - x_{ij}^{(h)}\}, \min_{(i,j) \in \Psi^-} \{x_{ij}^{(h)}\} \right\}. \tag{6.15}$$

Let (p, q) be the arc outgoing the basis according to (6.15). Then, $x_{pq}^{(h+1)} = u_{pq}$, if $(p, q) \in \Psi^+$, or $x_{pq}^{(h+1)} = 0$, if $(p, q) \in \Psi^-$. Observe that the outgoing arc (p, q) may be the same as the outgoing (v, w) if $t = u_{vw}$.

Boscheim is a German company manufacturing electronics convenience goods. Its VCR-12 video recorder is specifically designed for the British market. The VCR-12 is assembled in a plant near Rotterdam (The Netherlands), then stocked in two warehouses located in Bristol and Middlesborough and finally transported to the retailer outlets. The British market is divided into four sales districts whose centres of gravity are in London, Birmingham, Leeds and Edinburgh. Yearly demands amount to 90 000, 80 000, 50 000 and 70 000 items, respectively. The transportation costs per item from the assembly plant of Rotterdam to the warehouses of Bristol and Middlesborough are €24.5 and €26.0, respectively, whereas the transportation costs per item from the warehouses to the sales districts are reported in Table 6.3. Both warehouses have an estimated capacity of 15 000 items and are supplied 10 times a year. Consequently their maximum yearly throughput is 150 000 items.

The annual minimum cost distribution plan can be obtained by solving the following LMCF problem (see Figure 6.12).

Minimize

$$24.5x_{12} + 26.0x_{13} + 9.6x_{24} + 7.0x_{25} + 15.2x_{26}$$
$$+ 28.5x_{27} + 19.5x_{34} + 13.3x_{35} + 5.0x_{36} + 11.3x_{37}$$

subject to

$$x_{12} + x_{13} = 290\,000,$$
$$x_{24} + x_{25} + x_{26} + x_{27} - x_{12} = 0,$$
$$x_{34} + x_{35} + x_{36} + x_{37} - x_{13} = 0,$$
$$-x_{24} - x_{34} = -90\,000,$$
$$-x_{25} - x_{35} = -80\,000,$$
$$-x_{26} - x_{36} = -50\,000,$$
$$-x_{27} - x_{37} = -70\,000,$$

$$x_{12} \leqslant 150\,000,$$
$$x_{13} \leqslant 150\,000,$$
$$x_{12}, x_{13}, x_{24}, x_{25}, x_{26}, x_{27}, x_{34}, x_{35}, x_{36}, x_{37} \geqslant 0.$$

By using the network simplex method, the optimal solution is determined:

$$x_{12}^* = 150\,000, \quad x_{13}^* = 140\,000, \quad x_{24}^* = 90\,000, \quad x_{25}^* = 60\,000,$$
$$x_{35}^* = 20\,000, \quad x_{36}^* = 50\,000, \quad x_{37}^* = 70\,000$$

(as usual, only nonzero variables are reported). It is worth noting that the district of London will be entirely served by the warehouse of Bristol, while the sales districts of Leeds and Edinburgh will be served by the Middlesborough warehouse. The sales district of Birmingham is supplied by both the warehouse of Bristol (75%) and the warehouse of Middlesborough (25%). The total transportation cost is €9 906 000 per year.

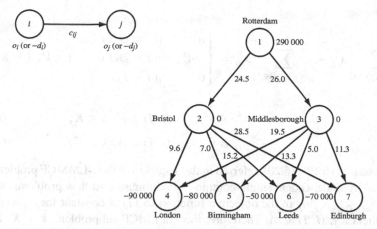

Figure 6.12   Graph representation of the Boscheim problem.

6.5.3   Linear multicommodity minimum-cost flow problems

The linear multicommodity minimum-cost flow (LMMCF) problem can be formulated as the following LP model.

Minimize

$$\sum_{k \in K} \sum_{(i,j) \in A} c_{ij}^k x_{ij}^k$$

subject to

$$\sum_{\{j \in V:(i,j) \in A\}} x_{ij}^k - \sum_{\{j \in V:(j,i) \in A\}} x_{ji}^k = \begin{cases} o_i^k, & \text{if } i \in O(k), \\ -d_i^k, & \text{if } i \in D(k), \qquad i \in V, \ k \in K, \\ 0, & \text{if } i \in T(k), \end{cases}$$

$$x_{ij}^k \leqslant u_{ij}^k, \qquad (i,j) \in A, \ k \in K,$$

$$\sum_{k \in K} x_{ij}^k \leqslant u_{ij}, \qquad (i,j) \in A, \qquad\qquad (6.16)$$

$$x_{ij}^k \geqslant 0, \qquad (i,j) \in A, \ k \in K.$$

The LMMCF problem can be solved efficiently through a tailored Lagrangian procedure. Let λ_{ij} ($\geqslant 0$) be the multipliers attached to constraints (6.16). The Lagrangian relaxation of the LMMCF problem is as follows.

Minimize

$$\sum_{k \in K} \sum_{(i,j) \in A} c_{ij}^k x_{ij}^k + \sum_{(i,j) \in A} \lambda_{ij} \left(\sum_{k \in K} x_{ij}^k - u_{ij} \right) \qquad (6.17)$$

subject to

$$\sum_{\{j \in V:(i,j) \in A\}} x_{ij}^k - \sum_{\{j \in V:(j,i) \in A\}} x_{ji}^k = \begin{cases} o_i^k, & \text{if } i \in O(k), \\ -d_i^k, & \text{if } i \in D(k), \\ 0, & \text{if } i \in T(k), \end{cases} \quad i \in V, \ k \in K,$$

(6.18)

$$x_{ij}^k \leqslant u_{ij}^k, \quad (i,j) \in A, \ k \in K,$$ (6.19)

$$x_{ij}^k \geqslant 0, \quad (i,j) \in A, \ k \in K.$$ (6.20)

The relaxation (6.17)–(6.20), referred in the sequel to as the r-LMMCF problem, is made up of $|K|$ independent single-commodity minimum-cost flow problems, since the sum $\sum_{(i,j) \in A} \lambda_{ij} u_{ij}$ in the objective function (6.17) is constant for a given set of multipliers λ_{ij}, $(i,j) \in A$. Therefore, the kth LMCF subproblem, $k \in K$, is as follows.

Minimize

$$\sum_{(i,j) \in A} (c_{ij}^k + \lambda_{ij}) x_{ij}^k$$ (6.21)

subject to

$$\sum_{\{j \in V:(i,j) \in A\}} x_{ij}^k - \sum_{\{j \in V:(j,i) \in A\}} x_{ji}^k = \begin{cases} o_i^k, & \text{if } i \in O(k), \\ -d_i^k, & \text{if } i \in D(k), \\ 0, & \text{if } i \in T(k), \end{cases} \quad i \in V,$$ (6.22)

$$x_{ij}^k \leqslant u_{ij}^k, \quad (i,j) \in A,$$ (6.23)

$$x_{ij}^k \geqslant 0, \quad (i,j) \in A,$$ (6.24)

can be solved through the network simplex algorithm. Let $\text{LB}_{\text{LMCF}}^k(\lambda)$, $k \in K$, be the optimal objective function value of the kth subproblem (6.21)–(6.24) and let $\text{LB}_{\text{r-LMMCF}}(\lambda)$ be the lower bound provided by solving the r-LMMCF problem. For a given set of multipliers, $\text{LB}_{\text{r-LMMCF}}(\lambda)$ is given by

$$\text{LB}_{\text{r-LMMCF}}(\lambda) = \sum_{k \in K} \text{LB}_{\text{LMCF}}^k(\lambda) - \sum_{(i,j) \in A} \lambda_{ij} u_{ij}.$$

Of course, $\text{LB}_{\text{r-LMMCF}}(\lambda)$ varies as the multiplier λ changes. The Lagrangian relaxation attaining the maximum lower bound value $\text{LB}_{\text{r-LMMCF}}(\lambda)$ as λ varies is called the *dual Lagrangian relaxation*. The following property follows from LP theory.

Property. The lower bound provided by the dual Lagrangian relaxation is equal to the optimal objective function value of the LMMCF model, i.e.

$$\max_{\lambda \geqslant 0} \{\text{LB}_{\text{r-LMMCF}}(\lambda)\} = z_{\text{LMMCF}}^*.$$

Moreover, the dual Lagrangian multipliers λ_{ij}^*, $(i, j) \in A$, are equal to the optimal dual variables π_{ij}^*, $(i, j) \in A$, associated with the relaxed constraints (6.16).

In order to compute the dual Lagrangian multipliers, or at least a set of multipliers associated with a 'good' lower bound, the classical *subgradient procedure*, already illustrated in Chapter 3, can be used.

Step 0. *(Initialization)*. Let H be a pre-established maximum number of subgradient iterations. Set LB $= -\infty$, $h = 1$ and $\lambda_{ij}^{(h)} = 0$, $(i, j) \in A$. Set UB equal to the cost of the best feasible solution if any is available, or set UB $= \infty$, otherwise.

Step 1. *(Calculation of a new lower bound)*. Solve the r-LMMCF problem using $\lambda^{(h)}$ as a vector of multipliers. If $\mathrm{LB_{r\text{-}LMMCF}}(\lambda^{(h)}) > \mathrm{LB}$, set LB $= \mathrm{LB_{r\text{-}LMMCF}}(\lambda^{(h)})$.

Step 2. *(Checking the stopping criterion)*. If solution $x_{ij}^{k,(h)}$, $(i, j) \in A$, $k \in K$, of the r-LMMCF problem satisfies the relaxed constraints (6.16), and

$$\lambda_{ij}^{(h)} \left(\sum_{k \in K} x_{ij}^{k,(h)} - u_{ij} \right) = 0, \quad (i, j) \in A,$$

STOP, the solution found is optimal ($z_{\mathrm{LMMCF}}^* = \mathrm{LB}$). If $x_{ij}^{k,(h)}$, $(i, j) \in A$, $k \in K$, satisfies the relaxed constraints (6.16), update UB if necessary. If UB $= $ LB, *STOP*, the feasible solution attaining UB is proved to be optimal. If $h = H$, *STOP*, LB represents the best lower bound available for z_{LMMCF}^*.

Step 3. *(Updating the multipliers)*. Determine, for each $(i, j) \in A$, the subgradient of the relaxed constraint:

$$s_{ij}^{(h)} = \sum_{k \in K} x_{ij}^{k,(h)} - u_{ij}, \quad (i, j) \in A.$$

Then set

$$\lambda_{ij}^{(h+1)} = \max(0, \lambda_{ij}^{(h)} + \beta^{(h)} s_{ij}^{(h)}), \quad (i, j) \in A,$$

where $\beta^{(h)}$ is a suitable coefficient

$$\beta^{(h)} = \frac{\alpha}{h}, \quad (i, j) \in A, \tag{6.25}$$

with α arbitrarily chosen in the interval $(0, 2]$. Alternatively, if a feasible solution the problem is available, set

$$\beta^{(h)} = \frac{\alpha(\mathrm{UB} - \mathrm{LB_{LMCF}}(\lambda^{(h)}))}{\sum_{(i,j) \in A} (s_{ij}^{(h)})^2}, \quad (i, j) \in A,$$

with UB equal to the objective function value of the best feasible solution available. Set $h = h + 1$ and go back to Step 1.

The subgradient method converges to z_{LMMCF}^* provided that the variations of the multipliers are 'small enough' (this assumption is satisfied if, for example, Equation (6.25) is used). However, this assumption makes the algorithm very slow.

Table 6.4 Demand (in tons) d_{kt}, $k = 1, 2$, $t = 1, 2$, in the Exofruit problem.

	$t = 1$	$t = 2$
$k = 1$	18 000	18 000
$k = 2$	12 000	14 000

Table 6.5 Maximum amounts available (in tons) o_{kt}, $k = 1, 2$, $t = 1, 2$, in the Exofruit problem.

	$t = 1$	$t = 2$
$k = 1$	26 000	20 000
$k = 2$	14 000	13 000

Table 6.6 Purchase prices (in euros/ton) p_{kt}, $k = 1, 2$, $t = 1, 2$, in the Exofruit problem.

	$t = 1$	$t = 2$
$k = 1$	500	700
$k = 2$	600	400

If the solution $x_{ij}^{k,(h)}$, $(i, j) \in A$, $k \in K$, of the r-LMMCF problem satisfies the relaxed constraints (6.16), it is not necessarily the optimal solution for the LMMCF problem. A *sufficient* (but *not necessary*) condition for it to be optimal is given by the *complementary slackness conditions*:

$$\lambda_{ij}^{(h)} \left(\sum_{k \in K} x_{ij}^{k,(h)} - u_{ij} \right) = 0, \quad (i, j) \in A. \qquad (6.26)$$

If relations (6.26) are not satisfied, then the feasible solution $x_{ij}^{k,(h)}$, $(i, j) \in A$, $k \in K$, is a simply a *candidate optimal* solution.

If $h = H$, the solution attaining LB could be infeasible for the LMMCF, or, if feasible, may not satisfy the complementarity slackness conditions. In any case, if subproblems (6.21)–(6.24) are solved by means of the network simplex method, a basic (feasible or infeasible) solution for the LMMCF model is available. In fact, the basic variables of the $|K|$ subproblems (6.21)–(6.24) make up a basis of the LMMCF problem. The basic solution obtained this way can be used as the starting solution for the primal or dual simplex method depending on whether the solution is feasible for the LMMCF problem or not. If initialized this way, the simplex method is particularly efficient since the initial basic solution provided by the subgradient algorithm is a good approximation of the optimal solution.

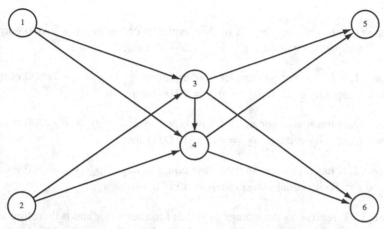

Figure 6.13   Graph representation of the Exofruit problem.

   Exofruit Ltd imports to the EU countries several varieties of tropical fruits, mainly coming from Northern Africa, Mozambique and Central America. The company purchases the products directly from farmers and transports them by sea to its warehouses in Marseille (France). The goods are then stored in refrigerated cells or at room temperature. As purchase and selling prices varies during the year, Exofruit has to decide when and how much to buy in order to satisfy demand over the year. The problem can be modelled as an LMMCF problem. In what follows, a simplified version of the problem is examined. It is assumed that a single source exists, products are grouped into two homogeneous groups (*macro-products*), and the planning horizon is divided into two semesters. Let d_{kt} and o_{kt}, $k = 1, 2$, $t = 1, 2$, be the demand and the maximum amount (in tons) of macro-product k available in semester t, respectively (see Tables 6.4 and 6.5). Purchase prices (in euros/tons) p_{kt}, $k = 1, 2$, $t = 1, 2$, of macro-product k in semester t are reported in Table 6.6.
   The transportation cost v of 1 ton of a macro-product is equal to €100, while the stocking cost w of 1 ton of a macro-product is €100 per semester. Finally, the maximum quantity q of goods that can be stored in a semester is 8000 tons.
   The problem can be formulated as an LMMCF problem with two commodities (one for each macro-product) on the directed graph shown in Figure 6.13. In such a representation,

- vertices 1 and 5 represent the source and the destination of macro-product 1, respectively;

- vertices 2 and 6 represent the source and the destination of macro-product 2, respectively;

- vertices 3 and 4 represent the warehouse in the first and in the second semester, respectively;

- arc $(1,3)$ has a cost per unit of flow equal to $c_{13}^1 = p_{11} + v = 600$ euros/ton and a capacity equal to $u_{13} = o_{11} = 26\,000$ tons;

- arc $(1,4)$ has a cost per unit of flow equal to $c_{14}^1 = p_{12} + v = 800$ euros/ton and a capacity equal to $u_{14} = o_{12} = 20\,000$ tons;

- arc $(2,3)$ has a cost per unit of flow equal to $c_{23}^2 = p_{21} + v = 700$ euros/ton and a capacity equal to $u_{23} = o_{21} = 14\,000$ tons;

- arc $(2,4)$ has a cost per unit of flow equal to $c_{24}^2 = p_{22} + v = 500$ euros/ton and a capacity equal to $u_{24} = o_{22} = 13\,000$ tons;

- arc $(3,4)$ represents the storage of goods for a semester and is therefore associated with a cost per unit of flow equal to $c_{34}^1 = c_{34}^2 = w = 100$ euros/ton and a capacity of $u_{34} = q = 8000$ tons;

- arc $(3,5)$ has a zero cost per unit of flow and a capacity equal to $u_{35} = d_{11} = 18\,000$ tons;

- arc $(4,5)$ has a zero cost per unit of flow and a capacity equal to $u_{45} = d_{12} = 18\,000$ tons;

- arc $(3,6)$ has a zero cost per unit of flow and a capacity equal to $u_{36} = d_{21} = 12\,000$ tons;

- arc $(4,6)$ has a zero cost per unit of flow and a capacity equal to $u_{46} = d_{22} = 14\,000$ tons.

The problem is formulated as follows.

Minimize

$$600x_{13}^1 + 800x_{14}^1 + 100x_{34}^1 + 700x_{23}^2 + 500x_{24}^2 + 100x_{34}^2$$

subject to

$$x_{35}^1 + x_{45}^1 = 36\,000,$$
$$x_{13}^1 - x_{34}^1 - x_{35}^1 = 0,$$
$$x_{14}^1 + x_{34}^1 - x_{45}^1 = 0,$$
$$x_{36}^2 + x_{46}^2 = 26\,000,$$
$$x_{23}^2 - x_{36}^2 - x_{34}^2 = 0,$$
$$x_{24}^2 + x_{34}^2 - x_{46}^2 = 0,$$

$$x_{13}^1 \leqslant 26\,000,$$

$$x_{14}^1 \leqslant 20\,000,$$

$$x_{35}^1 \leqslant 18\,000,$$

$$x_{45}^1 \leqslant 18\,000,$$

$$x_{23}^2 \leqslant 14\,000,$$

$$x_{24}^2 \leqslant 13\,000,$$

$$x_{36}^2 \leqslant 12\,000,$$

$$x_{46}^2 \leqslant 14\,000,$$

$$x_{34}^1 + x_{34}^2 \leqslant 8000,$$

$$x_{13}^1, \; x_{14}^1, \; x_{34}^1, \; x_{35}^1, \; x_{45}^1, \; x_{23}^2, \; x_{24}^2, \; x_{34}^2, \; x_{36}^2, \; x_{46}^2 \geqslant 0.$$

By relaxing in a Lagrangian fashion the constraint $x_{34}^1 + x_{34}^2 \leqslant 8000$ with a multiplier λ_{34}, the problem decomposes into two single-commodity linear minimum-cost flow problems. By initializing the subgradient algorithm with $\lambda_{34}^{(0)} = 0$ and using the updating formula (6.25) with $\alpha = 0.05$, the procedure provides, after 20 iterations, a lower bound LB $= €40\,197\,387$ ($\lambda_{34}^{(20)} = 99.887$). At the end, the procedure converges to $\lambda_{34}^* = 100$, which corresponds to an optimal objective value z_{LMMCF}^* equal to $€40\,200\,000$. Subproblem $k = 1$ has, for λ_{34}^*, two optimal basic solutions

$$x_{13}^{1,*} = x_{14}^{1,*} = x_{35}^{1,*} = x_{45}^{1,*} = 18\,000, \quad x_{34}^{1,*} = 0$$

and

$$x_{13}^{1,*} = 26\,000, \quad x_{14}^{1,*} = 10\,000, \quad x_{34}^{1,*} = 8000, \quad x_{35}^{1,*} = 18\,000, \quad x_{45}^{1,*} = 18\,000,$$

while subproblem $k = 2$ has a single optimal solution equal to

$$x_{23}^{2,*} = x_{24}^{2,*} = 13\,000, \quad x_{34}^{2,*} = 1000, \quad x_{36}^{2,*} = 12\,000, \quad x_{46}^{2,*} = 14\,000.$$

By combining the two partial solutions, the following two solutions are obtained:

$$x_{13}^{1,*} = x_{14}^{1,*} = x_{35}^{1,*} = x_{45}^{1,*} = 18\,000, \quad x_{34}^{1,*} = 0,$$

$$x_{23}^{2,*} = x_{24}^{2,*} = 13\,000, \quad x_{34}^{2,*} = 1000, \quad x_{36}^{2,*} = 12\,000, \quad x_{46}^{2,*} = 14\,000$$

and

$$x_{13}^{1,*} = 26\,000, \qquad x_{14}^{1,*} = 10\,000, \qquad x_{34}^{1,*} = 8000,$$

$$x_{35}^{1,*} = 18\,000, \qquad x_{45}^{1,*} = 18\,000, \qquad x_{23}^{2,*} = x_{24}^{2,*} = 13\,000,$$

$$x_{34}^{2,*} = 1000, \qquad x_{36}^{2,*} = 12\,000, \qquad x_{46}^{2,*} = 14\,000.$$

The former solution has an objective function value equal to $€41\,000\,000$ and is feasible, but does not satisfy the complementarity slackness conditions (6.26), while

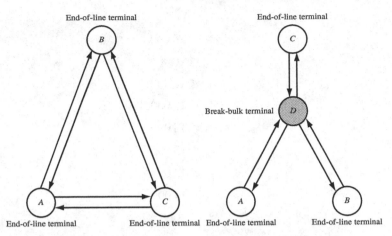

Figure 6.14   Two alternative service networks for a three end-of-line terminal transportation system.

the latter is infeasible. It can be easy verified that the optimal solution is a convex combination of the two previous solutions and corresponds to

$$x_{13}^{1,*} = 25\,000, \quad x_{14}^{1,*} = 11\,000, \quad x_{34}^{1,*} = 7000, \quad x_{35}^{1,*} = 18\,000, \quad x_{45}^{1,*} = 18\,000,$$
$$x_{23}^{2,*} = 12\,000, \quad x_{24}^{2,*} = 13\,000, \quad x_{34}^{2,*} = 1000, \quad x_{36}^{2,*} = 12\,000, \quad x_{46}^{2,*} = 14\,000.$$

6.6   Service Network Design Problems

The design of a network of transportation services is a tactical/operational decision particularly relevant to consolidation-based carriers. Given a set of terminals, the service network design problem consists of deciding on the characteristics (frequency, number of intermediate stops, etc.) of the routes to be operated, the traffic assignment along these routes, the operating rules at each terminal, and possibly the relocation of empty vehicles and containers. The objective is the minimization of a generalized cost taking into account a combination of carrier's operating costs and customers' expectations. Figure 6.14 shows two alternative service networks for a three-terminal transportation system in which it is assumed that each arc is associated with a line operated once a day. In the former network, each terminal is connected directly to every other terminal (so that each shipment takes one day) but this comes at the expense of a higher operating cost. In the latter network, operating costs are lower but the transportation between certain origin–destination pairs may require two days (unless all lines are synchronized).

Classification. Service network design models can be classified into two main categories: *frequency-based* and *dynamic* models. In frequency-based models the decision variables express how often each transportation service is operated in a given time horizon, while in dynamic models a time-expanded network is used to provide a more detailed description of the system. In the remainder of this chapter, the focus is on basic network design problems (namely, *fixed-charge network design* (FCND) problems) and on dynamic service network design models. For a description of frequency-based models, the reader should consult the references listed at the end of the chapter.

6.6.1 Fixed-charge network design models

FCND problems may be viewed as a generalization of NF problems in which a fixed cost f_{ij} has to be paid for using each arc $(i, j) \in A$. Therefore, FCND problems amount to determining

- which arcs have to be employed;

- how to transport the commodities on the selected arcs.

Let x_{ij}^k, $(i, j) \in A$, $k \in K$, be the flow of commodity k on arc (i, j), and let y_{ij}, $(i, j) \in A$, be a binary design variable, equal to 1 if arc (i, j) is used, 0 otherwise. A quite general formulation of the FCND problem is as follows.

Minimize

$$\sum_{k \in K} \sum_{(i,j) \in A} C_{ij}^k(x_{ij}^k) + \sum_{(i,j) \in A} f_{ij} y_{ij} \tag{6.27}$$

subject to

$$\sum_{\{j \in V:(i,j) \in A\}} x_{ij}^k - \sum_{\{j \in V:(j,i) \in A\}} x_{ji}^k = \begin{cases} o_i^k, & \text{if } i \in O(k), \\ -d_i^k, & \text{if } i \in D(k), \\ 0, & \text{if } i \in T(k), \end{cases} \quad i \in V, \ k \in K, \tag{6.28}$$

$$x_{ij}^k \leqslant u_{ij}^k, \quad (i, j) \in A, \ k \in K, \tag{6.29}$$

$$\sum_{k \in K} x_{ij}^k \leqslant u_{ij} y_{ij}, \quad (i, j) \in A, \tag{6.30}$$

$$x_{ij}^k \geqslant 0, \quad (i, j) \in A, \ k \in K,$$

$$y_{ij} \in \{0, 1\}, \quad (i, j) \in A.$$

The objective function (6.27) is the total transportation cost. Constraints (6.28) correspond to the flow conservation constraints holding at each vertex $i \in V$ and for each commodity $k \in K$; the constraints (6.29) impose that the flow of each commodity $k \in K$ does not exceed capacity u_{ij}^k on each arc $(i, j) \in A$; the constraints (6.30) (*bundle constraints*) require that, for each $(i, j) \in A$, the total flow on arc (i, j) is zero if the arc is not used, or not greater than capacity u_{ij}, otherwise.

In practice, some side constraints may be needed to represent economic and topological restrictions. For example, when several links share a common resource, the following *budget* constraint has to be added to the FCND model,

$$\sum_{(i,j)\in A} h_{ij} y_{ij} \leqslant H,$$

where h_{ij}, $(i, j) \in A$, is the consumption of resource made by arc $(i, j) \in A$, and H is the total amount of resource available.

6.6.2　The linear fixed-charge network design model

The *linear fixed-charge network design* (LFCND) problem is a particular FCND problem in which the transportation costs per flow unit c_{ij}^k are constant (hence the objective function (6.27) is linear).

More formally, the LFCND model can be formulated as follows.

Minimize

$$\sum_{k\in K} \sum_{(i,j)\in A} c_{ij}^k x_{ij}^k + \sum_{(i,j)\in A} f_{ij} y_{ij}$$

subject to

$$\sum_{\{j\in V:(i,j)\in A\}} x_{ij}^k - \sum_{\{j\in V:(j,i)\in A\}} x_{ji}^k = \begin{cases} o_i^k, & \text{if } i \in O(k), \\ -d_i^k, & \text{if } i \in D(k), \quad i \in V, \; k \in K, \\ 0, & \text{if } i \in T(k), \end{cases}$$

$$x_{ij}^k \leqslant u_{ij}^k, \quad (i, j) \in A, \; k \in K,$$

$$\sum_{k\in K} x_{ij}^k \leqslant u_{ij} y_{ij}, \quad (i, j) \in A,$$

$$x_{ij}^k \geqslant 0, \quad (i, j) \in A, \; k \in K,$$

$$y_{ij} \in \{0, 1\}, \quad (i, j) \in A.$$

The LFCND problem is NP-hard and branch-and-bound algorithms can hardly solve instances with a few hundreds of arcs and tens of commodities. Since instances arising in applications are much larger, heuristics are often used. To evaluate the quality of the solutions provided by heuristics, it is useful, as already observed in Chapter 3, to compute lower bounds on the optimal objective function value z_{LFCND}^*. In the following, two distinct continuous relaxations and a simple heuristic are illustrated.

The weak continuous relaxation

The weak continuous relaxation is obtained by relaxing the integrality requirement on the design variables.

Minimize

$$\sum_{k \in K} \sum_{(i,j) \in A} c_{ij}^k x_{ij}^k + \sum_{(i,j) \in A} f_{ij} y_{ij} \qquad (6.31)$$

subject to

$$\sum_{\{j \in V : (i,j) \in A\}} x_{ij}^k - \sum_{\{j \in V : (j,i) \in A\}} x_{ji}^k = \begin{cases} o_i^k, & \text{if } i \in O(k), \\ -d_i^k, & \text{if } i \in D(k), \qquad i \in V, \ k \in K, \\ 0, & \text{if } i \in T(k), \end{cases}$$

$$\qquad (6.32)$$

$$x_{ij}^k \leqslant u_{ij}^k, \quad (i,j) \in A, \ k \in K, \qquad (6.33)$$

$$\sum_{k \in K} x_{ij}^k \leqslant u_{ij} y_{ij}, \quad (i,j) \in A, \qquad (6.34)$$

$$x_{ij}^k \geqslant 0, \quad (i,j) \in A, \ k \in K, \qquad (6.35)$$

$$0 \leqslant y_{ij} \leqslant 1, \quad (i,j) \in A. \qquad (6.36)$$

It is easy to verify that every optimal solution of such a relaxation satisfies each constraint (6.34) as an equality since fixed costs f_{ij}, $(i, j) \in A$, are nonnegative. Therefore, design variables y_{ij}, $(i, j) \in A$, can be expressed as a function of flow variables x_{ij}^k, $(i, j) \in A, k \in K$:

$$y_{ij} = \frac{\sum_{k \in K} x_{ij}^k}{u_{ij}}, \quad (i, j) \in A.$$

Hence, the constraints (6.36) can be replaced by the following conditions:

$$\sum_{k \in K} x_{ij}^k \leqslant u_{ij}, \quad (i, j) \in A.$$

The relaxed problem (6.31)–(6.36) can be therefore equivalently formulated as follows.

Minimize

$$\sum_{k \in K} \sum_{(i,j) \in A} \left(c_{ij}^k + \frac{f_{ij}}{u_{ij}} \right) x_{ij}^k \qquad (6.37)$$

subject to

$$\sum_{\{j \in V : (i,j) \in A\}} x_{ij}^k - \sum_{\{j \in V : (j,i) \in A\}} x_{ji}^k = \begin{cases} o_i^k, & \text{if } i \in O(k), \\ -d_i^k, & \text{if } i \in D(k), \qquad i \in V, \ k \in K, \\ 0, & \text{if } i \in T(k), \end{cases}$$

$$\qquad (6.38)$$

$$x_{ij}^k \leqslant u_{ij}^k, \quad (i, j) \in A, \ k \in K, \tag{6.39}$$

$$\sum_{k \in K} x_{ij}^k \leqslant u_{ij}, \quad (i, j) \in A, \tag{6.40}$$

$$x_{ij}^k \geqslant 0, \quad (i, j) \in A, \ k \in K. \tag{6.41}$$

The model (6.37)–(6.41) is a minimum-cost flow problem with $|K|$ commodities. Let LB_w^* be the lower bound on z_{LFCND}^* given by the optimal objective function value of the above relaxation.

The strong continuous relaxation

The strong continuous relaxation is obtained by adding the following *valid* inequalities

$$x_{ij}^k \leqslant u_{ij}^k y_{ij}, \quad (i, j) \in A, \ k \in K, \tag{6.42}$$

to the LFCND model and removing the integrality constraints on the design variables y_{ij}, $(i, j) \in A$. Taking into account the fact that constraints (6.6.2) are dominated by constraints (6.42), and can therefore eliminated, the relaxed problem is as follows.

Minimize

$$\sum_{k \in K} \sum_{(i,j) \in A} c_{ij}^k x_{ij}^k + \sum_{(i,j) \in A} f_{ij} y_{ij} \tag{6.43}$$

subject to

$$\sum_{\{j \in V : (i,j) \in A\}} x_{ij}^k - \sum_{\{j \in V : (j,i) \in A\}} x_{ji}^k = \begin{cases} o_i^k, & \text{if } i \in O(k), \\ -d_i^k, & \text{if } i \in D(k), \\ 0, & \text{if } i \in T(k), \end{cases} \quad i \in V, \ k \in K, \tag{6.44}$$

$$x_{ij}^k \leqslant u_{ij}^k y_{ij}, \quad (i, j) \in A, \ k \in K, \tag{6.45}$$

$$\sum_{k \in K} x_{ij}^k \leqslant u_{ij} y_{ij}, \quad (i, j) \in A, \tag{6.46}$$

$$x_{ij}^k \geqslant 0, \quad (i, j) \in A, \ k \in K, \tag{6.47}$$

$$0 \leqslant y_{ij} \leqslant 1, \quad (i, j) \in A. \tag{6.48}$$

Let LB_s^* be the lower bound on z_{LFCND}^* given by the optimal objective function value of the relaxation (6.43)–(6.48). Such problem has no special structure and can therefore be solved by using any general purpose LP algorithm. By comparing the two continuous relaxations, it is clear that LB_s^* is always better than, or at least equal to, LB_w^*, i.e.

$$LB_s^* \geqslant LB_w^*.$$

This observation leads us to label the former relaxation as *weak*, and the latter as *strong*. Computational experiments have shown that LB_w^* can be as much as 40% lower than LB_s^*.

Table 6.7 Forecasted transportation demand of refrigerated goods (pallets per day) in the FHL problem.

	Bologna	Genoa	Milan	Padua
Bologna	—	3	8	2
Genoa	0	—	1	2
Milan	4	2	—	1
Padua	3	1	1	—

Table 6.8 Forecasted transportation demand of goods at room temperature (pallets per day) in the FHL problem.

	Bologna	Genoa	Milan	Padua
Bologna	—	3	4	2
Genoa	1	—	1	0
Milan	6	2	—	2
Padua	1	1	1	—

FHL is an Austrian fast carrier located in Lienz, whose core business is the transportation of small-sized and high-valued refrigerated goods (such as chemical reagents used by hospitals and laboratories). Goods are picked up from manufacturers' warehouses by small vans and carried to the nearest transportation terminal operated by the carrier. These goods are packed onto pallets and transported to destination terminals by means of large trucks. Then, the merchandise is unloaded and delivered to customers by small vans (usually the same vans employed for pick-up). In order to make capital investment in equipment as low as possible, FHL makes use of one-way rentals of trucks. Recently, the company has decided to enter the Italian fast parcel transportation market by opening four terminals in the cities of Bologna, Genoa, Padua and Milan. This choice made necessary a complete revision of the service network. The decision was complicated by the need to transport the refrigerated goods by special vehicles equipped with refrigerators, while parcels can be transported by any vehicle. The forecasted daily average demand of the two kinds of products in the next semester is reported in Tables 6.7 and 6.8.

Between each pair of terminals, the company can operate one or more lines (see Figure 6.15). Vehicles are of two types:

- trucks with refrigerated compartments, having a capacity of 12 pallets and a cost (inclusive of all charges) of €0.4 per kilometre;

- trucks with room-temperature compartments, having a capacity of 18 pallets and a cost (inclusive of all charges) of €0.5 per kilometre.

In addition, the company considers the possibility of transporting goods at room temperature through another carrier, by paying €0.1 per kilometre for each pallet. A directed graph representation of the problem is given in Figure 6.15.

Distances between terminals are reported in Table 6.9. The least-cost service network can be obtained as the solution of an LFCND model with $|K| = 22$ commodities (one for each combination of an origin–destination pair with positive demand and a kind of product). Let A_1 and A_2 be the set of lines operated by means of trucks having capacity equal to 12 pallets and 18 pallets, respectively, and let A_3 be the set of lines operated by an external carrier. Arc parameters are

$$c_{ij}^k = 0, \quad (i, j) \in A_1, \ k \in K,$$

$$f_{ij} = 0.4 \quad d_{ij}, \ (i, j) \in A_1,$$

$$u_{ij} = 12, \quad (i, j) \in A_1,$$

$$c_{ij}^k = 0, \quad (i, j) \in A_2, \ k \in K,$$

$$f_{ij} = 0.5, \quad d_{ij}, \ (i, j) \in A_2,$$

$$u_{ij} = 18, \quad (i, j) \in A_2,$$

$$c_{ij}^k = 0.1, \quad d_{ij}, \ (i, j) \in A_3, \ k \in K,$$

$$f_{ij} = 0, \quad (i, j) \in A_3,$$

$$u_{ij} = \infty, \quad (i, j) \in A_3,$$

where d_{ij} is the distance between terminals i and j.

The LFCND formulation is as follows.

Minimize

$$\sum_{k \in K} \sum_{(i,j) \in A_1 \cup A_2 \cup A_3} c_{ij}^k x_{ij}^k + \sum_{(i,j) \in A_1 \cup A_2 \cup A_3} f_{ij} y_{ij}$$

subject to

$$\sum_{\{j \in V:(i,j) \in A_1 \cup A_2 \cup A_3\}} x_{ij}^k - \sum_{\{j \in V:(j,i) \in A_1 \cup A_2 \cup A_3\}} x_{ji}^k = \begin{cases} o_i^k, & \text{if } i \in O(k), \\ -d_i^k, & \text{if } i \in D(k), \\ 0, & \text{if } i \in T(k), \end{cases}$$

$$i \in V, \ k \in K,$$

$$\sum_{k \in K} x_{ij}^k, \quad (i, j) \in A_1 \cup A_2 \cup A_3,$$

$$x_{ij}^k \geqslant 0, \quad (i, j) \in A_1 \cup A_2 \cup A_3, \ k \in K$$

$$y_{ij} \in \{0, 1\}, \ (i, j) \in A_1 \cup A_2 \cup A_3.$$

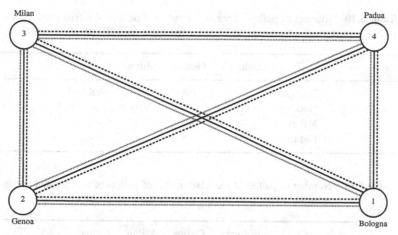

Figure 6.15 Graph representation of the FHL service network design problem (in order to make the picture clean, a single edge is drawn for each pair of opposite arcs).

Table 6.9 Distances (in kilometres) between terminals in the FHL problem.

	Bologna	Genoa	Milan	Padua
Bologna	0	225	115	292
Genoa	225	0	226	166
Milan	115	226	0	362
Padua	292	166	362	0

The strong continuous relaxation gives a lower bound $LB_s^* = €534.60$ per day. A branch-and-bound algorithm based on the strong continuous relaxation generates 696 nodes. The least-cost solution is reported in Figure 6.16 and has a cost of €886.70 per day. In the optimal solution, 7 lines are operated by 12-pallet trucks while a single line is operated by an 18-pallet truck (travelling from Bologna to Milan with 5 pallets of parcels). The number of pallets transported between each pair of terminals by means of 12-pallet trucks and by the external carrier is reported in Tables 6.10 and 6.11.

Add–drop heuristics

Add–drop heuristics are simple constructive procedures in which at each step one decides whether a new arc has to be used (*add procedure*) or an arc previously used has to be left out (*drop procedure*). Several criteria can be employed to choose which arc has to be added or dropped. In the following, a very simple drop procedure is illustrated. In order to describe such an algorithm, it is worth noting that a candidate optimal solution is characterized by the set $A' \subseteq A$ of selected arcs. A solution

Table 6.10 Number of pallets of parcels/number of pallets of refrigerated goods transported by means of 12-pallet trucks in the FHL problem.

	Bologna	Genoa	Milan	Padua
Bologna	—	4/8	—	4/8
Genoa	0/7	—	2/10	—
Milan	—	5/7	—	—
Padua	8/3	4/5	—	—

Table 6.11 Number of pallets of parcels/number of pallets of refrigerated goods transported by an external carrier in the FHL problem.

	Bologna	Genoa	Milan	Padua
Bologna	—	1/0	—	—
Genoa	—	—	4/0	9/0
Milan	—	—	—	—
Padua	—	—	—	—

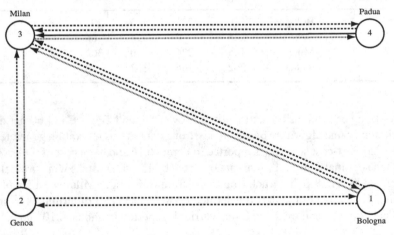

Figure 6.16 Transportation lines used in the optimal solution of the FHL problem.

is feasible if the LMMCF problem on the directed graph $G(V, A')$ induced by A' is feasible. If so, the solution cost is made up of the sum of the fixed costs $f_{ij}, (i, j) \in A'$, plus the optimal solution cost of the LMMCF problem. Moreover, it is worth noting that the LFCND solution associated with $A' = A$, if feasible, is characterized by a large fixed cost and by a low transportation cost. On the other hand, a feasible solution associated with a set A' with a few arcs is expected to be characterized by a low fixed cost and by a high variable cost. Consequently, an improved LFCND solution can be

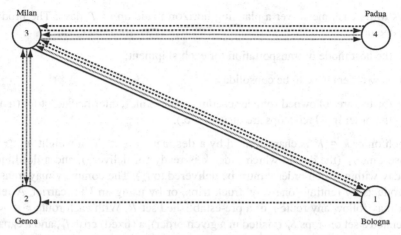

Figure 6.17 Solution provided by the drop heuristic in the FHL problem.

obtained by iteratively removing arcs from the set $A' = A$, while the current solution is still feasible and the total cost decreases. The drop procedure is as follows.

Step 1. Set $h = 0$ and $A^{(h)} = A$. Let $x_{ij}^{k,(h)}$, $(i, j) \in A$, $k \in K$, be the optimal solution (if any exists) of the LMMCF problem on the directed graph $G(V, A^{(h)})$ and let $z_{\text{LFCND}}^{(h)}$ be the cost of the associated LFCND problem. If the LMMCF problem is infeasible, *STOP*, the LFCND problem is also infeasible.

Step 2. For each arc $(i, j) \in A^{(h)}$, set $A'^{(h)} = A^{(h)} \setminus \{(i, j)\}$ and solve the LMMCF problem on the directed graph $(G(V, A'^{(h)})$. If all the LMMCF problems are infeasible, *STOP*, the set of the arcs $A^{(h)}$ and the flow pattern $x_{ij}^{k,(h)}$, $(i, j) \in A^{(h)}, k \in K$, are associated with the best feasible solution found; otherwise, let (v, w) be the arc whose removal from $A^{(h)}$ allows us to attain the least-cost LFCND feasible solution.

Step 3. Set $A^{(h+1)} = A^{(h)} \setminus (v, w)$, $h = h + 1$ and go back to Step 2.

The number of iterations of the algorithm is no more than the number of arcs and, at each iteration, Step 2 requires the solution of $O(|A|)$ LMMCF problems.

By applying the drop heuristic to the FHL problem, a solution having a cost equal to €899 per day is obtained (see Figure 6.17).

6.7 Shipment Consolidation and Dispatching

In this section, we examine a consolidation and dispatching problem often faced by manufacturers. A producer has to choose the best way of delivering timely a set of orders to its customers over a planning horizon made up of T days. The producer must decide

- the best mode of transportation for each shipment;

- how orders have to be consolidated;

- the features of owned vehicle schedules (start times, intermediate stops (if any), the order in which stops are visited, etc.).

Each order $k \in K$ is characterized by a destination $i_k \in N$, a weight $w_k \geqslant 0$, a release time r_k (the day in which order k is ready for delivery), and a deadline d_k (the day within which order k must be delivered to i_k). The company may transport its products by renting 'one-way' truck trips, or by using an LTL carrier. A rented truck may follow any route r of a pre-established set R. With each route $r \in R$ are associated a set of stops S_r (visited in a given order), a (fixed) cost f_r, and a capacity q_r (the maximum weight that the vehicle operating route r can carry). Moreover, let τ_{kr}, $k \in K$, $r \in R$, be the number of travel days needed to deliver order k on route r. Transporting order k to its destination by common carrier costs g_k and takes τ'_k days. The decision variables are x_{krt}, $k \in K$, $r \in R$, $t = 1, \ldots, T$, of a binary type, having a value equal to 1 if order k is assigned to route r starting on day t, 0 otherwise; y_{rt}, $r \in R$, $t = 1, \ldots, T$, a binary variable equal to 1 if route r is operated on day t, 0 otherwise; w_k, a binary variable equal to 1 if order k is transported by the common carrier, 0 otherwise (such a variable is defined only if $r_k + \tau'_k \leqslant d_k$).

Minimize

$$\sum_{r \in R} \sum_{t=1}^{T} f_r y_{rt} + \sum_{k \in K} g_k w_k \tag{6.49}$$

subject to

$$\sum_{k: r_k \leqslant t \leqslant d_k - \tau_{kr}, i_k \in S_r} w_k x_{krt} \leqslant q_r y_{rt}, \quad r \in R, \ t = 1, \ldots, T, \tag{6.50}$$

$$\sum_{r: i_k \in S_r} \sum_{t: r_k \leqslant t \leqslant d_k - \tau_{kr}} x_{krt} + w_k = 1, \quad k \in K, \tag{6.51}$$

$$x_{krt} \in \{0, 1\}, \quad k \in K, \ r \in R, \ t = 1, \ldots, T, \tag{6.52}$$

$$y_{rt} \in \{0, 1\}, \quad r \in R, \ t = 1, \ldots, T, \tag{6.53}$$

$$w_k \in \{0, 1\}, \quad k \in K. \tag{6.54}$$

The objective function (6.49) is the total cost paid to transport orders. Constraints (6.50) state that, for each route $r \in R$ and for each day $t = 1, \ldots, T$ the total weight carried on route r, on day t, must not exceed capacity q_r if y_{rt} is equal to 1, and is equal to 0, otherwise. Constraints (6.51) impose that each order is assigned to a route operated by a rented truck or to the common carrier. It is easy to show that formulation (6.49)–(6.54) can be transformed into a network design model on a time-expanded directed graph.

Table 6.12 Trip fixed costs (in euros) in the Oxximet problem.

Trip	Fixed cost
Bari–Marseille	800
Bari–Milan	750
Bari–Ancona	400
Bari–Ancona–Milan	780
Bari–Ancona–Marseille	830
Bari–Milan–Marseille	830
Bari–Ancona–Milan–Marseille	860

Table 6.13 Customers orders (in hundreds of kilograms) in the Oxximet problem.

Order	Customer location	Release time	Deadline	Weight
1	Marseille	16 June	3 July	15.44
2	Ancona	14 June	15 June	242.65
3	Ancona	14 June	16 June	102.54
4	Milan	13 June	13 June	100.46
5	Marseille	14 June	15 June	154.79
6	Marseille	14 June	14 June	78.53
7	Marseille	13 June	13 June	56.89
8	Marseille	14 June	14 June	45.42
9	Marseille	13 June	13 June	39.55
10	Marseille	11 June	15 June	207.34
11	Marseille	11 June	16 June	19.05
12	Marseille	11 June	16 June	19.59
13	Marseille	11 June	16 June	35.23
14	Marseille	11 June	16 June	61.54
15	Milan	11 June	16 June	38.31

Oxximet manufactures semi-finished chemical products in a plant located close to Bari (Italy). The main customers are located in Marseille (France), Milan (Italy) and Ancona (Italy). Oxximet rents 'one-way' truck trips visiting one or more customers. No common carrier is used. Trip fixed costs are reported in Table 6.12. Truck capacity is 260 hundred kilograms. Last 11 June, 15 customer orders were waiting to be satisfied (Table 6.13). Oxximet solved model (6.49)–(6.54) with a six-day horizon ($T = 6$) and $g_k = \infty$. No trips were allowed on the subsequent Saturday and Sunday. The optimal solution is reported in Table 6.14. It is worth noting that no truck trip was rented on 12 June.

Table 6.14 Optimal trip schedule in the Oxximet problem.

Day	Trip	Orders
11 June	Bari–Marseille	10, 13
12 June	—	—
13 June	Bari–Milan–Marseille	4, 7, 9, 11, 15
	Bari–Marseille	5, 6, 12
14 June	Bari–Ancona	2
	Bari–Ancona–Marseille	3, 8, 14
17 June	Bari–Marseille	1

6.8   Freight Terminal Design and Operations

Freight terminals are places where shipments are classified, consolidated, possibly stored for a short time, and moved from incoming to outgoing vehicles. These include LTL terminals, crossdocks, package-handling terminals (such as those of UPS and FedEx) as well as rail, port and airport terminals. These facilities share several features and consequently their design and management can be approached using common methodologies. However, there are also significant differences among them. In this section the analysis is restricted to terminals where material handling is labour intensive (e.g. loads are moved by forklifts). This is the case, e.g. of LTL terminals and crossdocks, where, as packages often have different sizes, it is difficult to make use of automatic equipment.

6.8.1   Design issues

The first design decision that has to be made is how many *doors* (or *gates*) a terminal should have. Crossdocks have two kinds of doors: *receiving* doors and *shipping* doors (see Chapter 1). In both cases, a door is assigned a share of floor space (see Figure 6.18). The numbers of receiving and shipping doors are customarily set equal to the number of destinations that have to be served, although high-volume destinations may be assigned more than one door in order to accommodate multiple trucks at the same time. In LTL terminals, the number of doors may be estimated by means of simple formulae like those used in Chapter 5 for calculating the number of docks of a warehouse. A better evaluation can be done through simulation models once other design variables have been set.

Another key design decision is related to the choice of the terminal shape. Terminals may have hundreds of doors so that distances travelled by workers in a trip may exceed several hundred metres. It is then crucial to choose a shape that minimizes the expected overall workload for a given number of doors. In practice, docks in the shape of an I, L, T and H are quite common. In order to compare the different shapes, two performance measures (named the *diameter* and the *centrality* of a terminal) have been introduced. The diameter of a terminal is the largest distance between any pair

Table 6.15 Characteristics of the most common terminal shapes.

Shape	Centrality	Number of corners (Inside − Outside)
I	$4/2 = 2$	$0 - 4$
L	$4/2 = 2$	$1 - 5$
T	$6/2 = 3$	$2 - 6$
H	$8/2 = 4$	$4 - 8$

of doors. The centrality is the rate at which the diameter grows as the number of door increases. For an I-shaped terminal the diameter is 2. In fact, if two doors are added at each end, the diameter increases by two doors so that the centrality is $4/2 = 2$. The centrality of the most common shapes are reported in Table 6.15. Larger values of centrality are better for two reasons: the workload is smaller and the forklift traffic congestion is lighter. For example, for an I-shape, forklift traffic may be very heavy in the middle of the terminal (in fact, it varies with the square of the number of the doors) while for an H-shape it is much lighter *ceteris paribus*. On the other hand, designs different from the I-shape show a deterioration in efficiency due to the higher number of *corners*. *Corners* reduce the potential number of doors along the perimeter for two reasons (see Figures 6.18 and 6.19). First, each door needs a suitable amount of floor space, otherwise there is an interference•between adjacent doors. Second, where orthogonal segments of a terminal join, doors are unusable because vehicles would overlap otherwise. As a result, for a given number of doors, the terminal has to be larger if the number of corners increases.

The previous considerations suggest discarding L-shape designs because they have the same value of centrality as the I-shape but more corners. Simulation-based studies have shown that the I-shape is best for small to mid-sized terminals. The T-shape is best for terminals with about 150 to 250 doors, while for larger terminals the H-shape should be selected. Figure 6.20 illustrates qualitatively the expected workloads for the I, T and H-shapes. The exact breakpoints depend on the material flow pattern and on how vehicles are assigned to doors.

6.8.2 Tactical and operational issues

At a tactical or operational level, one must decide how vehicles should be assigned to gates. Such a decision is influenced mostly by the type of terminal. In rail terminals and airports, vehicles are dynamically assigned to doors, while in crossdocks doors are permanently labelled as receiving or shipping doors. Moreover, in a crossdock, doors do not usually change designations because this allows workers to be more efficient when handling freight.

For a crossdock, a good quality assignment can be obtained through the following constructive heuristic (*alternating heuristic*), possibly followed by a local search procedure.

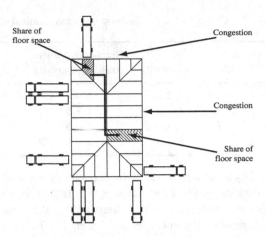

Figure 6.18 I-shaped terminal.

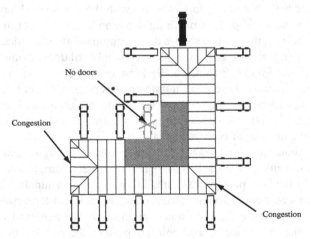

Figure 6.19 L-shaped terminal (unusable doors are shadowed).

Step 1. Sort doors by nondecreasing average distance to all other doors.

Step 2. Sort outbound trailers by nonincreasing freight flows.

Step 3. Assign alternatively the busiest inbound vehicle (trailer) and the busiest outbound trailer to the best locations still available. If there exist some more trailers to allocate, repeat Step 3, otherwise *STOP*.

> Blue Freight has a T-shaped crossdock with 200 doors in Denver (Colorado, USA). By applying the alternating heuristic the solution shown in Figure 6.21 is obtained.

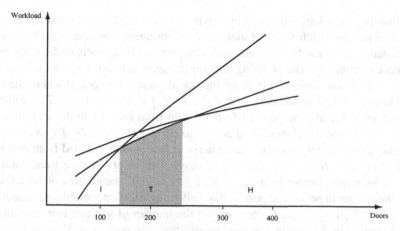

Figure 6.20   Expected workloads for the I, T and H-shapes.

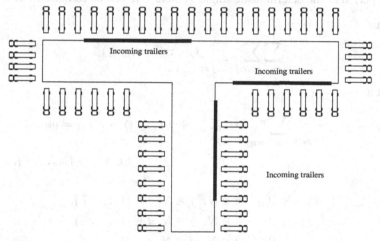

Figure 6.21   Trailer allocation provided by the alternating heuristic in
the Blue Freight crossdock (receiving doors are bold).

6.9   Vehicle Allocation Problems

Vehicle allocation problems (VAPs) are faced by carriers that generate revenue by
transporting *full loads* over long distances, as in TL trucking and container shipping.
Once a vehicle delivers a load, it becomes empty and has to be moved to the pick-
up point of another load, or has to repositioned in anticipation of future demands.
A VAP amounts to deciding which loads have to be accepted and which ones have
to be rejected, as well as repositioning empty vehicles. In what follows, the VAP is
modelled as a minimum-cost flow problem on a time-expanded directed graph for
the case where all demands are known in advance. For the sake of simplicity, we

examine the case where a single vehicle type exists, while the extension to the multi-vehicle type case is left to the reader as an exercise (see Problem 6.9). The case in which demands are random is much more complex and is currently under study by the scientific community. The planning horizon is assumed to comprise a finite number $\{1, \ldots, T\}$ of time periods. Let N be the set of points (e.g. cities) where the (full) loads have to picked up and delivered; $d_{ijt}, i \in N, j \in N, t = 1, \ldots, T$, the number of loads available at time period t to be moved from origin i to destination j; τ_{ij}, $i \in N, j \in N$, the travel time from point i to point j; $p_{ij}, i \in N, j \in N$, the profit (revenue minus direct operating costs) derived from moving a load from point i to point j; $c_{ij}, i \in N, j \in N$, the cost of moving an empty vehicle from point i to point j. Moreover, denote by $m_{it}, i \in N, t = 1, \ldots, T$, the number of vehicles that enter the system in period t at point i. The following decision variables are used: x_{ijt}, $i \in N, j \in N, t = 1, \ldots, T$, representing the number of vehicles that start moving a load from point i to point j at time period t; $y_{ijt}, i \in N, j \in N, t = 1, \ldots, T$, representing the number of vehicles that start moving empty from point i to point j at time period t. The deterministic single-vehicle VAP can be formulated as follows.

Maximize

$$\sum_{t=1}^{T} \sum_{i \in N} \sum_{j \in N, j \neq i} (p_{ij} x_{ijt} - c_{ij} y_{ijt}) \tag{6.55}$$

subject to

$$\sum_{j \in N} (x_{ijt} + y_{ijt}) - \sum_{k \in N, k \neq i : t > \tau_{ki}} (x_{ki(t - \tau_{ki})} + y_{ki(t - \tau_{ki})}) - y_{iit-1} = m_{it},$$

$$i \in V, t \in \{1, \ldots, T\}, \tag{6.56}$$

$$x_{ijt} \leqslant d_{ijt}, \quad i \in N, \ j \in N, \ t \in \{1, \ldots, T\}, \tag{6.57}$$

$$x_{ijt} \geqslant 0, \quad i \in N, \ j \in N, \ t \in \{1, \ldots, T\},$$

$$y_{ijt} \geqslant 0, \quad i \in N, \ j \in N, \ t \in \{1, \ldots, T\}.$$

The objective function (6.55) is the total discounted profit over the planning horizon. Constraints (6.56) impose flow conservation at the beginning of each time period, while constraints (6.57) state that the number of loaded movements is bounded above by the demand. It is worth noting that the $d_{ijt} - x_{ijt}$ differences, $i \in N, j \in N, t = 1, \ldots, T$, represent the loads that should be rejected, while y_{iit} variables, $i \in N, t = 1, \ldots, T$, represent vehicles staying idle (the so-called *inventory movements*). It is easy to recognize that the VAP can be modelled as a minimum-cost flow problem on a time-expanded directed graph in which vertices are associated with (i, t) pairs, $i \in N, j \in N$, and arcs represent loaded, empty and inventory movements. Therefore, integrality constraints on x_{iit} and y_{iit} variables, $i \in N, j \in N, t = 1, \ldots, T$, are implicitly satisfied. Since there exists a pair of arcs between each pair of distinct nodes, such a network is a directed multigraph.

Table 6.16 Travel times (in days) between terminals in the Murty problem.

	Ananthapur	Chittoor	Ichapur	Khammam	Srikakulam
Ananthapur	0	1	2	2	2
Chittoor	1	0	2	2	2
Ichapur	2	2	0	2	1
Khammam	2	2	2	0	2
Srikakulam	2	2	1	2	0

Murty is a motor carrier operating in the Andhraachuki region (India). Last 11 July, four TL transportation requests were made: from Chittoor to Khammam on 11 July, from Srikakulam to Ichapur on 11 July, from Ananthapur to Chittoor on 13 July (two loads). On 11 July, one vehicle was available in Chittoor and one in Khammam. A further vehicle was currently transporting a previously scheduled shipment and would be available in Chittoor on 12 July. Transportation times between terminals are shown in Table 6.16. The revenue provided by a truck carrying a full load is 1.8 times the transportation cost of a deadheading truck. Let $T = \{11 \text{ July}, 12 \text{ July}, 13 \text{ July}\} = \{1, 2, 3\}$ and $N = \{\text{Ananthapur, Chittoor, Ichapur, Khammam, Srikakulam}\} = \{1, 2, 3, 4, 5\}$. The optimal VAP solution is $x_{141} = 1$, $x_{123} = 1$, $y_{441} = 1$, $y_{443} = 2$, while the remaining variables are zero. It is worth noting that the requests from Srikakulam to Ichapur on 11 July and from Ananthapur to Chittoor on 13 July are not satisfied.

6.10 The Dynamic Driver Assignment Problem

The *dynamic driver assignment problem* (DDAP) arises in TL trucking where full-load trips are assigned to drivers in an on-going fashion. In TL trucking a trip may take several days (a four-day duration is not unusual both in Europe and in North America) and customer service requests arrive at random. Consequently, each driver is assigned a single trip at a time. Let $D = \{1, \ldots, n\}$ be the set of drivers waiting to be assigned a task and let $L = \{1, \ldots, n\}$ be the current set of full-load trips to be performed. When the number of drivers exceed the number of loads, a dummy load 0 is inserted in L, while if the number of loads exceed the number of drivers, a dummy driver 0 is inserted in D. The DDAP can be formulated as a particular single-commodity uncapacitated minimum-cost flow problem (a classical transportation problem) and can be solved efficiently through the network simplex method or a tailored algorithm. The cost c_{ij}, $i \in D$, $j \in L$, of assigning driver i to load j is set equal to the cost of moving empty from the current location of driver i to the pick-up point of load j. Let x_{ij}, $i \in D$, $j \in L$, be a binary variable equal to 1 if driver i is assigned to load j, 0 otherwise. If the number of drivers exceed the number of loads, the DDAP can be formulated as follows (see Figure 6.22).

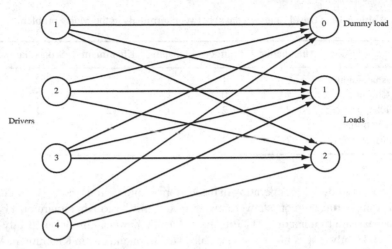

Figure 6.22   Driver assignment network.

Minimize

$$\sum_{i \in D} \sum_{j \in L} c_{ij} x_{ij}$$

subject to

$$\sum_{j \in L} x_{ij} = 1, \quad i \in D,$$

$$\sum_{i \in D} x_{ij} = 1, \quad j \in L \setminus \{0\},$$

$$x_{ij} \in \{0, 1\}, \quad i \in D, \ j \in L.$$

Costs $c_{i0}, i \in D$, can be set equal to 0. In practice, the previous model is reoptimized as vehicle locations and customer requests are revealed during the planning horizon (this explains the dynamic component of the model). In addition, penalties or bonuses may have added to arc costs $c_{ij}, i \in D, j \in L$, to reflect the cost of taking drivers home after a given number of weeks. A dispatcher who wants to take a driver $i \in D$ home at a given point in time can simply reduce the cost of the assignment of that particular driver to loads $j \in L$ whose delivery points are close to the driver home location. This can be accomplished by subtracting a suitable quantity from such c_{ij} costs.

Planet Transport Ltd is an Illinois (USA) motor carrier specializing in TL trucking. Last 26 January, the company had to solve the problem of assigning four full-load trips each three days long. The pick-up points were in Champaign, Danville, Peoria and Springfield. At that time six drivers were available. The first four were located

Table 6.17 Distance (in miles) between driver locations and trip pick-up points in the Planet Transport problem.

	Champaign	Danville	Peoria	Springfield
Bloomington	51.9	84.1	42.0	67.4
Decatur	46.8	83.3	107.0	40.0
Mason City	97.0	129.1	51.2	47.5
Pekin	95.4	127.6	13.1	69.4
Springfield	85.5	121.9	74.3	0.0

Table 6.18 Optimal driver assignment in the Planet Transport problem.

Driver	Location driver	Trip	Trip pick-up point
1	Bloomington	2	Danville
2	Decatur	1	Champaign
3	Mason City	—	—
4	Pekin	3	Peoria
5	Springfield	4	Springfield
6	Springfield	—	—

in Bloomington, Decatur, Mason City and Pekin, the last two in Springfield. The company formulated and solved a DDAP model with $|D| = 6$, and a dummy load 0. Costs $c_{ij}, i \in D, j \in L \setminus \{0\}$, were assumed to be proportional to distances between driver locations i and full-load trip pick-up points j (see Table 6.17). The optimal driver assignment is reported in Table 6.18. It is worth noting that the driver located in Mason City and one of the two drivers situated in Springfield were not assigned.

6.11 Questions and Problems

6.1 Assuming that the transportation rates for parcel, LTL trucking and TL trucking are those depicted in Figures 6.1, 6.2, 6.3, respectively, draw the most convenient rate as a function of load weight.

6.2 Canberra Freight is in charge of transporting auto parts for a US car manufacturer in Australia. Every week a tractor and one or two trailers move from the port of Melbourne to a warehouse located 430 km away. A tractor costs $75 per day, a trailer $30, a driver $7.5 per hour while running costs are $0.75 per kilometre. A trailer can contain 36 pallets. Derive the transportation cost per pallet as a function of shipment size for the case where one or two trailers are used.

6.3 Truckload trucking rates from Boston (Massachusetts, USA) to Miami (Florida, USA) are usually higher than those from Miami to Boston. Why?

6.4 Class 55 rates is cheaper than analogous class 70 rates (see Table 6.1). Why?

6.5 Extend Equation (6.2) to the case where two types of vehicles are available.

6.6 Devise a local search heuristic for the service network design problem in which at each step an existing arc is removed or a new arc is added.

6.7 Why should U-shapes be avoided when designing a crossdock?

6.8 Apply the alternating heuristic to an H-shaped terminal. How are shipping doors located?

6.9 Show that the deterministic VAP with multiple vehicle types can be modelled as an LMMCF problem on a time-expanded directed graph.

6.10 Show that, in the DDAP, costs c_{i0}, $i \in D$, can be set equal equivalently to ∞.

6.12   Annotated Bibliography

An broad introduction to operations research methods for planning freight transportation is:

1. Crainic TG and Laporte G 1997 Planning models for freight transportation. *European Journal of Operational Research* **97**, 409–438.

For a survey on long-haul freight transportation, including a review of strategic planning activities performed by governments and international organizations, see:

2. Crainic TG 1998 A survey of optimization models for long-haul freight transportation. Technical Report CRT-98-67, Centre de recherche sur les transports, Montréal, Canada.

A recommended textbook on network optimization problems is:

3. Ahuja RK, Magnanti TL and Orlin JB 1993 *Network Flows: Theory, Algorithms, and Applications*. Prentice Halll, Englewood Cliffs, New Jersey.

An introductory article to network design problems is:

4. Magnanti TL and Wong RT 1984 Network design and transportation planning: models and algorithms. *Transportation Science* **18**, 1–55.

A good description of a number of issues related to dynamic models in transportation and logistics is provided in:

5. Powell WB, Jaillet P and Odoni AR 1995 Stochastic and dynamic networks and routing. In *Handbooks in Operations Research and Management Science, 8: Network Routing* (ed. Ball MO, Magnanti TL, Monma CL and Nemhauser GL), pp. 141–295. Elsevier Science, Amsterdam.

Several aspects related to the design and operation of a crossdock are illustrated in the following articles:

6. Bartholdi III JJ and Gue KR 2000 The best shape for a crossdock. Technical Report, Department of Systems Management Naval Postgraduate School Monterey, California, USA.

7. Gue KR 1999 The effects of trailer scheduling on the layout of freight terminals. *Transportation Science* **33**, 419–428.

8. Bartholdi III JJ and Gue KR 2000 Reducing labour costs in an LTL crossdocking terminal. *Operations Research* **48**, 823–832.

7

Planning and Managing Short-Haul Freight Transportation

7.1 Introduction

Short-haul freight transportation concerns the pick-up and delivery of goods in a relatively small area (e.g. a city or a county) using a fleet of trucks. As a rule, vehicles are based at a single depot, and vehicle tours are performed in a single work shift and may include several pick-up and delivery points.

Classification of short-haul transportation services.   Short-haul transportation is relevant to *distribution companies* that have to supply retail outlets or customer orders from a warehouse using small vans (see Figure 7.1a). It is also crucial to *local fast couriers* transporting loads between origin–destination pairs situated in the same area. Similarly, *long-haul carriers* need to collect locally outbound parcels before sending them to a remote terminal as a consolidated load, and to locally distribute loads coming from remote terminals (see Figure 7.1b). Short-haul transportation problems also arise in *garbage collection*, *mail delivery*, appliance *repair services*, *dial-a-ride* systems (providing transportation services to the elderly and the handicapped), and *emergency services* (including fire fighting and ambulance services).

Decision problems.   Short-haul transportation often involves a large number of users. For instance, in soft drink and beer distribution, the average number of customers visited daily can be up to 600, while in sanitation applications the number of sites visited daily is often between 200 and 1000. At a strategic level, the main decision is related to warehouse location. For this purpose, the methods illustrated in Chapter 3 can be used, although some adaptations are sometimes needed in order to take vehicle routes into account explicitly (see Section 7.8). At a tactical level, the main issue is fleet sizing (see Section 6.4). Finally, at an operational level, the

---

Introduction to Logistics Systems Planning and Control   G. Ghiani, G. Laporte and R. Musmanno
© 2004 John Wiley & Sons, Ltd   ISBN: 0-470-84916-9 (HB) 0-470-84917-7 (PB)

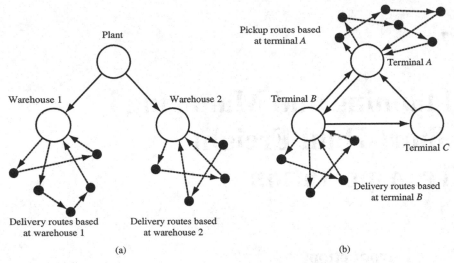

Figure 7.1   (a) Delivery routes based at warehouses. (b) Pickup and
delivery routes based at terminals of a long-haul carrier.

main problem (usually referred to as the *vehicle routing and scheduling problem*
(VRSP)) is to build vehicle routes in order to satisfy user requests. At this stage, a
number of operational constraints have to be taken into account. In some settings,
vehicle routes can be planned on a regular basis as all data are known beforehand.
This is the case of a company devising daily its distribution plan on the basis of the
customer orders collected the day before. Another example arises in sanitation applications, where vehicle routes are designed two or three times a year as the amount
of garbage to be collected daily is the same over several months. In contrast to such
static environments, there are settings where vehicle routes are built in an on-going
fashion as customer requests arrive. This is the case, for example, of long-distance
carriers whose pick-up requests arise during the day of operations and have to be
serviced the same day whenever possible. Due to recent advances in information
and communication technologies, vehicle fleets can now be managed easily in real-
time. When jointly used, devices like GIS, GPS, traffic flow sensors and cellular
telephones are able to provide relevant real-time data, such as current vehicle locations, new customer requests and up-to-date estimates of road travel times. If suitably
processed, these data can be used to devise revised vehicle routes. The main features
of such *vehicle routing and dispatching problems* (VRDPs) are briefly illustrated in
Section 7.7.

A slightly different class of operational problems arises in *vendor-managed* distribution systems which are quite common in the petrochemical and gas industry as
well as in the soft drink business. *Vendor-managed resupplying* (VMR) requires that
the distribution companies estimate customer inventory levels so that replenishment
can occur before they run out of stock. Such systems are discussed in Section 7.9.

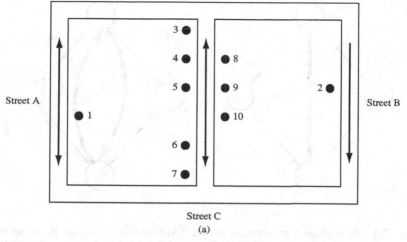

Figure 7.2 (a) A road network where 10 customers (represented by black dots) are to be served. Streets A and C are two-way. Street B is one-way.

7.2 Vehicle Routing Problems

VRPs consist of determining the routes to be used by a fleet of vehicles to serve a set of users. VRPs can be defined on a mixed graph $G = (V, A, E)$, where V is a set of vertices, A is a set of arcs and E is a set of edges. A vertex 0 represents the depot at which m vehicles are based, while a subset $U \subseteq V$ of *required vertices* and a subset $R \subseteq A \cup E$ of *required arcs* and *required edges* represent the users. VRPs amount to determining a least-cost set of m tours based at a depot, and including the required vertices, arcs and edges.

In this graph representation, arcs and edges correspond to road segments, and vertices correspond to road intersections. Isolated users are represented by required vertices, whereas subsets of customers distributed almost continuously along a set of customers are modelled as required arcs or edges (this is often the case of mail delivery and solid waste collection in urban areas). See Figures 7.2 and 7.3 for an example. If $R = \emptyset$, the VRP is called a *node routing problem* (NRP), while if $U = \emptyset$ it is called an *arc routing problem* (ARP). NRPs have been studied more extensively than ARPs and are usually referred to simply as VRPs. However, for the sake of clarity, in this textbook we use the appellation NRPs. If $m = 1$ and there are no side constraints, the NRP is the classical *travelling salesman problem*, which consists of determining a single circuit spanning the vertices of G, whereas the ARP is the *rural postman problem* (RPP), which amounts to designing a single circuit including the arcs and edges of R. The RPP reduces to the *Chinese postman problem* (CPP) if every arc and edge has to be serviced ($R = A \cup E$).

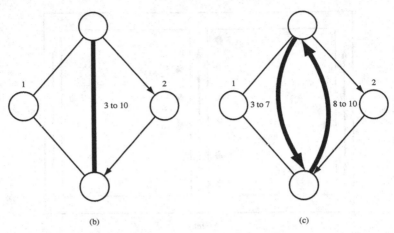

Figure 7.3   (b) A graph representation in case a vehicle traversing road B can serve the customers on both sides. (c) A graph representation in case a vehicle traversing road B can serve the customers on a single side. Bold vertices, arcs and edges are required.

Operational constraints.   The most common operational constraints are

- the number of vehicles m can be fixed or can be a decision variable, possibly subject to an upper bound constraint;

- the total demand transported by a vehicle at any time must not exceed its capacity;

- the duration of any route must not exceed a work shift duration;

- customers require to be served within pre-established *time windows*;

- some customers must be served by specific vehicles;

- the service of a customer must be performed by a single vehicle or may be shared by several vehicles;

- customers are subject to precedence relations.

When customers impose service time windows or when travel times vary during the day, time issues have to be considered explicitly in the design of vehicle routes, in which case VRPs are often referred to as VRSPs.

Precedence constraints arise naturally whenever some goods have to be transported between specified pairs of pick-up and delivery points. In such problems, a pick-up and delivery pair is to be serviced by the same vehicle (no transshipment is allowed at the depot) and each pick-up point must be visited before the associated delivery point. Another kind of precedence relation has to be imposed whenever vehicles have first to perform a set of deliveries (*linehaul customers*) and then a set of pick-ups (*backhaul customers*), as is customary in some industries (VRPs with backhauls, see Figure 7.4).

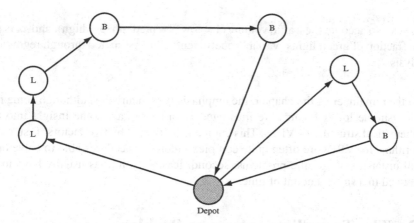

Figure 7.4 Vehicle routing with backhauls (L, linehaul customer; B, backhaul customers).

Objective. With each arc and edge $(i, j) \in A \cup E$ are associated a travel time t_{ij} and a travel cost c_{ij}. In addition, with each vehicle may be associated a fixed cost. The most common objective is to minimize the cost of traversing the arcs and edges of the graph plus the sum of the fixed costs associated with using the vehicles.

Travel time estimate. While the computation of distances in a road network is straightforward, the accurate estimation of travel times is often difficult. A rough evaluation of the travel time along a road segment can be obtained by dividing road length by the average speed at which the road segment can normally be traversed. This method is fairly accurate for inter-city roads, for which a constant speed can be kept for a long time, but performs poorly for intra-city streets. In such a case, average travel times can be estimated by using a regression method. To this end, the factors affecting travel time along a street are identified, and then a regression equation is used to forecast the average travel time as a function of these factors. The most relevant factors are the number of lanes, street width, whether the street is one-way or two-way, parking regulations, traffic volume, the number of traffic lights, the number of stop signs and the quality of the road surface.

In a school bus routing and scheduling study, the traversal times of the streets and avenues of Manhattan (New York, USA) were computed by estimating vehicle speed v through the following formula,

$$v = \bar{v} + 2.07x_1 + 7.52x_2 + 1.52x_3 + 1.36x_4 - 3.26x_5 + 4.04x_6,$$

where $\bar{v} = 7.69$ miles per hour is the average bus speed in normal conditions, x_1 is the total number of street lanes, x_2 is the number of street lanes available for buses, x_3 is a binary constant, equal to 1 in case of one-way street, 0 otherwise, x_4 is equal to 1 in case of bad road surface conditions, 2 in case of good road surface conditions,

x_5 takes into account the traffic volume ($1 = $ low, $2 = $ medium, $3 = $ high), and x_6 is the time fraction of green lights. Variable coefficients were estimated through regression analysis.

In the remainder of the chapter, the emphasis is on heuristics, although some relatively simple lower bounds are illustrated in order to gain some insight into the mathematical structure of VRPs. This choice is motivated by two factors: first, exact algorithms for VRPs are often quite complex tailored procedures able to solve only small instances of specific problems; second, feasible solutions usually have to be generated in a small amount of time.

7.3 The Travelling Salesman Problem

In the absence of operational constraints, there always exists an optimal NRP solution in which a single vehicle is used (see Problem 7.3). Hence, the NRP reduces to a TSP which consists of finding a least-cost tour including all the required vertices and the depot. In any TSP feasible solution on graph G, each vertex of $U \cup \{0\}$ appears at least once and two successive vertices of $U \cup \{0\}$ are linked by a least-cost path. As a consequence, the TSP can be reformulated on an auxiliary complete directed graph $G' = (V', A')$, where $V' = U \cup \{0\}$ is the vertex set and A' is the arc set. With each arc $(i, j) \in A'$ is associated a cost c_{ij} equal to that of a least-cost path from i to j in G. These costs satisfy the *triangle inequality*:

$$c_{ij} \leqslant c_{ik} + c_{kj}, \quad \forall (i, j) \in A', \ \forall k \in V', \ k \neq i, j.$$

Because of this property, there exists a TSP optimal solution which is a Hamiltonian tour in G', i.e. a cycle in which each vertex in V' appears exactly once. In what follows, the search for an optimal or suboptimal TSP solution is restricted to Hamiltonian tours.

If $c_{ij} = c_{ji}$ for each pair of distinct vertices $i, j \in V'$, the TSP is said to be *symmetric* (STSP), otherwise it is called *asymmetric* (ATSP). The STSP is suitable for inter-city transportation, while the ATSP is recommended in urban settings because of one-way streets. Of course, the solution techniques developed for the ATSP can also be applied to the STSP. This approach could, however, be very inefficient, as explained later. It is therefore customary to deal with the two cases separately.

7.3.1 The asymmetric travelling salesman problem

The ATSP can be formulated as follows. Let x_{ij}, $(i, j) \in A'$, be a binary decision variable equal to 1 if arc (i, j) is part of the solution, 0 otherwise.

Minimize

$$\sum_{(i, j) \in A'} c_{ij} x_{ij}$$

subject to

$$\sum_{i \in V' \setminus \{j\}} x_{ij} = 1, \quad j \in V', \tag{7.1}$$

$$\sum_{j \in V' \setminus \{i\}} x_{ij} = 1, \quad i \in V', \tag{7.2}$$

$$x_{ij} \in X, \qquad (i, j) \in A', \tag{7.3}$$

$$x_{ij} \in \{0, 1\}, \quad (i, j) \in A'.$$

Equations (7.1) and (7.2) are referred to as *degree* constraints. Constraints (7.1) mean that a unique arc enters each vertex $j \in V'$. Similarly, constraints (7.2) state that a single arc exits each vertex $i \in V'$. Constraints (7.3) specify that the x_{ij} values must lie in a set X that will yield a feasible solution consisting of a single directed tour (circuit). They can be formulated in two alternative ways, which are algebraically equivalent (see Problem 7.10):

$$\sum_{i \in S} \sum_{j \notin S} x_{ij} \geqslant 1, \qquad S \subset V', |S| \geqslant 2, \tag{7.4}$$

$$\sum_{i \in S} \sum_{j \in S} x_{ij} \leqslant |S| - 1, \quad S \subset V', |S| \geqslant 2. \tag{7.5}$$

Inequalities (7.4) guarantee that the circuit has at least one arc coming out from each proper and nonempty subset S of vertices in V' (*connectivity constraints*). Inequalities (7.5) prevent the formation of subcircuits containing less than $|S|$ vertices (*subcircuit elimination constraints*). It is worth noting that the number of constraints (7.4) (or, equivalently, (7.5)) is $2^{|V'|} - |V'| - 2$. Constraints (7.4) and (7.5) are redundant for $|S| = 1$ because of constraints (7.2).

A lower bound. The ATSP has been shown to be NP-hard. A good lower bound on the ATSP optimal solution cost z^*_{ATSP} can be obtained by removing constraints (7.3) from ATSP formulation. The relaxed problem is the following linear *assignment problem* (AP).

Minimize

$$\sum_{i \in V'} \sum_{j \in V'} c_{ij} x_{ij}$$

subject to

$$\sum_{i \in V'} x_{ij} = 1, \quad j \in V',$$

$$\sum_{j \in V'} x_{ij} = 1, \quad i \in V',$$

$$x_{ij} \in \{0, 1\}, \quad i, j \in V',$$

where $c_{ii} = \infty$, $i \in V'$, in order to force $x^*_{ii} = 0$, for all $i \in V'$.

The optimal AP solution x_{AP}^* corresponds to a collection of p directed subcircuits $C_1, \ldots, C_p$, spanning all vertices of the directed graph G'. If $p = 1$, the AP solution is feasible (and hence optimal) for the ATSP.

As a rule, z_{AP}^* is a good lower bound on z_{ATSP}^* if the cost matrix is strongly asymmetric (in this case, it has been empirically demonstrated that the deviation from the optimal solution cost $(z_{ATSP}^* - z_{AP}^*)/z_{AP}^*$ is often less than 1%). On the contrary, in the case of symmetric costs, the deviation is typically 30% or more. The reason of this behaviour can be explained by the fact that for symmetric costs, if the AP solution contains arc $(i, j) \in A'$, then the AP optimal solution is likely to include arc $(j, i) \in A'$ too. As a result, the optimal AP solution usually shows several small subcircuits and is quite different from the ATSP optimal solution.

Bontur is a pastry producer founded in Prague (Czech Republic) in the 19th century. The firm currently operates, in addition to four modern plants, a workshop in Gorazdova street, where the founder began the business. The workshop serves Prague and its surroundings. Every day at 6:30 a.m. a fleet of vans carries the pastries from the workshop to several retail outlets (small shops, supermarkets and hotels). In particular, all outlets of the Vltava river district are usually served by a single vehicle. For the sake of simplicity, arc transportation costs are assumed to be proportional to arc lengths. In Figure 7.5 the road network is modelled as a mixed graph $G(V, A)$, where a length is associated with each arc/edge (i, j). The workshop and the vehicle depot are located in vertex 0. Last 23 March, seven shops (located at vertices 1, 3, 9, 18, 20 and 22) needed to be supplied. The problem can be formulated as an ATSP on a complete directed graph $G' = (V', A')$, where V' is formed by the seven vertices associated with the customers and by vertex 0. With each arc $(i, j) \in A'$ is associated a cost c_{ij} corresponding to the length of the shortest path from i to j on G (see Table 7.1). The optimal AP solution x_{AP}^* is made up of the following three subcircuits (see Figure 7.6):

$$C_1 = \{(1, 4), (4, 3), (3, 9), (9, 1)\},$$

of cost equal to 11.0 km;

$$C_2 = \{(0, 18), (18, 0)\},$$

of cost equal to 5.2 km;

$$C_3 = \{(20, 22), (22, 20)\},$$

of cost equal to 4.6 km. Therefore, the AP lower bound z_{AP}^* on the objective function value of ATSP is equal to

$$z_{AP}^* = 11.0 + 5.2 + 4.6 = 20.8 \text{ km.}$$

Patching heuristic.   The patching heuristic works as follows. First, the AP relaxation is solved. If a single circuit is obtained, the procedure stops (the AP solution is

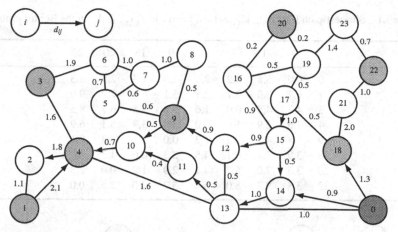

Figure 7.5 A graph representation of Bontur distribution problem (one-way street segments are represented by arcs, while two-way street segments are modelled through edges).

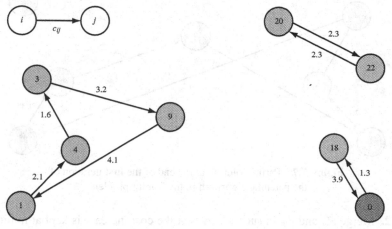

Figure 7.6 Optimal solution of the AP relaxation in the Bontur problem.

the optimal ATSP solution). Otherwise, a feasible ATSP solution is constructed by merging the subcircuits of the AP solution. When merging two subcircuits, one arc is removed from each subcircuit and two new arcs are added in such a way a single connected subcircuit is obtained.

Step 0 (*Initialization*). Let $C = \{C_1, \dots, C_p\}$ be the set of the p subcircuits in the AP optimal solution. If $p = 1$, *STOP*. The AP solution is feasible (and hence optimal) for the ATSP.

Step 1. Identify the two subcircuits $C_h, C_k \in C$ with the largest number of vertices.

Table 7.1 Shortest-path length (in kilometres) from i to j, $i, j \in V'$ in the Bontur problem.

	0	1	3	4	9	18	20	22
0	0.0	5.5	4.2	2.6	2.4	1.3	2.5	4.3
1	4.7	0.0	3.7	2.1	5.1	6.0	7.2	9.0
3	4.2	4.5	0.0	1.6	3.2	5.5	6.7	8.5
4	2.6	2.9	1.6	0.0	3.0	3.9	5.1	6.9
9	3.8	4.1	2.8	1.2	0.0	5.1	6.3	8.1
18	3.9	7.4	6.1	4.5	3.3	0.0	1.2	3.0
20	3.5	7.0	5.7	4.1	2.9	1.2	0.0	2.3
22	5.8	9.3	8.0	6.4	5.2	3.0	2.3	0.0

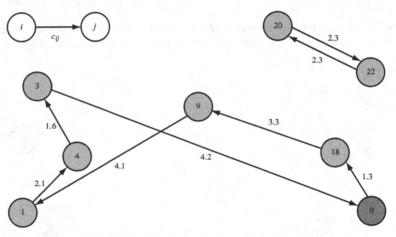

Figure 7.7 Partial solution at the end of the first iteration of the patching algorithm in the Bontur problem.

Step 2. Merge C_h and C_k in such a way that the cost increase is kept at minimum. Update C and let $p = p - 1$. If $p = 1$, *STOP*, a feasible solution ATSP has been determined, otherwise go back to Step 1.

In order to find a feasible solution $\bar{x}_{ATSP}$ to the Bontur distribution problem, the patching algorithm is applied to the AP solution shown in Figure 7.6. At the first iteration, C_1 and C_2 are selected to be merged (alternatively, C_3 could have been used instead of C_2). By merging C_1 and C_2 at minimum cost (through the removal of arcs (3,9) and (18,0) and the insertion of arcs (3,0) and (18,9)), the following subcircuit (having length equal to 16.6 km) is obtained (see Figure 7.7):

$$C_4 = \{(0, 18), (18, 9), ((9, 1), (1, 4), (4, 3), (3, 0))\}.$$

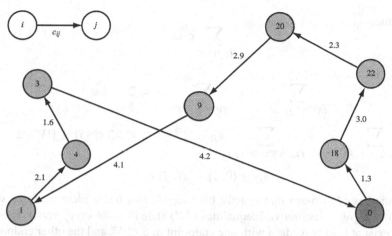

Figure 7.8 ATSP feasible solution generated by the patching algorithm in the Bontur problem.

The partial solution, formed by the two subcircuits C_3 and C_4, is depicted in Figure 7.7. The total length increases by 0.4 km with respect to the initial solution. At the end of the second iteration, the two subcircuits in Figure 7.7 are merged at the minimum cost increase of 0.3 km through the removal of arcs (18,9) and (20,22) and the insertion of arcs (18,22) and (20,9). This way, a feasible ATSP solution of cost $\bar{z}_{ATSP} = 21.5$ km is obtained (see Figure 7.8). In order to evaluate the quality of the heuristic solution, the following deviation from the AP lower bound can be computed,

$$\frac{\bar{z}_{ATSP} - z^*_{AP}}{z^*_{AP}} = \frac{21.5 - 20.8}{20.8} = 0.0337,$$

which corresponds to a percentage deviation of 3.37%.

7.3.2 The symmetric travelling salesman problem

As explained in the previous section, the ATSP lower and upper bounding procedures perform poorly when applied to the symmetric TSP. For this reason, several STSP tailored procedures have been developed.

The STSP can be formulated on a complete undirected graph $G' = (V', E')$, in which with each edge $(i, j) \in E'$ is associated a transportation cost c_{ij} equal to that of a least-cost path between i and j in G. Hence, c_{ij} costs satisfy the triangle inequality, and there exists an optimal solution which is a Hamiltonian cycle in G'. Let x_{ij}, $(i, j) \in E'$, be a binary decision variable equal to 1 if edge $(i, j) \in E'$ belongs to the least-cost Hamiltonian cycle, and to 0 otherwise. The formulation of the STSP is as follows (recall that $i < j$ for each edge $(i, j) \in E'$).

Minimize

$$\sum_{(i,j)\in E'} c_{ij}x_{ij}$$

subject to

$$\sum_{i\in V':(i,j)\in E'} x_{ij} + \sum_{i\in V':(j,i)\in E'} x_{ji} = 2, \quad j \in V', \qquad (7.6)$$

$$\sum_{(i,j)\in E':i\in S, j\notin S} x_{ij} + \sum_{(j,i)\in E':i\in S, j\notin S} x_{ji} \geqslant 2, \quad S \subset V', 2 \leqslant |S| \leqslant \lceil|V'|/2\rceil, \quad (7.7)$$

$$x_{ij} \in \{0, 1\}, \quad (i, j) \in E'.$$

Equations (7.6) mean that exactly two edges must be incident to every vertex $j \in V'$ (*degree constraints*). Inequalities (7.7) state that, for every vertex subset S, there exist at least two edges with one endpoint in $S \subset V'$ and the other endpoint in $V' \setminus S$ (*connectivity constraints*). Since the connectivity constraints of a subset S and that of its complement $V' \setminus S$ are equivalent, one has to consider only inequalities (7.7) associated with subsets $S \subset V'$ such that $|S| \leqslant \lceil|V'|/2\rceil$. Constraints (7.7) are redundant if $|S| = 1$ because of (7.6). Alternatively, the connectivity constraints (7.7) can be replaced with the following equivalent *subcycle elimination constraints*:

$$\sum_{(i,j)\in E':i\in S, j\in S} x_{ij} \leqslant |S| - 1, \quad S \subset V', 2 \leqslant |S| \leqslant \lceil|V'|/2\rceil.$$

A lower bound. The STSP is an NP-hard problem. A lower bound on the optimal solution cost z^*_{STSP} can be obtained by solving the following problem (see Exercise 7.5).

Minimize

$$\sum_{(i,j)\in E'} c_{ij}x_{ij} \qquad (7.8)$$

subject to

$$\sum_{i\in V':(i,r)\in E'} x_{ir} + \sum_{i\in V':(r,i)\in E'} x_{ri} = 2, \qquad (7.9)$$

$$\sum_{(i,j)\in E':i\in S, i\neq r, j\notin S, j\neq r} x_{ij} + \sum_{(j,i)\in E':i\in S, i\neq r, j\notin S, j\neq r} x_{ji} \geqslant 1,$$

$$S \subset V', 1 \leqslant |S| \leqslant \lceil|V'|/2\rceil, \quad (7.10)$$

$$x_{ij} \in \{0, 1\}, \quad (i, j) \in E', \qquad (7.11)$$

where $r \in V'$ is arbitrarily chosen (*root* vertex). Model (7.8)–(7.11) corresponds to a minimum-cost spanning r-tree problem (MSrTP), for which the optimal solution is a least-cost connected subgraph spanning G' and such that vertex $r \in V'$ has degree 2. The MSrTP can be solved in $O(|V'|^2)$ steps with the following procedure:

Table 7.2 Shortest distances (in kilometres) between terminals in the Saint-Martin problem.

	Betteville	Bolbec	Dieppe	Fécamp	Le Havre	Luneray	Rouen	Valmont
Betteville	0.0	27.9	54.6	42.0	56.5	37.0	30.9	34.1
Bolbec	27.9	0.0	67.2	25.6	28.8	48.4	57.4	21.6
Dieppe	54.6	67.2	0.0	60.5	95.8	18.8	60.4	52.1
Fécamp	42.0	25.6	60.5	0.0	39.4	43.1	70.2	12.2
Le Havre	56.5	28.8	95.8	39.4	0.0	77.2	84.5	44.4
Luneray	37.0	48.4	18.8	43.1	77.2	0.0	51.6	34.0
Rouen	30.9	57.4	60.4	70.2	84.5	51.6	0.0	59.3
Valmont	34.1	21.6	52.1	12.2	44.4	34.0	59.3	0.0

Step 1. Determine a minimum-cost tree $T^*(V' \setminus \{r\}, E_T)$ spanning $V' \setminus \{r\}$.

Step 2. Insert in T^* vertex r as well as the two least-cost edges incident to vertex r.

Saint-Martin distributes fresh fishing products in Normandy (France). Last 7 June, the company received seven orders from sales points all located in northern Normandy. It was decided to serve the seven requests by means of a single vehicle sited in Betteville. The problem can be formulated as an STSP on a complete graph $G'(V', E')$, where V' is composed of eight vertices corresponding to the sales points and of vertex 0 associated with the depot. With each edge $(i, j) \in E'$, is associated a cost c_{ij} equal to the shortest distance between vertices i and j (see Table 7.2). The minimum-cost spanning r-tree is depicted in Figure 7.9, to which corresponds a cost $z^*_{MSrTP} = 225.8$ km.

The MSrTP lower bound can be improved in two ways. In the former approach, the MSrTP relaxation is solved for more choices of the roots $r \in V'$ and then the largest MSrTP lower bound is selected. In the latter method, $r \in V'$ is fixed but each constraint (7.6) with the only exception of $j = r$ is relaxed in a Lagrangian fashion. Let $\lambda_j, j \in V' \setminus \{r\}$, be the Lagrangian multiplier attached to vertex $j \in V' \setminus \{r\}$. A Lagrangian relaxation of the STSP is as follows.

Minimize

$$\sum_{(i,j)\in E'} c_{ij}x_{ij} + \sum_{j\in V'\setminus\{r\}} \lambda_j \left(\sum_{i\in V':(i,j)\in E'} x_{ij} + \sum_{i\in V':(j,i)\in E'} x_{ji} - 2 \right) \qquad (7.12)$$

subject to

$$\sum_{i\in V':(i,r)\in E'} x_{ir} + \sum_{i\in V':(r,i)\in E'} x_{ri} = 2, \qquad (7.13)$$

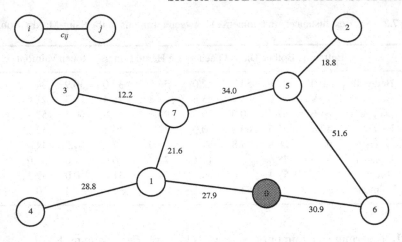

Figure 7.9 Minimum-cost spanning r-tree in the Saint-Martin problem.

$$\sum_{(i,j)\in E':i\in S,i\neq r,j\notin S,j\neq r} x_{ij} + \sum_{(j,i)\in E':i\in S,i\neq r,j\notin S,j\neq r} x_{ji} \geqslant 1,$$

$$S \subset V', \ 1 \leqslant |S| \leqslant \lceil |V'|/2 \rceil, \quad (7.14)$$

$$x_{ij} \in \{0, 1\}, \quad (i, j) \in E'. \quad (7.15)$$

Setting arbitrarily $\lambda_r = 0$, the objective function (7.12) can be rewritten as

$$\sum_{(i,j)\in E'} (c_{ij} + \lambda_i + \lambda_j)x_{ij} - 2\sum_{j\in V'} \lambda_j. \quad (7.16)$$

To determine the optimal multipliers (or at least a set of 'good' multipliers), a suitable variant of the subgradient method illustrated in Section 3.3.1 can be used. In particular, at the kth iteration the updating formula of the Lagrangian multipliers is the following,

$$\lambda_j^{k+1} = \lambda_j^k + \beta^k s_j^k, \quad j \in V' \setminus \{r\},$$

where

$$s_j^k = \sum_{i\in V':(i,j)\in E'} x_{ij}^k + \sum_{i\in V':(j,i)\in E'} x_{ji}^k - 2, \quad j \in V' \setminus \{r\},$$

x_{ij}^k, $(i, j) \in E'$, is the optimal solution of the Lagrangian relaxation MSrTP(λ) (7.16), (7.13)–(7.15) at the kth iteration, and β^k can be set equal to

$$\beta^k = \frac{1}{k}, \quad k = 1, \ldots.$$

The results of the first three iterations of the subgradient method in the Saint-Martin problem ($r = 0$) are

$$\lambda_j^1 = 0, \quad j \in V' \setminus \{r\}; \quad z_{\text{MS}_r\text{TP}(\lambda^1)}^* = 225.8; \quad s^1 = [1, -1, -1, -1, 1, 0, 1]^T;$$
$$\beta^1 = 1;$$
$$\lambda^2 = [1, -1, -1, -1, 1, 0, 1]^T; \quad z_{\text{MS}_r\text{TP}(\lambda^2)}^* = 231.8;$$
$$s^2 = [1, -1, -1, -1, 1, 0, 1]^T; \quad \beta^2 = 0.5;$$
$$\lambda^3 = [\tfrac{3}{2}, -\tfrac{3}{2}, -\tfrac{3}{2}, -\tfrac{3}{2}, \tfrac{3}{2}, 0, \tfrac{3}{2}]^T; \quad z_{\text{MS}_r\text{TP}(\lambda^3)}^* = 234.8.$$

Nearest-neighbour heuristic. The nearest-neighbour heuristic is a simple constructive procedure that builds a Hamiltonian path by iteratively linking the vertex inserted at the previous iteration to its nearest unrouted neighbour. Finally, a Hamiltonian cycle is obtained by connecting the two endpoints of the path. The nearest-neighbour heuristic often provides low-quality solutions, since the edges added in the final iterations may be very costly.

Step 0. Set $C = \{r\}$, where $r \in V'$, is a vertex chosen arbitrarily, and set $h = r$.

Step 1. Identify the vertex $k \in V' \setminus S$ such that $c_{hk} = \min_{j \in V' \setminus C} \{c_{hj}\}$. Add k at the end of C.

Step 2. If $|C| = |V'|$, add r at the end of C, STOP (C corresponds to a Hamiltonian cycle), otherwise let $h = k$ and go back to Step 1.

In order to find a feasible solution $\bar{x}_{\text{STSP}}$ to the Saint-Martin distribution problem, the nearest-neighbour heuristic is applied ($r = 0$), and the following Hamiltonian cycle is obtained (see Figure 7.10),

$$C = \{(0, 1), (1, 7), (3, 7), (3, 4), (4, 5), (2, 5), (2, 6), (0, 6)\},$$

whose cost is 288.4 km. The deviation of this solution cost from the available lower bound LB $= 234.8$ is

$$\frac{\bar{z}_{\text{STSP}} - \text{LB}}{\text{LB}} = \frac{288.4 - 234.8}{234.8} = 22.8\%.$$

The Christofides heuristic. The Christofides heuristic is a constructive procedure that works as follows.

Step 1. Compute a minimum-cost tree $T = (V', E_T')$ spanning the vertices of $G'(V', E')$. Let z_{MSTP}^* be the cost of this tree.

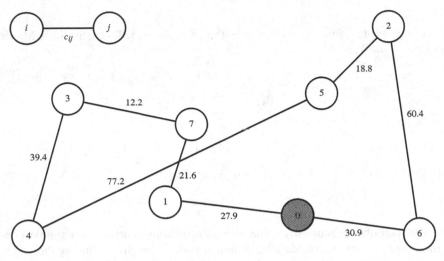

Figure 7.10   STSP feasible solution generated by the nearest-neighbour
algorithm in the Saint-Martin problem.

Step 2. Compute a least-cost perfect matching $M(V'_D, E'_D)$ among the vertices of
odd degree in the tree T ($|V'_D|$ is always an even number). Let z^*_M be the optimal
matching cost. Let $H(V', E'_H)$ be the Eulerian subgraph (or multigraph) of G'
induced by the union of the edges of T and M ($E'_H = E'_T \cup E'_D$).

Step 3. If there is a vertex $j \in V'$ of degree greater than 2 in subgraph H, eliminate
from E'_S two edges incident in j and in vertices $h \in V'$ and $k \in V'$, with $h \neq k$.
Insert in E'_H edge $(h, k) \in E'$ (or the edge $(k, h) \in E'$ if $(h, k) \notin E'$) (the *shortcuts
method*, see Figure 7.11). Repeat Step 3 until all vertices in V' have a degree of 2
in subgraph H.

Step 4. *STOP*, the set E'_H is a Hamiltonian cycle.

It is worth stressing that the substitution of a pair of edges (h, j) and (j, k) with
edge (h, k) (Step 3 of the algorithm) generally involves a cost reduction since the
triangle inequality holds. It can be shown that the cost of the Christofides solution is
at most 50% higher than the optimal solution cost. The proof is omitted for brevity.

The Saint-Martin distribution problem is solved by means of the Christofides algorithm. The minimum-cost spanning tree is made up of edges {(0,1), (0,6), (1,4), (1,7),
(2,5), (3,7), (5,7)}, and has a cost of 174.2 km. The optimal matching of the odd-
degree vertices (1, 2, 3, 4, 6 and 7) is composed of edges (1,4), (2,6) and (3,7), and
has a cost of 101.4 km. The Eulerian multigraph generated at the end of Step 2 is
illustrated in Figure 7.12. Edges (1,7) and (3,7) are substituted for edge (1,3), edges
(0,1) and (1,4) are substituted for edge (0,4). At this stage a Hamiltonian cycle of

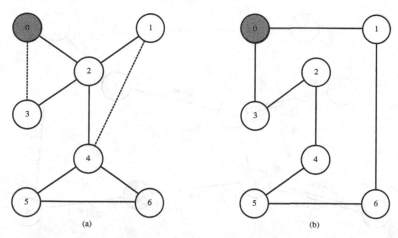

Figure 7.11 The Christofides algorithm. (a) Partial solution at the end of Step 2 (minimum-cost spanning tree edges are continuous lines while matching edges are dashed lines). (b) The Hamiltonian cycle obtained after Step 3 (where the degree of vertex 2 is reduced by removing edges (0,2) and (2,1) and by inserting edge (0,1), and the degree of vertex 4 is reduced by removing edges (1,4) and (4,6) and by inserting edge (1,6)).

cost equal to 267.2 km is obtained (see Figure 7.13). The deviation from the available lower bound LB is

$$\frac{\bar{z}_{STSP} - LB}{LB} = \frac{267.2 - 234.8}{234.8} = 13.8\%.$$

Local search algorithms. Local search algorithms are iterative procedures that try to improve an initial feasible solution $x^{(0)}$. At the kth step, the solutions contained in a 'neighbourhood' of the current solution $x^{(k)}$ are enumerated. If there are feasible solutions less costly than the current solution $x^{(k)}$, the best solution of the neighbourhood is taken as the new current solution $x^{(k+1)}$ and the procedure is iterated. Otherwise, the procedure is stopped (the last current solution is a *local optimum*).

Step 0. (*Initialization*). Let $x^{(0)}$ be the initial feasible solution and let $N(x^{(0)})$ be its neighbourhood. Set $h = 0$.

Step 1. Enumerate the feasible solutions belonging to $N(x^{(h)})$. Select the best feasible solution $x^{(h+1)} \in N(x^{(h)})$.

Step 2. If the cost of $x^{(h+1)}$ is less than that of $x^{(h)}$, set $h = h + 1$ and go back to Step 1; otherwise, *STOP*, $x^{(h)}$ is the best solution found.

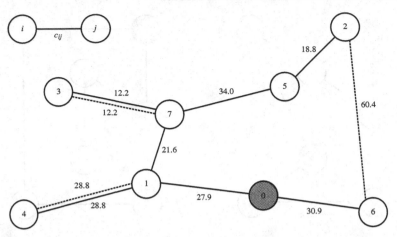

Figure 7.12 Eulerian multigraph generated by the Christofides heuristic in the Saint-Martin problem (minimum cost spanning tree edges are full lines and minimum-cost matching edges are dashed lines).

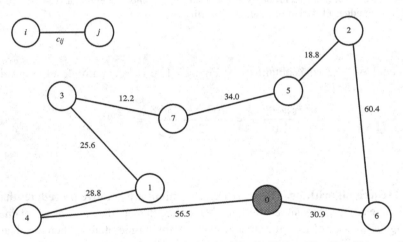

Figure 7.13 Hamiltonian cycle provided by the Christofides algorithm in the Saint-Martin problem.

For the STSP, $N(x^{(h)})$ is commonly defined as the set of all Hamiltonian cycles that can be obtained by substituting k edges ($2 \leqslant k \leqslant |V'|$) of $x^{(h)}$ for k other edges in E' (k-*exchange*) (Figure 7.14).

Step 0. Let $C^{(0)}$ be the initial Hamiltonian cycle and let $z_{STSP}^{(0)}$ be the cost of $C^{(0)}$. Set $h = 0$.

Step 1. Identify the best feasible solution $C^{(h+1)}$ that can be obtained through a k-exchange. If $z_{STSP}^{(h+1)} < z_{STSP}^{(h)}$, *STOP*, $C^{(h)}$ is a Hamiltonian cycle for the STSP.

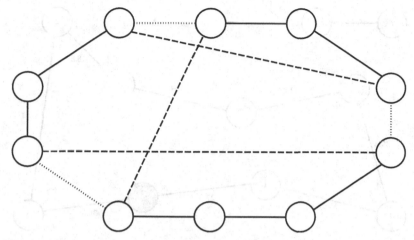

Figure 7.14 A feasible 3-exchange (dotted edges are removed, dashed edges are inserted).

Step 2. Let $h = h + 1$ and go back to Step 1.

In a local search algorithm based on k-exchanges, k can be constant or can be dynamically increased in order to intensify the search when improvements are likely to occur. If k is constant, each execution of Step 1 requires $O(|V'|^k)$ operations. In general, k is set equal to 2 or 3 at most, in order to limit the computational effort.

If a 2-exchange procedure is applied to the solution provided by the Christofides heuristic in the Saint-Martin problem, a less costly Hamiltonian cycle (see Figure 7.15) is obtained at the first iteration by replacing edges (0,4) and (1,3) with edges (0,1) and (3,4). As a consequence, the solution cost decreases by 14.8 km.

7.4 The Node Routing Problem with Capacity and Length Constraints

As illustrated in Section 7.2, in several settings operational constraints come into play when designing vehicle routes. These restrictions lead to a large number of variants and the algorithms described in the literature are often dependent on the type of constraint. For this reason, in the remainder of the chapter the most important constrained NRPs are examined and a limited number of techniques, representative of the most-used approaches, are described. As usual, the interested reader should consult the references listed at the end of the chapter for further information.

The *node routing problem with capacity and length constraints* (NRPCL) consists of designing a set of m least-cost vehicle routes starting and ending at the depot, such that

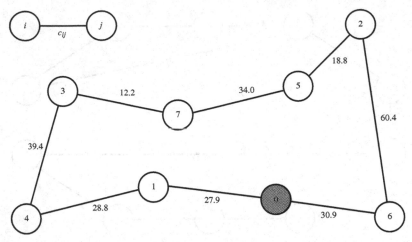

Figure 7.15   Best 2-exchange neighbour of the STSP feasible solution in Figure 7.13.

- each customer is visited exactly once;

- with each customer i is associated a demand p_i (demands are either collected or delivered, but not both); then the total demand of each vehicle cannot exceed a given vehicle capacity q (vehicles are assumed to be identical) (*capacity constraints*);

- with each customer is associated a service time s_i; then the total duration of each route, including service and travel times, may not exceed a given work shift duration T.

This problem can be formulated on a complete directed graph $G' = (V', A')$ or on a complete undirected graph $G' = (V', E')$ depending on whether the cost matrix is asymmetric or symmetric. In both cases, the vertex set V' is composed of the depot 0 and the customers in U. In what follows, the focus is on the symmetric version of the problem.

The NRPCL can be formulated by suitably modifying the STSP model.

Minimize

$$\sum_{(i,j)\in E'} c_{ij}x_{ij}$$

subject to

$$\sum_{i\in V':(i,j)\in E'} x_{ij} + \sum_{i\in V':(j,i)\in E'} x_{ji} = 2, \quad j \in U, \tag{7.17}$$

$$\sum_{i\in V':(0,i)\in E'} x_{0i} = 2m, \tag{7.18}$$

$$\sum_{(i,j)\in E':i\in S, j\notin S} x_{ij} + \sum_{(j,i)\in E':i\in S, j\notin S} x_{ji} \geqslant 2\alpha(S), \quad S \subseteq V' \setminus \{0\}, \ |S| \geqslant 2, \tag{7.19}$$

$$\sum_{(i,j)\in E':i\in S, j\notin S} x_{ij} + \sum_{(j,i)\in E':i\in S, j\notin S} x_{ji} \geqslant 4, \quad S \subseteq V' \setminus \{0\}, \ |S| \geqslant 2, \ t^*_{STSP}(S) > T,$$

(7.20)

$$x_{ij} \in \{0, 1\}, \quad (i, j) \in E'.$$

Constraints (7.17) state that two edges are incident to each customer $j \in U$ (*customer degree constraints*). Similarly, constraints (7.18) guarantee that $2m$ edges are incident to vertex 0 (*depot degree constraints*). Capacity constraints (7.19) impose that the number of vehicles serving customers in S is at least a lower bound $\alpha(S)$ on the optimal solution value of a 1-BP problem with $|S|$ items having weights p_i, $i \in S$ and bins of capacity q. In practice, it is common to use $\alpha(S) = \lceil (\sum_{i\in S} p_i)/q \rceil$. Length constraints (7.20) state that a single route is not sufficient to serve all the customers in S whenever the duration $t^*_{STSP}(S)$ of a least-cost Hamiltonian cycle spanning $S \cup \{0\}$ exceeds T.

Set partitioning formulation. An alternative formulation of the NRPCL can be obtained as follows. Let K be the set of routes in G' satisfying the capacity and length constraints and let c_k, $k \in K$, be the cost of route k. Define a_{ik}, $i \in V'$, $k \in K$, as a binary constant equal to 1 if vertex i is included into route k, and to 0 otherwise. Let y_k, $k \in K$, be a binary decision variable equal to 1 if route k used in an optimal solution, and to 0 otherwise. The NRPCL can be reformulated as a *set partitioning problem* (NRPSP) in the following way.

Minimize

$$\sum_{k\in K} c_k y_k$$

subject to

$$\sum_{k\in K} a_{ik} y_k = 1, \quad i \in V',$$

(7.21)

$$\sum_{k\in K} y_k = m,$$

(7.22)

$$y_k \in \{0, 1\}, \quad k \in K.$$

Constraints (7.21) establish that each customer $i \in V'$ must be served, while constraint (7.22) requires that exactly m vehicles are used.

The NRPSP is very flexible as it can be easily modified in order to include additional operational constraints. Its main weakness is the large number of variables, especially for 'weakly constrained' problems. For example, if $p_i = 1$, $i \in U$, and the length constraints are not binding, $|K| = O(\sum_{h=1}^{q} |V'|/h)$. Consequently, even if $|U| = 50$ and $q = 10$, there can be several billion variables. However, in some applications the characteristics of the operational constraints can considerably reduce $|K|$. This happens very often, for example, in fuel distribution where the demand of a user $i \in U$ (a gas pump) is customarily a small part (usually a half or a third) of a vehicle

Table 7.3 Orders (in hectolitres) received by Bengalur Oil.

Gas station	Orders
1	50
2	75
3	50
4	50
5	75

Table 7.4 Distance (in kilometres) between the gas stations and the firm's depot in the Bengalur Oil problem (depot corresponds to vertex 0).

	0	1	2	3	4	5
0	0	90	100	90	80	80
1	90	0	10	20	10	30
2	100	10	0	10	20	40
3	90	20	10	0	10	30
4	80	10	20	10	0	20
5	80	30	40	30	20	0

capacity. Therefore, the customers visited in each route can be three at most. As a consequence,

$$|K| = O\left(\binom{|V'|}{3} + \binom{|V'|}{2} + \binom{|V'|}{1}\right) = O(|V'|^3).$$

It is easy to show (see Problem 7.6) that constraints (7.21) can be replaced with the following relations,

$$\sum_{k \in K} a_{ik} y_k \geq 1, \quad i \in V',$$

in which case an easier-to-solve *set covering formulation* (NRPSC) is obtained.

Finally, both the NRPSP and NRPSC models can be used to generate a heuristic solution by including only a limited number $K' \subset K$ of feasible routes in the model.

Bengalur Oil manufactures and distributes fuel to filling stations in the Karnataka region (India). Last 2 July, the firm received five orders (see Table 7.3). The distances between the gas stations and the firm's depot are reported in Table 7.4. The vehicles have a capacity of 150 hectolitres. In order to formulate the problem as an NRPSC, the feasible routes are enumerated (Table 7.5).

The NRPSC model is as follows.

Table 7.5 Feasible routes in the Bengalur Oil problem (costs are expressed in kilometres).

Route	Vertices	Cost	Route	Vertices	Cost
1	{0,1,0}	180	9	{0,1,5,0}	200
2	{0,2,0}	200	10	{0,2,3,0}	200
3	{0,3,0}	180	11	{0,2,4,0}	200
4	{0,4,0}	160	12	{0,2,5,0}	220
5	{0,5,0}	160	13	{0,3,4,0}	180
6	{0,1,2,0}	200	14	{0,3,5,0}	200
7	{0,1,3,0}	200	15	{0,4,5,0}	180
8	{0,1,4,0}	180	16	{0,1,3,4,0}	200

Minimize

$$180y_1 + 200y_2 + 180y_3 + 160y_4 + 160y_5 + 200y_6 + 200y_7 + 180y_8$$
$$+ 200y_9 + 200y_{10} + 200y_{11} + 220y_{12} + 180y_{13} + 200y_{14} + 180y_{15} + 200y_{16}$$

subject to

$$y_1 + y_6 + y_7 + y_8 + y_9 + y_{16} \geqslant 1,$$
$$y_2 + y_6 + y_{10} + y_{11} + y_{12} \geqslant 1,$$
$$y_3 + y_7 + y_{10} + y_{13} + y_{14} + y_{16} \geqslant 1,$$
$$y_4 + y_8 + y_{11} + y_{13} + y_{15} + y_{16} \geqslant 1,$$
$$y_5 + y_9 + y_{12} + y_{14} + y_{15} \geqslant 1,$$
$$y_1, y_2, y_3, y_4, y_5, y_6, y_7, y_8, y_9, y_{10}, y_{11}, y_{12}, y_{13}, y_{14}, y_{15}, y_{16} \in \{0, 1\}.$$

In the optimal solution, the 12th and the 16th routes are used ($y_{12}^* = y_{16}^* = 1$) (see Figure 7.16), and the total distance covered by the vehicles is 420 km.

7.4.1 Constructive heuristics

In the remainder of this section, some constructive procedures for the NRPCL are illustrated.

'Cluster first, route second' heuristics. *Cluster first, route second* heuristics attempt to determine a good NRPCL solution in two steps. First, customers are partitioned into subsets $U_k \subset V' \setminus \{0\}$, each of which is associated with a vehicle $k = 1, \ldots, m$. Second, for each vehicle $k = 1, \ldots, m$, the STSP on the complete subgraph induced by $U_k \cup \{0\}$ is solved (exactly or heuristically). The partitioning of the customer set can be made visually or by more formalized procedures (as the one

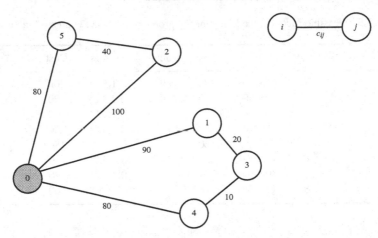

Figure 7.16 Optimal solution of the Bengalur Oil problem.

proposed by Fisher and Jaikumar). Readers interested in a deeper examination of this subject are referred to the references listed at the end of the chapter.

In the Bengalur Oil problem, customers can be partitioned into two clusters:

$$U_1 = \{2, 5\},$$
$$U_2 = \{1, 3, 4\}.$$

Then, two STSPs are solved on the complete subgraphs induced by $U_1 \cup \{0\}$ and $U_2 \cup \{0\}$, respectively. The result of this 'cluster first, route second' procedure is shown in Figure 7.16 (the total transportation cost is equal to 420 km).

'Route first, cluster second' heuristics. *Route first, cluster second* heuristics attempt to determine an NRPCL solution in two stages. First, a single Hamiltonian cycle (generally infeasible for the NRPCL) is generated through an exact or heuristic STSP algorithm. Then, the cycle is decomposed into m feasible routes, originating and terminating at the depot. The route decomposition can be performed visually or by means of formalized procedures, like the one proposed by Beasley. Readers interested in this method should consult the references listed at the end of the chapter.

Applying a 'route first, cluster second' procedure to the Bengalur Oil problem, a Hamiltonian cycle, having cost equal to 240 km, is generated (Figure 7.17). At the second stage, the cycle is decomposed into two feasible routes, which are the same as those illustrated in Figure 7.16.

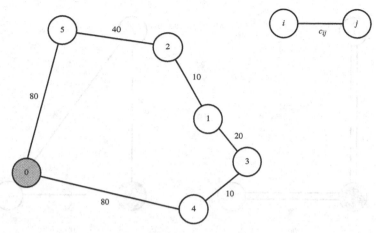

Figure 7.17 Hamiltonian cycle generated at the first step of the 'route first, cluster second' procedure in the Bengalur Oil problem.

Savings heuristic. The *savings* heuristic is an iterative procedure that initially generates $|U|$ distinct routes each of which serves a single customer. At each subsequent iteration, the algorithm attempts to merge a pair of routes in order to obtain a cost reduction (a *saving*). The cost saving s_{ij} achieved when servicing customers i and j, $i, j \in U$ on one route, as opposed to servicing them individually, is (see Figure 7.18)

$$s_{ij} = c_{0i} + c_{0j} - c_{ij}, \quad i, j \in V' \setminus \{0\}\}, \quad i \neq j. \tag{7.23}$$

It is worth noting that $s_{ij}, i, j \in V' \setminus \{0\}$, are nonnegative since the triangle inequality holds for all costs $c_{ij}, i, j \in E'$. The savings formula still holds if $i \in U$ is the last customer of the first route involved in a merge, and $j \in U$ is the first customer of the second route. The algorithm stops when it is no more possible to merge feasibly a pair of routes.

Step 0. (*Initialization*). Let C be the set of $|U|$ initial routes $C_i = \{0, i, 0\}, i \in V' \setminus \{0\}$. For each pair of vertices $i, j \in V' \setminus \{0\}, i \neq j$, compute the saving s_{ij} by using Equations (7.23). Let L be the list of savings sorted in a nonincreasing fashion (since $s_{ij} = s_{ji}, i, j \in V' \setminus \{0\}, i \neq j$, then the list L contains only one saving value for each pair of different customers).

Step 1. Extract from the top of list L a saving s_{ij}. If vertices i and j belong to two separate routes of C in which i and j are directly linked to the depot (Figure 7.19), and if the route obtained by replacing edges $(0, i)$ and $(0, j)$ with edge (i, j) is feasible, then merge the two routes and update C.

Step 2. If $L = \emptyset$, STOP, it is not possible to merge further pairs of routes. If $L \neq \emptyset$, go back to Step 1.

The computational complexity of the algorithm is determined by the saving sorting phase and is therefore $O(|V'|^2 \log |V'|^2) = O(|V'|^2 \log |V'|)$. In practice, the algo-

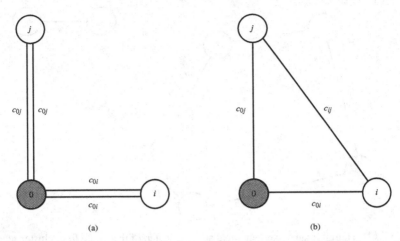

Figure 7.18   Computation of saving s_{ij}: (a) the cost of the two individual routes is $2c_{0i} + 2c_{0j}$; (b) the cost of the merged route is $c_{0i} + c_{0j} - c_{ij}$.

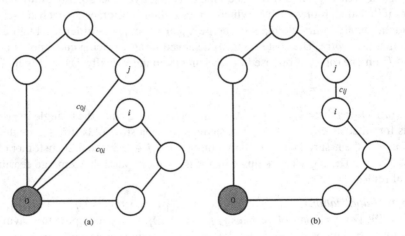

Figure 7.19   Merging two routes in a single route (the saving $s_{ij} = c_{0i} + c_{0j} - c_{ij}$).

rithm is very fast as it takes less than a second on the most common computers to solve a problem with hundreds of customers. However, the quality of the solutions can be poor. According to extensive computational experiments, the error made by the savings algorithm is typically in the 5–20% range.

The savings method is very flexible since it can easily be modified to take into account additional operational constraints (such as customer time window restrictions). However, with such variations, solution quality can be very poor. For this reason, tailored procedures are usually employed when dealing with constraints different from capacity and length constraints.

Table 7.6 Savings s_{ij}, $i, j \in V' \setminus \{0\}$, $i \neq j$, in the Bengalur Oil problem.

	1	2	3	4	5
1	—	180	160	160	140
2		—	180	160	140
3			—	160	140
4				—	140
5					—

By applying the savings algorithm to the Bengalur Oil problem, five individual routes are initially generated:

$$C_1 = \{0, 1, 0\},$$
$$C_2 = \{0, 2, 0\},$$
$$C_3 = \{0, 3, 0\},$$
$$C_4 = \{0, 4, 0\},$$
$$C_5 = \{0, 5, 0\}.$$

Then savings s_{ij}, $i, j \in V' \setminus \{0\}$, $i \neq j$, are calculated (see Table 7.6) and list L is initialized:

$$L = \{s_{12}, s_{23}, s_{13}, s_{14}, s_{24}, s_{34}, s_{15}, s_{25}, s_{35}, s_{45}\}.$$

Subsequently, routes C_1 and C_2 are merged while savings $s_{23}, s_{13}, s_{14}, s_{24}$ are discarded. Then, saving s_{34} is implemented by merging routes C_3 and C_4. At this stage there are no further feasible route merges. The final solution has a cost of 540 km (see Figure 7.20).

7.5 The Node Routing and Scheduling Problem with Time Windows

In several settings, customers need to be serviced within specified time windows. This is the case, for example, with retail outlets that cannot be replenished during busy periods. In the simplest version of the *node routing and scheduling problem with time windows* (NRSPTW), each customer specifies a single time window, while in other variants each customer can set multiple time windows (e.g. a time window in the morning and one in the afternoon). Let e_i, $i \in U$, be the *earliest time* at which service can start at customer i, and let l_i, $i \in U$, be the *latest time* (or *deadline*) at which service must start at customer i. Similarly, let e_0 be the earliest time at which

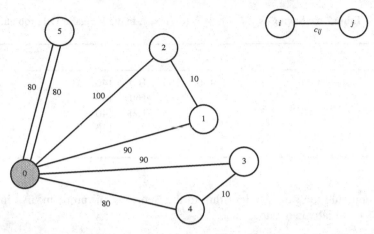

Figure 7.20 Solution provided by the savings algorithm for the Bengalur Oil problem.

vehicles can leave the depot, and let l_0 be the deadline within which vehicles must return to the depot. In the NRSPTW, the service starting time at each customer $i \in U$, is a decision variable b_i. If a vehicle arrives too early at a customer $j \in U$, it has to wait. Therefore, b_j, $j \in U$, is given by

$$b_j = \max\{e_j, b_i + s_i + t_{ij}\}, \quad j \in U,$$

where i is the customer visited just before j, t_{ij} is the quickest travel time between customers i and j and s_i is the service time of customer i.

It is worth noting, that, even though travel costs and times are symmetric, a solution is made up of a set of circuits, because of the time windows that do not allow reversals of route orientations.

In the remainder of this section, a constructive procedure is illustrated for the NRSPTW, while in Subsection 7.5.2 a tabu search procedure capable of providing high-quality solutions to a large number of constrained NRPs is described.

7.5.1 An insertion heuristic

Insertion-type heuristics are among the most efficient for the NRSPTW. In this section, the I1 Solomon heuristic is described. The procedure builds a feasible solution by constructing one route at a time. At each iteration the procedure decides which new customer $u^* \in U$ has to be inserted in the current solution, and between which adjacent customers $i(u^*)$ and $j(u^*)$ the new customer u^* has to be inserted on the current route. When choosing u^*, the algorithm takes into account both the cost increase associated with the insertion of u^*, and the delay in service time at customers following u^* on the route.

Step 0. (*Initialization*). The first route is initially $C_1 = \{0, \bar{i}, 0\}$, where $\bar{i}$ is the customer with the earliest deadline. Set $k = 1$.

Step 1. Let $C_k = \{i_0, i_1, \ldots, i_m\}$ be the current route, where $i_0 = i_m = 0$. Set

$$f_1(i_{p-1}, u, i_p) = \alpha(c_{i_{p-1}u} + c_{ui_p} - \mu c_{i_{p-1}i_p}) + (1 - \alpha)(b^u_{i_p} - b_{i_p}), \quad (7.24)$$

where $0 \leqslant \alpha \leqslant 1$, $\mu \geqslant 0$ and $b^u_{i_p}$ is the time when service begins at customer i_p provided that customer u is inserted between i_{p-1} and i_p. For each unrouted customer u, compute its best feasible insertion position in route C_k as

$$f_1(i(u), u, j(u)) = \min_{p=1,\ldots,m} f_1(i_{p-1}, u, i_p),$$

where $i(u)$ and $j(u)$ are the two adjacent vertices of the current route between which u should to be inserted. Determine the best unrouted customer u^* to be inserted yielding

$$f_2(i(u^*), u^*, j(u^*)) = \max_u \{f_2(i(u), u, j(u))\},$$

where

$$f_2(i(u), u, j(u)) = \lambda c_{0u} - f_1(i(u), u, j(u)) \quad (7.25)$$

with $\lambda \geqslant 0$.

Step 2. Insert customer u^* in route C_k between $i(u^*)$ and $j(u^*)$ and go back to Step 1. If u^* does not exist, but there are still unrouted customers, set $k = k + 1$, initialize a new route C_k (as in Step 0) and go back to Step 1. Otherwise, *STOP*, a feasible solution of the NRSPTW has been found.

The insertion heuristic tries to maximize the benefit obtained when servicing a customer on the current route rather than on an individual route. For example, when $\mu = \alpha = \lambda = 1$, Equation (7.25) corresponds to the saving in distance from servicing customer u on the same route as customers i and j rather than using an individual route. The best feasible insertion place of an unrouted customer is determined by minimizing a measure, defined by Equation (7.24), of the extra distance and the extra time required to visit it. Different values of the parameters μ, α and λ lead to different possible criteria for selecting the customer to be inserted and its best position in the current route.

McNish is a chain of supermarkets located in Scotland. Last 13 October, the warehouse situated in Aberdeen was required to serve 12 sales points located in Banchory, Clova, Cornhill, Dufftown, Fyvie, Huntly, Newbyth, Newmill, Peterhead, Strichen, Towie and Turriff. The number of requested pallets and the time windows within which service was allowed are reported in Table 7.7, whereas distances and travel times on the fastest routes between the cities are reported in Tables 7.8–7.11. Each vehicle had a capacity of 30 pallets.

Each vehicle leaves the warehouse at 9:00 a.m. and must return to the warehouse by 2:00 p.m. Service time (time needed for unloading a vehicle) can be assumed to

Table 7.7   Number of pallets and time windows for the sales point in the McNish problem.

Sales point	Vertex	Number of pallets	Time window
Banchory	1	9	9:00 a.m. – 11:30 a.m.
Clova	2	7	9:00 a.m. – 12:30 a.m.
Cornhill	3	5	9:00 a.m. – 12:30 a.m.
Dufftown	4	4	11:00 a.m. – 2:30 a.m.
Fyvie	5	8	9:00 a.m. – 12:30 a.m.
Huntly	6	8	11:00 a.m. – 1:30 p.m.
Newbyth	7	7	9:00 a.m. – 12:30 a.m.
Newmill	8	6	10:00 a.m. – 12:30 a.m.
Peterhead	9	6	9:00 a.m. – 10:30 a.m.
Strichen	10	6	9:00 a.m. – 11:15 a.m.
Towie	11	4	10:00 a.m. – 12:45 a.m.
Turriff	12	6	10:00 a.m. – 12:30 a.m.

Table 7.8   Distance (in kilometres) between cities computed on the fastest route in the McNish problem (Part I).

	Aberdeen	Banchory	Clova	Cornhill	Dufftown	Fyvie	Huntly
Aberdeen	0.0	28.4	58.1	84.7	83.4	41.1	63.0
Banchory	28.4	0.0	48.1	89.7	76.1	52.7	67.9
Clova	58.1	48.1	0.0	46.3	34.1	48.6	24.5
Cornhill	84.7	89.7	46.3	0.0	36.8	34.9	23.2
Dufftown	83.4	76.1	34.1	36.8	0.0	53.8	21.8
Fyvie	41.1	52.7	48.6	34.9	53.8	0.0	33.4
Huntly	63.0	67.9	24.5	23.2	21.8	33.4	0.0
Newbyth	59.5	75.3	66.1	35.1	63.4	22.7	42.9
Newmill	31.8	36.8	48.6	55.2	64.4	21.0	43.9
Peterhead	51.5	82.5	95.3	70.0	100.4	50.2	79.9
Strichen	56.7	87.8	80.0	47.0	77.3	34.8	56.8
Towie	62.1	47.5	13.0	53.3	41.0	55.5	31.4
Turriff	55.1	66.7	51.9	21.0	49.2	14.2	28.7

be 15 min on average for every sales point, regardless of demand. For the sake of simplicity, time is computed in minutes starting at 9:00 a.m. (for example, 11:00 a.m. corresponds to 120 min). The I1 insertion procedure (with parameters $\alpha = 0.9$; $\mu = \lambda = 1$) gave the following results. At the first iteration,

$$C_1 = \{0, 9, 0\}.$$

Table 7.9 Distance (in kilometres) between cities computed on the fastest route in the McNish problem (Part II).

	Newbyth	Newmill	Peterhead	Strichen	Towie	Turriff
Aberdeen	59.5	31.8	51.5	56.7	62.1	55.1
Banchory	75.3	36.8	82.5	87.8	47.5	66.7
Clova	66.1	48.6	95.3	80.0	13.0	51.9
Cornhill	35.1	55.2	70.0	47.0	53.3	21.0
Dufftown	63.4	64.4	100.4	77.3	41.0	49.2
Fyvie	22.7	21.0	50.2	34.8	55.5	14.2
Huntly	42.9	43.9	79.9	56.8	31.4	28.7
Newbyth	0.0	43.5	39.8	16.9	72.9	15.1
Newmill	43.5	0.0	50.3	43.9	52.7	34.9
Peterhead	39.8	50.3	0.0	24.7	100.6	47.8
Strichen	16.9	43.9	24.7	0.0	86.9	28.4
Towie	72.9	52.7	100.6	86.9	0.0	58.8
Turriff	15.1	34.9	47.8	28.4	58.8	0.0

Table 7.10 Travel times (in minutes) on the fastest routes between cities in the McNish problem (Part I).

	Aberdeen	Banchory	Clova	Cornhill	Dufftown	Fyvie	Huntly
Aberdeen	0	34	74	89	87	51	65
Banchory	34	0	58	96	87	63	72
Clova	74	58	0	56	42	70	31
Cornhill	89	96	56	0	43	42	27
Dufftown	87	87	42	43	0	65	26
Fyvie	51	63	70	42	65	0	43
Huntly	65	72	31	27	26	43	0
Newbyth	70	90	80	38	74	27	51
Newmill	37	44	61	68	67	28	44
Peterhead	58	88	117	80	113	63	90
Strichen	62	92	97	52	91	43	68
Towie	67	54	19	61	47	75	36
Turriff	67	79	64	27	58	16	35

For each unrouted customer u, its best feasible insertion position in route C_1 is computed, as reported in Table 7.12. Based on these evaluations, it is decided to insert vertex 10 in C_1 as follows:

$$C_1 = \{0, 9, 10, 0\}.$$

Table 7.11   Travel times (in minutes) on the fastest routes between cities in the McNish problem (Part II).

	Newbyth	Newmill	Peterhead	Strichen	Towie	Turriff
Aberdeen	70	37	58	62	67	67
Banchory	90	44	88	92	54	79
Clova	80	61	117	97	19	64
Cornhill	38	68	80	52	61	27
Dufftown	74	67	113	91	47	58
Fyvie	27	28	63	43	75	16
Huntly	51	44	90	68	36	35
Newbyth	0	54	49	21	84	18
Newmill	54	0	63	57	63	44
Peterhead	49	63	0	30	118	59
Strichen	21	57	30	0	102	35
Towie	84	63	118	102	0	69
Turriff	18	44	59	35	69	0

Then, two more customers are accommodated in route C_1:

$$C_1 = \{0, 9, 10, 7, 12, 0\}.$$

At this stage the length of C_1 is 163.3 km. Hence, a new route C_2 is initialized:

$$C_2 = \{0, 1, 0\}.$$

After three more iterations, C_2 is

$$C_2 = \{0, 1, 8, 5, 3, 0\},$$

with a distance of 205.8 km. Finally, route C_3 is constructed,

$$C_3 = \{0, 11, 2, 4, 6, 0\},$$

with a distance of 194.0 km. The final solution covers 563.1 km and corresponds to the schedule reported in Tables 7.13–7.15.

7.5.2   A unified tabu search procedure for constrained node routing problems

In recent years the field of heuristics has been transformed by the development of tabu search (TS). This is essentially a local search method that generates a sequence of solutions in the hope of generating better local optima. TS differs from classical

Table 7.12 Best feasible insertion positions in route C_1 for each unrouted customer in the McNish problem at the first iteration of the I1 algorithm.

u	$f_1(i, u, j)$	$i(u)$	$j(u)$	$f_2(i, u, j)$
2	106.51	9	0	−48.41
3	105.48	9	0	−20.78
4	134.77	9	0	−51.37
5	42.92	9	0	−1.82
6	93.46	9	0	−30.46
7	50.62	9	0	8.88
8	33.24	9	0	−1.44
9	—	—	—	—
10	31.81	9	0	24.89
11	114.28	9	0	−52.18
12	54.56	9	0	0.54

Table 7.13 Schedule of the first route in the McNish problem.

City	Arrival	Departure	Cumulated load
Aberdeen	—	9:00 a.m.	0
Peterhead	9:58 a.m.	10:13 a.m.	6
Strichen	10:43 a.m.	10:58 a.m.	12
Newbyth	11:19 a.m.	11:34 a.m.	19
Turriff	11:52 a.m.	12:07 p.m.	25
Aberdeen	1:14 p.m.	—	

Table 7.14 Schedule of the second route in the McNish problem.

City	Arrival	Departure	Cumulated load
Aberdeen	—	9:00 a.m.	0
Banchory	9:34 a.m.	9:49 a.m.	9
Newmill	10:33 a.m.	10:48 a.m.	15
Fyvie	11:16 a.m.	11:31 a.m.	23
Cornhill	12:13 a.m.	12:28 a.m.	28
Aberdeen	1:57 p.m.	—	

methods in that the successive solutions it examines do not necessarily improve upon each other. A key concept at the heart of TS is that of *neighbourhood*. The neighbourhood $N(s)$ of a solution s is the set of all solutions that can be reached from s by performing a simple operation. For example, in the context of the NRP, two common

Table 7.15   Schedule of the third route determined for the McNish problem.

City	Arrival	Departure	Cumulated load
Aberdeen	—	9:00 a.m.	0
Towie	10:07 a.m.	10:22 a.m.	4
Clova	10:41 a.m.	10:56 a.m.	11
Dufftown	11:38 a.m.	11:53 a.m.	15
Huntly	12:19 a.m.	12:34 a.m.	23
Aberdeen	1:39 p.m.	—	

neighbourhood structures are obtained by moving a customer from its current route to another route or by swapping two customers between two different routes. The standard TS mechanism is to move from s to the best neighbour in $N(s)$. This way of proceeding may, however, induce cycling. For example, s' may be the best neighbour of s which, in turn, is the best neighbour of s'. To avoid cycling the search process is prevented from returning to solutions processing some attributes of solutions already considered. Such solutions are declared *tabu* for a number of iterations. For example, if a customer v is moved from route r to route r' at iteration t, then moving v back to route r will be declared tabu until iteration $t + \theta$, where θ is called the length of the tabu tenure (typically θ is chosen between 5 and 10). When the tabu tenure has expired, v may be moved back to route r at which time the risk of cycling will most likely have disappeared because of changes that have occurred elsewhere in the solution.

Not only is it possible to accept deteriorating solutions in TS, but it may also be interesting to consider infeasible solutions. For example, in the sequence of solutions s, s', s'', both s and s'' may be feasible while s' is infeasible. If s'' cannot be reached directly from s, but only from s', and if it improves upon s, then it pays to go through the infeasible solution s'. This can occur if, for example, s' contains a route r that violates vehicle capacity due to the inclusion of a new customer in that route. Feasibility may be restored at the next iteration if a customer is removed from route s'. A practical way of handling infeasible solutions in TS is to work with a penalized objective function. If $f(s)$ is the actual cost of solution s, then the penalized objective is defined as

$$f'(s) = f(s) + \alpha Q(s) + \beta D(s) + \gamma W(s), \qquad (7.26)$$

where $Q(s)$, $D(s)$, $W(s)$ are the total violations of the vehicle capacity constraints, route duration constraints and time window constraints, respectively. Other types of constraints can of course be handled in the same way. The parameters α, β and γ are positive weights associated with constraint violations. These parameters are initially set equal to 1 and self-adjust during the course of the search to produce a mix of feasible and infeasible solutions. For example, if at a given iteration s is feasible with respect to the vehicle capacity constraint, then dividing α by a factor $1 + \delta$ (where $\delta > 0$) will increase the likelihood of generating an infeasible solution at the next iteration. Conversely, if s is infeasible, multiplying α by $1 + \delta$ will help the search move to a feasible solution. A good choice of δ is typically 0.5. The same principle applies to β and γ.

The algorithm repeatedly performs these operations starting from an initial solution which may be infeasible. It stops after a preset number of iterations. As is common in TS, this number can be very large (e.g. several thousands).

The article by Cordeau, Laporte and Mercier quoted at the end of this chapter illustrates how this implementation of TS has been employed to generate high-quality solutions for a large class of difficult NRPs.

7.6   Arc Routing Problems

According to the definition given in Section 7.2, an ARP consists of designing a least-cost set of vehicles routes in a graph $G(V, A, E)$, such that each arc and edge in a subset $R \subseteq A \cup E$ should be visited. Unlike NRPs (which are formulated and solved on an auxiliary complete graph G'), ARPs are generally modelled directly on G.

In this section, unconstrained ARPs, namely the CPP and the RPP, are examined. Constrained ARPs can be approached using the algorithmic ideas employed for the constrained NRPs along with the solution procedures for the CPP and the RPP. For example, the ARP with capacity and length constraints, whose applications arise in garbage collection and mail delivery, can be heuristically solved using a 'cluster first, route second' approach: in a first stage, the required arcs and edges are divided into clusters, each of which is assigned to a vehicle, while at a second stage an RPP is solved for each cluster. In this textbook, constrained ARPs are not tackled for the sake of brevity.

7.6.1   The Chinese postman problem

The CPP is to determine a minimum-cost route traversing all arcs and all edges of a graph at least once. Its main applications arise in garbage collection, mail delivery, network maintenance, snow removal and meter reading in urban areas.

The CPP is related to the problem of determining whether a graph $G(V, A, E)$ is Eulerian, i.e. whether it contains a tour traversing each arc and each edge of the graph exactly once. Obviously, in an Eulerian graph with nonnegative arc and edge costs, each Eulerian tour constitutes an optimal CPP solution. In a nonEulerian graph, an optimal CPP solution must traverse at least one arc or edge twice.

Necessary and sufficient conditions for the existence of an Eulerian tour depend on the type of graph G considered (directed, undirected or mixed), as stated in the following propositions, whose proofs are omitted for brevity.

Property.   A directed and strongly connected graph G is Eulerian if and only if it is *symmetric*, i.e. for any vertex the number of incoming arcs (*incoming semi-degree*) is equal to the number of outgoing arcs (*outgoing semi-degree*) (*symmetric* vertex).

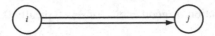

Figure 7.21   A mixed Eulerian graph which is not symmetric.

Property.   An undirected and connected graph G is Eulerian if and only if it is *even*, i.e. each vertex has an even degree (*even* vertex).

Property.   A mixed and strongly connected graph G is Eulerian if and only if

(a) the total number of arcs and of edges incident to any vertex is even (*even* graph);

(b) for each set S of vertices ($S \subset V$ and $S \neq \emptyset$), the difference between the number of the arcs traversing the cut $(S, V \setminus S)$ in the two directions is less than or equal to the number of edges of the cut (*balanced* graph).

Furthermore, since an even and symmetric graph is balanced, the following proposition holds.

Property.   A mixed, strongly connected, even and symmetric graph G is Eulerian.
   This condition is sufficient, but not necessary, as illustrated in the example reported in Figure 7.21.
   The solution of the CPP can be decomposed in two steps.

Step 1.   Define a least-cost set of arcs $A^{(a)}$ and of edges $E^{(a)}$ such that the multigraph $G^{(a)} = (V, A \cup A^{(a)}, E \cup E^{(a)})$ is Eulerian (if G is itself Eulerian, then $A^{(a)} = \emptyset$ and $E^{(a)} = \emptyset$ and, therefore, $G = G^{(a)}$).

Step 2.   Determine an Eulerian tour in $G^{(a)}$.

The first step of the procedure can be executed in polynomial time if G is a directed or undirected graph, while it results in an NP-hard algorithm if G is mixed. The second step can be performed in $O(|A \cup E|)$ time with the following *end-pairing* procedure.

Step 1.   Determine a covering of the edges and arcs of $G^{(a)}$ through a set C of tours, in such a way that an edge/arc is traversed exactly once.

Step 2.   If $|C| = 1$, *STOP*, the tour obtained is Eulerian.

Step 3.   Determine two tours in C which contain at least one common vertex. Merge the two tours, update C and go back to step 1.

In step 1, a tour in $G^{(a)}$ can be obtained by visiting the multigraph randomly until the initial vertex is included twice. Then, the edges/arcs of the tour are removed. In the new multigraph a new tour can be looked for.
   In the sequel it is shown how to determine $G^{(a)}$ in the cases where G is directed or undirected.

Directed Chinese postman problem. When G is directed, in an optimal solution the arcs in $A^{(a)}$ form a set of least-cost paths connecting the asymmetric vertices. Therefore $A^{(a)}$ can be obtained by solving a minimum-cost flow problem (a transportation problem, see Section 3.4) on a bipartite directed graph suitably defined. Let V^+ and V^- be the subsets of V whose vertices have a positive and a negative difference between the incoming semi-degree and the outgoing semi-degree, respectively. The bipartite directed graph is $G_T(V^+ \cup V^-, A_T)$, where $A_T = \{(i, j) : i \in V^+, j \in V^-\}$.

With each arc $(i, j) \in A_T$ is associated a cost w_{ij} equal to that of a least-cost path in G from vertex i to vertex j. Let also $o_i (> 0)$, $i \in V^+$, be the supply of the vertex i, equal to the difference between its incoming semi-degree and its outgoing semi-degree. Similarly, let $d_i (> 0)$, $i \in V^-$, be the demand of vertex i, equal to the difference between its outgoing semi-degree and its incoming semi-degree. Furthermore, let s_{ij}, $(i, j) \in A_T$, be the decision variable associated with the flow along arc (i, j). The transportation problem is as follows.

Minimize

$$\text{minimize} \quad \sum_{(i,j)\in A_T} w_{ij} s_{ij} \tag{7.27}$$

subject to

$$\sum_{j\in V^-} s_{ij} = o_i, \qquad i \in V^+, \tag{7.28}$$

$$\sum_{i\in V^+} s_{ij} = d_j, \qquad j \in V^+, \tag{7.29}$$

$$s_{ij} \geqslant 0, \quad (i, j) \in A_T. \tag{7.30}$$

Of course, $\sum_{i\in V^+} o_i = \sum_{j\in V^-} d_j$, so that problem (7.27)–(7.30) is feasible. Let s_{ij}^*, $(i, j) \in A_T$, be an optimal (integer) solution of the transportation problem. $A^{(a)}$ is formed by the arcs $(r, s) \in A$ belonging to the least-cost paths associated with the arcs $(i, j) \in A_T$ such that $s_{ij}^* > 0$ ((r, s) is taken s_{ij} times).

In the directed graph $G(V, A)$ shown in Figure 7.22, the differences between the incoming and outgoing semi-degrees of vertices 0, 1, 2, 3, 4 and 5 are -1, 0, 1, 0, 1, -1, respectively. The least-cost paths from vertex 2 to vertex 0 and from vertex 2 to vertex 5 are given by the sequences of arcs $\{(2,3), (3,4), (4,5), (5,0)\}$ (of cost equal to 109) and $\{(2,3), (3,4), (4,5)\}$ (of cost equal to 86), respectively. Similarly, the least-cost path from vertex 4 to vertex 0 is formed by $\{(4,5), (5,0)\}$, of cost 51, while the least-cost path from vertex 4 to vertex 5 is given by arc (4,5) whose cost is equal to 28. We can therefore formulate the transportation problem on the bipartite directed graph $G_T(V^+ \cup V^-, A_T)$ represented in Figure 7.23. The optimal solution to the transportation problem is

$$s_{20}^* = 0, \qquad s_{25}^* = 1, \qquad s_{40}^* = 1, \qquad s_{45}^* = 0.$$

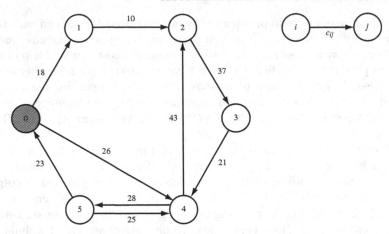

Figure 7.22 A directed graph $G(V, A)$.

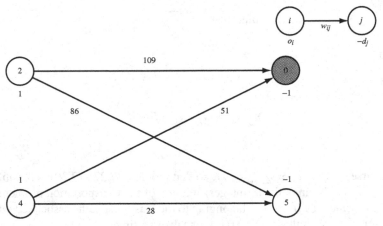

Figure 7.23 Bipartite directed graph $G_T(V^+ \cup V^-, A_T)$ associated
with graph G in Figure 7.22.

Therefore, set $A^{(a)}$ is formed by the arcs of the least-cost paths from vertex 2 to
vertex 5 and from vertex 4 to vertex 0. Adding such arcs to the directed graph G,
a least-cost Eulerian multigraph is obtained (Figure 7.24). Hence, an optimal CPP
solution of cost 368 is defined by the following arcs:

$$\{(0, 1), (1, 2), (2, 3), (3, 4), (4, 5), (5, 0), (0, 4),$$
$$(4, 2), (2, 3), (3, 4), (4, 5), (5, 4), (4, 5), (5, 0)\}.$$

In this solution some arcs are traversed more than once (for example, arc (4,5)), but
on these arcs it is performed only once.

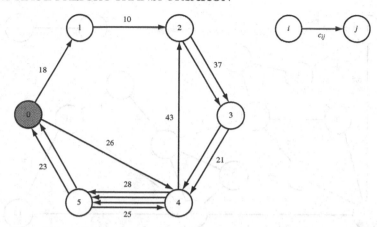

Figure 7.24 Least-cost Eulerian multigraph associated with the directed graph G in Figure 7.22.

Undirected Chinese postman problem. When G is undirected, the set $E^{(a)}$ can be obtained as a solution of a matching problem on an auxiliary graph $G_D(V_D, E_D)$. V_D is the set of odd vertices in G (V_D is formed by an even number of vertices) and $E_D = \{(i, j) : i \in V_D,\ j \in V_D,\ i \neq j\}$. With each edge $(i, j) \in E_D$ is associated a cost w_{ij} equal to that of a least-cost chain in G between vertices i and j. The set $E^{(a)}$ is therefore obtained as the union of the edges which are part of the least-cost chains associated with the edges of the optimal matching on G_D.

Welles is in charge of maintaining the road network of Wales (Great Britain). Among other things, the company has to monitor periodically the roads in order to locate craps and potholes in the asphalt. To this purpose, the road network has been divided into about 10 subnetworks, each of which has to be visited every 15 days by a dedicated vehicle. The graph representing one such subnetwork is shown in Figure 7.25. In order to determine the optimal undirected CPP solution, a minimum-cost matching problem between the odd-degree vertices (vertices 2, 3, 6, 7, 9 and 11) is solved. The optimal matching is 2–3, 6–7, 9–11. The associated set of chains in G is (2,3), (6,7) and (9,11) (total cost is 7.5 km). Adding these edges to G, the Eulerian multigraph in Figure 7.26 is obtained.

Finally, by using the end-pairing procedure, the optimal undirected CPP solution is obtained:

$$\{(0, 1), (1, 3), (3, 2), (2, 5), (5, 4), (4, 3), (3, 2), (2, 0),$$
$$(0, 6), (6, 7), (7, 8), (8, 5), (5, 6), (6, 7), (7, 9), (9, 11),$$
$$(11, 10), (10, 8), (8, 9), (9, 11), (11, 12), (12, 9), (9, 0)\}$$

(total cost is 52.8 km, the cost of the edges of G, plus 7.5 km). In this solution, edges (2, 3), (6, 7) and (9, 11) are traversed twice.

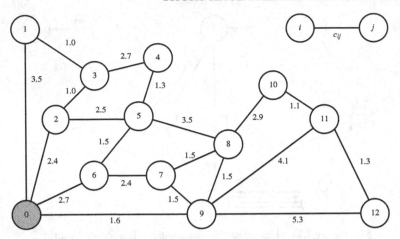

Figure 7.25   Graph representation used in the Welles problem.
Costs c_{ij}, $(i, j) \in E$, are in kilometres.

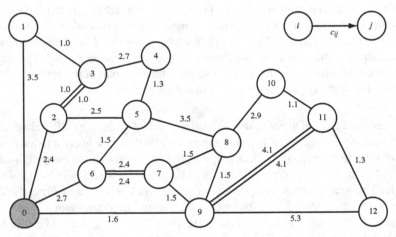

Figure 7.26   Least-cost Eulerian multigraph in the Welles problem.

7.6.2   The rural postman problem

The RPP is to determine in a graph $G(V, A, E)$ a least-cost route traversing a subset $R \subseteq A \cup E$ of required arcs and edges at least once. Its applications arise in garbage collection, mail delivery, network maintenance, snow removal and meter reading in scarcely populated areas.

Let $G_1(V_1, A_1, E_1), \ldots, G_p(V_p, A_p, E_p)$ be the p connected components of graph $G(V, R)$ induced by the required arcs and edges (see Figures (7.27) and (7.28)). The RPP solution can be obtained in two steps.

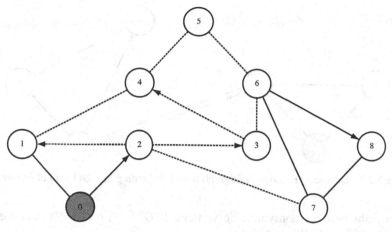

Figure 7.27 A mixed graph $G(V, A, E)$.

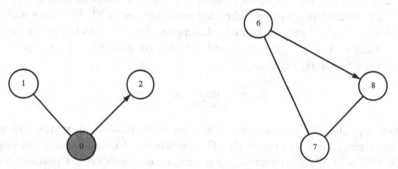

Figure 7.28 Connected components induced by the required arcs and edges of graph G in Figure 7.27.

Step 1. Determine a least-cost set of arcs $A^{(a)}$ and edges $E^{(a)}$ such that the multigraph $G^{(a)} = (V, (R \cap A) \cup A^{(a)}, (R \cap E) \cup E^{(a)})$ is Eulerian (see Figure 7.29).

Step 2. Determine an Eulerian tour in $G^{(a)}$.

The first step is NP-hard even for directed and undirected graphs, if $p > 1$. For $p = 1$, the RPP can be reduced to a CPP. The second step can be solved in polynomial time with the end-pairing procedure. In what follows, the first stage of two constructive heuristics is illustrated for directed and undirected graphs.

Directed rural postman problem. A heuristic solution to the directed RPP can be obtained through the 'balance-and-connect' heuristic.

Step 1. Using the procedure employed for the directed CPP, construct a directed symmetric graph $G'^{(a)}(V, R \cup A'^{(a)})$, by adding to $G(V, R)$ a suitable set of least-

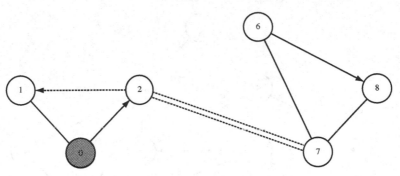

Figure 7.29   Least-cost Eulerian multigraph associated with graph $G(V, R)$ of Figure 7.27.

cost paths between nonsymmetric vertices. If $G'^{(a)}(V, R \cup A'^{(a)})$ is connected, *STOP*, $G'^{(a)} = G^{(a)}$ is Eulerian.

Step 2. Let p' ($1 < p' \leqslant p$) be the number of connected components of $G'^{(a)}(V, R \cup A'^{(a)})$. Construct an auxiliary undirected graph $G^{(c)} = (V^{(c)}, E^{(c)})$, in which there is a vertex $h \in V^{(c)}$ for each connected component of $G'^{(a)}$, and, between each pair of vertices $h, k \in V^{(c)}$, $h \neq k$, there is an edge $(h, k) \in E^{(c)}$. With edge (h, k) is associated a cost g_{hk} equal to

$$g_{hk} = \min_{i \in V_h, j \in V_k} \{w_{ij} + w_{ji}\},$$

where w_{ij} and w_{ji} are the costs of the least-cost paths from vertex i to vertex j and from vertex j to vertex i in G, respectively. Compute the minimum-cost tree $T^{(c)} = (V^{(c)}, E_T^{(c)})$ spanning the vertices of graph $G^{(c)}$. Construct a symmetric, connected and directed graph $G^{(a)}(V, R \cup A'^{(a)} \cup A''^{(a)})$ by adding to $G'^{(a)}(V, R \cup A'^{(a)})$ the set of arcs $A''^{(a)}$ belonging to the least-cost paths corresponding to the edges $E_T^{(c)}$ of the tree $T^{(c)}$.

Step 3. Apply, when possible, the shortcuts method (see the Christofides algorithm for the STSP) in order to reduce the solution cost.

The 'balance and connect' algorithm is applied to problem represented in Figure 7.30. The directed graph $G(V, R)$ has five connected components of required arcs. At the end of Step 1 (see Figure 7.31), $A'^{(a)}$ is formed by arcs (2,1) (the least-cost path from vertex 1 to vertex 2), (3,4) (the least-cost path from vertex 3 to vertex 4), (5,6) (the least-cost path from vertex 5 to vertex 6), (8,9) and (9,11) (the least-cost path from vertex 8 to vertex 11), and (10,7) (the least-cost path from vertex 10 to vertex 7). At Step 2, $V^{(c)} = \{1, 2, 3\}$, and the least-cost paths from vertex 1 to vertex 4 (arc (1,4) and vice versa (arc (4,1)), and from vertex 6 to vertex 7 (arc (6,7)) and vice versa (arcs (7,2), (2,3) and (3,6)) are added to the partial solution (see Figure 7.32).

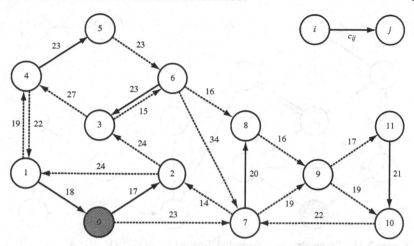

Figure 7.30 Directed graph $G(V, A)$.

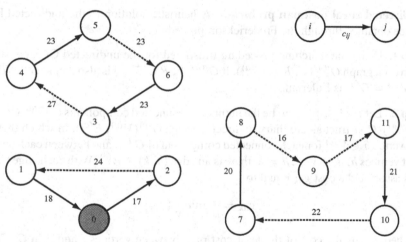

Figure 7.31 Directed graph $G'^{(a)}(V, R \cup A'^{(a)})$ obtained at the end of Step 1 of the 'balance and connect' algorithm.

Finally, using the end-pairing procedure, the following circuit of cost 379 is obtained:

$$\{(0, 2), (2, 1), (1, 4), (4, 5), (5, 6), (6, 7), (7, 8), (8, 9), (9, 11),$$
$$(11, 10), (10, 7), (7, 2), (2, 3), (3, 6), (6, 3), (3, 4), (4, 1), (1, 0)\}.$$

It can be shown that in this case the 'balance and connect' solution is optimal.

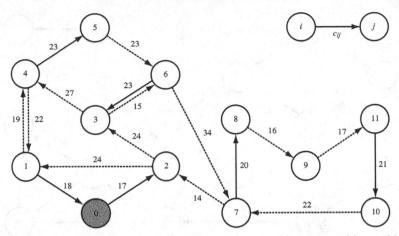

Figure 7.32 Symmetric and connected directed graph $G^{(a)}(V, R \cup A'^{(a)} \cup A''^{(a)})$ obtained at the end of Step 2 of the 'balance and connect' algorithm.

Undirected rural postman problem. A heuristic solution to the undirected RPP can be obtained through the Frederickson procedure.

Step 1. Using the matching procedure illustrated for the undirected CPP, construct an even graph $G'^{(a)}(V, R \cup E'^{(a)})$. If $G'^{(a)}(V, R \cup E'^{(a)})$ is also connected, *STOP*, $G'^{(a)} = G^{(a)}$ is Eulerian.

Step 2. Let p' $(1 < p' \leqslant p)$ be the number of connected components of $G'^{(a)}(V, R \cup E'^{(a)})$. Construct an auxiliary undirected graph $G^{(c)}(V^{(c)}, E^{(c)})$, in which there is a vertex $h \in V^{(c)}$ for each connected component of $G'^{(a)}$, and between each couple of vertices h, $k \in V^{(c)}, h \neq k$, there is an edge $(h, k) \in E^{(c)}$. With each edge (h, k) is associated a cost g_{hk} equal to

$$g_{hk} = \min_{i \in V_h, j \in V_k} \{w_{ij}\},$$

where w_{ij} is the cost of the least-cost path between vertices i and j in G. Compute a minimum-cost tree $T^{(c)} = (V^{(c)}, E_T^{(c)})$ spanning the vertices of graph $G^{(c)}$. Construct an even and connected graph $G^{(a)}(V, R \cup E'^{(a)} \cup E''^{(a)})$ by adding to $R \cup E'^{(a)}$ the set of edges $E''^{(a)}$ (each of which taken twice) belonging to the least-cost chains corresponding to the edges $E_T^{(c)}$ of tree $T^{(c)}$.

Step 3. Apply, if possible, the shortcuts method (see the Christofides algorithm for the STSP) in order to reduce solution cost.

Tracon distributes newspapers and milk door-to-door all over Wales. In the same road subnetwork as in the Welles problem, customers are uniformly distributed along some roads (represented by continuous lines in Figure 7.33). The entire demand of the

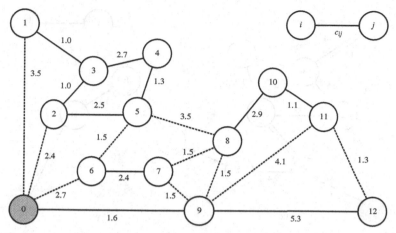

Figure 7.33 Graph $G(V, E)$ associated with the Tracon problem (costs are in kilometres).

subnetwork can be served by a single vehicle. By applying the Frederickson heuristic, the even and connected multigraph $G^{(a)}(V, R \cup E^{(a)})$ shown in Figure 7.34 is obtained. Finally, by using the end-pairing procedure, the following cycle is generated:

$$\{(0, 9), (9, 12), (12, 11), (11, 10), (10, 8), (8, 7), (7, 6),$$
$$(6, 5), (5, 4), (4, 3), (3, 1), (1, 3), (3, 2), (2, 5), (5, 6), (6, 0)\}$$

(total cost is 31.3 km). It is worth noting that edge (5,6) is traversed twice without being served. It can be shown that the Frederickson solution is optimal.

7.7 Real-Time Vehicle Routing and Dispatching

As pointed out in Section 7.2, there exist several important short-haul transportation problems that must be solved in real time. In this section, the main features of such problems are illustrated.

In *real-time* VRDPs, uncertain data are gradually revealed during the operational interval, and routes are constructed in an on-going fashion as new data arrive. The *events* that lead to route modifications can be

- the arrival of new user requests,

- the arrival of a vehicle at a destination,

- the update of travel times.

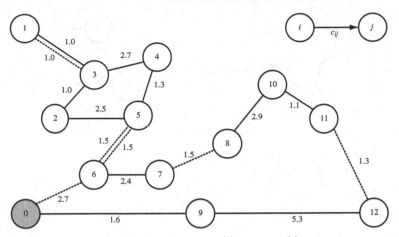

Figure 7.34 Even and connected multigraph $G^{(a)}(V, R \cup E^{(a)})$, obtained at the end of Step 2 of the Frederickson algorithm applied to the Tracon problem.

Every event must be processed according to the policies set by the company or organization operating the fleet of vehicles. As a rule, when a new request is received, one must decide whether it can be serviced on the same day, or whether it must be delayed or rejected. If the request is accepted, it is assigned temporarily to a position in a vehicle route. The request is effectively serviced as planned if no other event occurs in the meantime. Otherwise, it can be assigned to a different position of the same vehicle route, or even dispatched to a different vehicle. It is worth noting that at any time each driver just needs to know his next stop. Hence, when a vehicle reaches a destination it has to be assigned a new destination. Because of the difficulty of estimating the current position of a moving vehicle, reassignments could not easily made until quite recently. However, due to advances in vehicle positioning and communication technologies, route diversions and reassignments are now a feasible option and should take place if this results in a cost saving or in an improved service level. Finally, if an improved estimation of vehicle travel times is available, it may be useful to modify the current routes or even the decision of accepting a request or not. For example, if an unexpected traffic jam occurs, some user services can be deferred. If the demand rate is low, it is sometimes useful to relocate idle vehicles in order to anticipate future demands or to escape a forecasted traffic congestion.

Real-time problems possess a number of particular features, some of which have just been described. In the following, the remaining characteristics are outlined.

Quick response.   Algorithms for solving real-time VRDPs must provide a quick response so that route modifications can be transmitted timely to the fleet. To this end, two approaches can be used: simple policies (like the FCFS), or more involved algorithms running on parallel hardware. The choice between them depends mainly on the objective, the degree of dynamism and the demand rate.

Denied or deferred service. In some applications it is valid to deny service to some users, or to forward them to a competitor, in order to avoid excessive delays or unacceptable costs. For instance, requests that cannot be serviced within a given time windows are rejected.

Congestion. If the demand rate exceeds a given threshold, the system becomes saturated, i.e. the expected waiting time of a request goes to infinity.

The degree of dynamism. Designing an algorithm for solving real-time VRDPs depends to a large extent on how dynamic the problem is. To quantify this concept, the *degree of dynamism* of a problem has been defined. Let $[0, T]$ be the operational interval and let n_s and n_d be the number of static and dynamic requests, respectively ($n_s + n_d = |U|$). Moreover, let $t_i \in [0, T]$ be the *occurrence time* of service request of customer $i \in U$. Static requests are such that $t_i = 0, i \in U$, while dynamic ones have $t_i \in (0, T], i \in U$. The degree of dynamism δ can be simply defined as

$$\delta = \frac{n_d}{n_s + n_d}$$

and may vary between 0 and 1. Its meaning is straightforward. For instance, if δ is equal to 0.3, then 3 customers out of 10 are dynamic. This definition can be generalized in order to take into account both dynamic request occurrence times and possible time windows. For a given δ value, a problem is more dynamic if immediate requests occur at the end of the operational interval $[0, T]$. As a result, the measure of dynamism can be generalized as follows:

$$\delta' = \frac{\sum_{i=1}^{n_d} t_i / T}{n_s + n_d}.$$

Again δ' ranges between 0 and 1. It is equal to 0 if all user requests are known in advance while it is equal to 1 if all user requests occur at time T. Finally, the definition of δ' can be modified to take into account possible time windows on user service time. Let a_i and b_i be the *ready time* and *deadline* of customer $i \in U$, respectively. Then,

$$\delta'' = \frac{\sum_{i=1}^{n_d} [T - (b_i - t_i)] / T}{n_s + n_d}.$$

It can be shown that δ'' also varies between 0 and 1. Moreover, it is worth noting that if no time windows are imposed (i.e. $a_i = t_i$ and $b_i = T$ for each customer $i \in U$), then $\delta'' = \delta'$. As a rule, vendor-based distribution systems (such as those distributing heating oil) are weakly dynamic. Problems faced by long-distance couriers and appliance repair service companies are moderately dynamic. Finally, emergency services exhibit a strong dynamic behaviour.

Objectives. In real-time VRDPs the objective to be optimized is often a combination of different measures. In weakly dynamic systems the focus is on minimizing routing cost but, when operating a strongly dynamic system, minimizing the expected *response time* (i.e. the expected time lag between the instant a user service begins and its occurrence time) becomes a key issue. Another meaningful criterion which is often considered (alone or combined with other measures) is throughput optimization, i.e. the maximization of the expected number of requests serviced within a given period of time.

7.8 Integrated Location and Routing

Facility location and vehicle routing are two of the most fundamental decisions in logistics. Location decisions that are very costly and difficult to change are said to be *strategic*, e.g. those involving major installations such as factories, airports, fixed transportation links, etc. Others are said to be *tactical*, because, while still being relatively costly, they can still be modified after several years. Warehouse and store location fall in that category. Finally, *operational* location decisions involve easily movable facilities such as parking areas, mail boxes, etc. Once facilities are located, a routing plan must be put in place to link them together on a regular basis. All too often, facilities are first located without sufficient consideration of transportation costs, which may result in systemic inefficiencies. When planning to locate facilities it is preferable to integrate into the analysis the routing costs that these will generate. This applies equally well to strategic, tactical and operational decisions. A strategic location-routing decision is the location of airline hubs whose choice bears on routing costs. The location of depots and warehouses in a supply chain is a tactical decision influencing delivery costs to customers. A simple example arising at the operational level is mail box location. Locating a large number of mail boxes in a city will improve customer convenience since average walking distance to a mail box will be reduced. At the same time, the cost of emptying a larger number of mail boxes on a regular basis will be higher. Unfortunately, integrated location-routing mathematical models combining these two aspects will often contain too many integer variables and constraints to be solvable optimally. Heuristics based on a decomposition principle are often used instead. Facilities are first located, customers are assigned to facilities, and routing is then performed. These three decisions are then iteratively updated until no significant improvement can be reached. A detailed account of location-routing applications and methods can be found in the survey by Laporte.

7.9 Vendor-Managed Inventory Routing

VMR refers to situations where decisions about the timing and level of customer replenishment is determined not by the customer, but by the supplier. Traditionally, VMR has been the practice in the gas, petroleum and heating oil distribution but it is now becoming more frequent in automobile parts distribution and in the food and beverage industries. It requires the supplier to be aware of the customers' stock levels, which is not a major difficulty in cases where consumption is easily predictable, as in heating oil distribution. It is now possible to link vending machines to automatic dialing systems in order to inform suppliers of their stock level. There are two distinct advantages to VMR: by deciding when to deliver, the supplier can better plan its routes and delivery times by suitably combining deliveries to customers located in the same geographical zone; VMR also relieves customers of the costs associated with inventory monitoring and ordering.

The particular problem of combining routing and resupply decisions is known as the *inventory-routing problem* (IRP). The IRP consists of deciding which customers to visit during each period (e.g. one day) of a given time horizon (e.g. one week), and how much to deliver to each of them at each visit. In the simplest of cases, consider a planning horizon of t days, a single customer whose initial inventory is zero and whose usage rate (per day) is r. Denote by Q the customer capacity, by q the vehicle capacity, and by c the delivery cost. Then it is optimal to make deliveries only when customer stock is zero, which generates a cost of

$$z = c \max \left\{ \left\lceil \frac{tr}{Q} \right\rceil, \left\lceil \frac{tr}{q} \right\rceil \right\}.$$

The term within the braces is the number of deliveries per period of t days. It is driven by the smallest of the two capacities Q or q. To provide an idea of just how difficult the problem becomes when there is more than one customer, even with initial inventories of zero, consider two customers $i = 1, 2$, each of them with a capacity Q_i, usage rate r_i, and delivery cost c_i. Two extreme policies are possible here. Under the first policy, each customer is visited separately, generating a cost,

$$z_1 = c_1 \max \left\{ \left\lceil \frac{tr_1}{Q_1} \right\rceil, \left\lceil \frac{tr_1}{q} \right\rceil \right\} + c_2 \max \left\{ \left\lceil \frac{tr_2}{Q_2} \right\rceil, \left\lceil \frac{tr_2}{q} \right\rceil \right\}. \quad (7.31)$$

Under the second policy, the two customers are always visited jointly, yielding a cost,

$$z_2 = c_{1,2} \max \left\{ \left\lceil \frac{tr_1}{Q_1} \right\rceil, \left\lceil \frac{tr_2}{Q_2} \right\rceil, \left\lceil \frac{t(r_1 + r_2)}{q} \right\rceil \right\}, \quad (7.32)$$

where $c_{1,2}$ is the combined routing cost through the two customers. The first two terms in braces apply if the number of deliveries is driven by Q_1 or Q_2. In the third case, vehicle capacity is the most binding constraint. In this case, one must also decide how much to deliver to each customer. Matters complicate when the number of customers is larger than two since all possible ways of making joint deliveries must be considered and the joint routing cost through a given set of customers is the solution value of a TSP.

Another level of difficulty arises from the fact that usage rates are not constant in practice and must be treated as random variables. As a result, safety margins must be incorporated into the models and the cost of stockout must be explicitly accounted for.

With respect to the standard VRP, research on the IRP is still in its early stages and it is doubtful whether this problem can be solved exactly for any meaningful size. A practical and relatively simple heuristic is to select at each period a subset of customers having a high degree of urgency, i.e. a low inventory level, and solve a VRP associated with that set of customers, using one of the known VRP heuristics. The same process can be applied to each period. Solution improvements can be obtained by rescheduling some visits between sets of consecutive days and reoptimizing the individual VRP solutions.

An interesting account of VRM and IRP can be found in the article by Campbell et al. listed in the references.

Monsanto is a company based at Guimares (Portugal) distributing soft drinks in the Costa Verde and Montanhas regions. The main two customers are located in Porto and Braga and have demand rates equal to 42 and 26 cartons per day, respectively. The planning horizon consists of 30 days. The vehicle capacity is 100 cartons. Both a round trip from Guimares to Porto and a round trip from Guimares to Braga cost €180, while a tour Guimares–Porto–Braga–Guimares costs €210. The customer in Porto can hold at most 70 cartons while the customer in Braga can stock at most 85 cartons. If each customer is visited separately, the cost computed by using Equation 7.31 is $z_1 = 5040$ euros/month, while, if the two customers are visited jointly, the costs given by Equation 7.32 is $z_2 = 4410$ euros/month.

7.10 Questions and Problems

7.1 Show that if the costs associated with the arcs of a complete directed graph G satisfy the triangle inequality property, then there exists an ATSP optimal solution which is a Hamiltonian circuit in G'.

7.2 You have an algorithm capable of solving the capacitated NRP with no fixed vehicle costs and you would like to solve a problem where a fixed cost f is attached to each vehicle. Show how such a problem can be solved using the algorithm at hand.

7.3 Show that, if there are no operational constraints, there always exists an optimal NRP solution in which a single vehicle is used. (Hint: least-cost path costs satisfy the triangle inequality.)

7.4 Show that the two alternative subcircuit elimination constraints (7.4) and (7.5) are equivalent.

7.5 Demonstrate that the optimal solution value of MSrTP is a lower bound on the optimal solution value of STSP.

7.6 Show that the NRPSC formulation is correct.

7.7 Explain why the distances in Tables 7.8 and 7.9 do not necessarily satisfy the triangle inequality.

7.8 Modify the savings algorithm for the case in which the same vehicle may be assigned to several routes during a given planning period (*node routing problem with multiple use of vehicles*).

7.9 Devise a local search for the capacitated ARP.

7.10 Illustrate how the Christofides and Frederickson heuristics can be adapted to the *undirected general routing problem*, which consists of determining a least-cost cycle including a set of required vertices and edges.

7.11 Annotated Bibliography

Statistics reported in Section 7.1 are taken from:

1. Golden BL, Wasil EA, Kelly JA and Chao IM 1998 The impact of metaheuristics on solving the vehicle routing problem: algorithms, problem sets and computational results. In *Fleet Management and Logistics* (ed. Laporte G and Crainic TG). Kluwer, Boston.

A survey on node routing problems is:

2. Fisher ML 1995 Vehicle routing. In *Network Routing* (ed. Ball MO, Magnanti TL, Monma CL and Nemhauser GL). North-Holland, Amsterdam.

A description of a good tabu search heuristic for node routing can be found in:

3. Cordeau JF, Laporte G and Mercier A 2001 A unified tabu search heuristic for vehicle routing problems with time windows. *Journal of the Operational Research Society* **52**, 928–936.

Two reviews on arc routing problems are:

4. Assad A and Golden BL 1995 Arc routing methods and applications. In *Handbooks in Operations Research and Management Science, 8: Network Routing* (ed. Ball MO, Magnanti TL, Monma CL and Nemhauser GL), pp. 375–483. Elsevier Science, Amsterdam.

5. Eiselt HA, Gendreau M and Laporte G 1995 Arc routing problems. Part I. The Chinese postman problem. *Operations Research* **43**, 231–242.

6. Eiselt HA, Gendreau M and Laporte G 1995 Arc routing problems. Part II. The rural postman problem. *Operations Research* **43**, 399–414.

Two recent surveys on the real-time VRDP are:

7. Psaraftis HN 1995 Dynamic vehicle routing: status and prospects. *Annals of Operations Research* **61**, 143–164.

8. Gendreau M and Potvin JY 1998 Dynamic vehicle routing and dispatching. In *Fleet Management and Logistics* (ed. Laporte G and Crainic TG). Kluwer, Boston.

A survey of location-routing problems is provided in:

9. Laporte G 1988 Location-routing problems. In *Vehicle Routing: Methods and Studies* (ed. Assad and Golden BL). North-Holland, Amsterdam.

A survey on quantitative methods for operating vendor-managed distribution systems is:

10. Campbell A, Clarke L, Kleywegt A and Savelsberg M 1998 The inventory routing problem. In *Fleet Management and Logistics* (ed. Laporte G and Crainic TG). Kluwer, Boston.

8

Linking Theory to Practice

8.1 Introduction

Many real-world logistics problems possess slightly different features from those found in the problems we have discussed so far. Some of these features can easily be taken into account, while some others require more complex modifications, both in models and algorithms. Two examples will help illustrate this statement.

- When designing vehicle routes in practice, a meal break may have to be inserted in each route. In the single vehicle case, this can be simply accomplished by introducing a dummy customer with a service time equal to the required rest time, and a service time window equal to the interval within which meals are allowed.

- When designing vehicle routes in garbage collection applications, the street network is often modelled as a mixed graph. As a result, the single vehicle case amounts to solving a mixed RPP (see Section 7.6). This can be done, in principle, by suitably assigning a traversal direction to each edge and then applying the 'balance-and-connect' heuristic for the directed RPP. However, while this approach is easy to implement, it is not clear whether it yields a good quality solution for the mixed RPP. Devising a good heuristic for this problem may indeed require a substantial effort and is a research topic on its own.

The aim of this chapter is to link theory to practice in logistics management planning and control by providing supplementary material and cases. In Sections 8.2 to 8.6 a few real-world logistics systems are depicted while in Sections 8.7 to 8.15 some studies, taken from the scientific literature, illustrate the adaptation of basic techniques to more complex settings. Finally, further insightful case studies are listed in Section 8.17.

Introduction to Logistics Systems Planning and Control G. Ghiani, G. Laporte and R. Musmanno
© 2004 John Wiley & Sons, Ltd   ISBN: 0-470-84916-9 (HB) 0-470-84917-7 (PB)

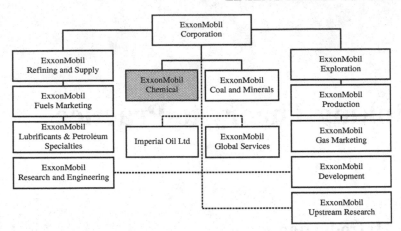

Figure 8.1   ExxonMobil corporation functional companies.

8.2   Shipment Consolidation and Dispatching at ExxonMobil Chemical

ExxonMobil Chemical is a functional company of ExxonMobil corporation. Exxon-Mobil is a US corporation formed in 1999 by the combination of Exxon and Mobil, two companies whose roots can be traced back to the late 19th century. ExxonMobil is an industry leader in almost every aspect of the energy and petrochemical business. Its activities range from the exploration and production of oil and gas to coal and copper mining, from the refining of petroleum products to the marketing of fuels (under the Exxon, Mobil and Esso brands), waxes, asphalt and chemicals. In addition, ExxonMobil is active in electric power generation. Figure 8.1 depicts the organization of ExxonMobil into several functional companies.

ExxonMobil Chemical is one of the largest petrochemical companies in the world. Its products include olefins, aromatics, synthetic rubber, polyethylene, polypropylene and oriented polypropylene packaging films. The company operates its 54 manufacturing plants in more than 20 countries and markets its products in more than 150 countries (see Table 8.1).

At many sites, the ExxonMobil Chemical operations are integrated with refining operations within a single complex. The ExxonMobil Chemical plant in Brindisi (Italy) is devoted to the manufacturing of oriented polypropylene packaging films for the European market. Oriented polypropylene (OPP) is a flexible material derived from melting and orienting (i.e. stretching) a polymer called polypropylene. This raw material is unaffected by most chemical agents encountered in everyday life. It meets the requirements of the US Food and Drug Administration and other relevant authorities throughout the world. By orienting polypropylene, one can improve its physical properties, such as water vapour impermeability, stiffness, dimensional stability and optics. OPP films are used as flexible packages for food (e.g. biscuits, bakery products and frozen food) and as high strength films for garbage bags and liners. Every

Table 8.1 ExxonMobil Chemical manufacturing plants.

Site	Country	Site	Country
Adelaide	Australia	Karlsruhe	Germany
Al-Jubayl	Saudi Arabia	Kashima	Japan
Altona	Australia	Kawasaki	Japan
Amsterdam	The Netherlands	Kerkrade*	The Netherlands
Antwerp	Belgium	LaGrange*	Georgia (USA)
Augusta	Italy	Managua	Nicaragua
Baton Rouge	Louisiana (USA)	Meerhout	Belgium
Baytown	Texas (USA)	Mont Belvieu	Texas (USA)
Bayway	New Jersey (USA)	Newport	United Kingdom
Beaumont	Texas (USA)	Notre-Dame-	
		de-Gravenchon	France
Belleville*	Canada	Panyu	China
Botany Bay	Australia	Paulina	Brazil
Brindisi*	Italy	Pensacola	Florida (USA)
Campana	Argentina	Plaquemine	Louisiana (USA)
Chalmette	Louisiana (USA)	Rotterdam	The Netherlands
Cologne	Germany	Sakai	Japan
Dartmouth	Canada	San Antonio	Chile
Edison	New Jersey (USA)	Sarnia	Canada
Fawley	United Kingdom	Shawnee*	Arkansas (USA)
Fife	United Kingdom	Singapore	Singapore
Fos-sur-Mer	France	Sriracha	Thailland
Geleen	The Netherlands	Stratford*	New Jersey (USA)
Harnes	France	Trecate	Italy
Houston	Texas (USA)	Virton*	Belgium
Ingolstadt	Germany	Wakayama	Japan
Jeffersonville	Indiana (USA)	Yanbu' al Bahr	Saudi Arabia
Jinshan	China	Yosu	South Korea

* = oriented polypropylene film plant

packaging application is different. For example, special OPP films are needed for complex products containing chocolate, sugar and cream that are more sensitive and need special protection, particularly against oxidation, odour loss and uptake of off-odours. OPP films may be transparent, opaque or metallized. ExxonMobil Chemical rigorously tests every packaging product before it is used commercially.

Although overall market growth is slow, indicating maturity across most sectors, there remain significant growth prospects within niche markets, such as those for individually wrapped biscuits. ExxonMobil Chemical produces more than 230 000 tonnes of OPP annually in seven plants in the USA, Canada, Belgium, Italy and The Netherlands (shown by asterisks in Table 8.1). The OPP process begins with pellets of polypropylene resin derived from crude oil or natural gas. Resins are transported to the Brindisi harbour by boat and then moved to the plant by a local dedicated train.

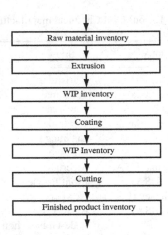

Figure 8.2   OPP film manufacturing process.

The manufacturing process is made up of three main stages (see Figure 8.2). First, the pellets are fed into an extruder, where they are melted by heat and friction from a continuously revolving screw. At the end of this stage the molten plastic is cast into a sheet form. Then, sheets are stretched lengthwise or crosswise and an acrylic coating is applied on one or both sides. Finally, large rolls are cut into smaller rolls to meet customers' specifications. At the end of this process, the custom slit film is shipped to an end-user or to third-party plants to be metallized or printed. Films manufactured in Brindisi needing to be metallized are sent either to the Metalvuoto plants in Termoli and Roncello (Italy), to the Neograf plant in Cuneo (Italy), or to the Metlux plant in Luxembourg, where a very thin coating of aluminium is applied to one side (see Figure 8.3).

As a rule, Italian end-users are supplied directly by the Brindisi plant, while customers and third-party plants outside Italy are replenished through the DC located in Milan (Italy). In particular, this warehouse supplies three DCs located in Herstal, Athus and Zeebrugge (Belgium), which in turn replenish customers in Eastern Europe, Central Europe and United Kingdom, respectively.

8.3   Distribution Management at Pfizer

The Pfizer Pharmaceuticals Group is the largest pharmaceutical corporation in the world. Its mission is 'to discover, develop, manufacture and market innovative, value-added products that improve the quality of life of people around the world and help them enjoy longer, healthier, and more productive lives'. The Pfizer range of products includes a broad portfolio of human pharmaceuticals meeting essential medical needs, a wide range of consumer products in the area of self-care and well-being, and health products for livestock and pets.

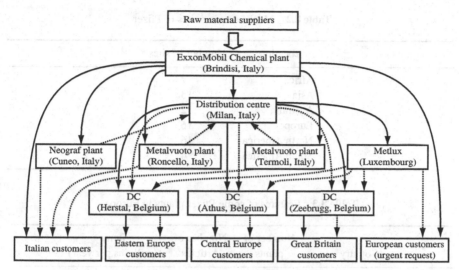

Figure 8.3   OPP film distribution patterns at ExxonMobil Chemical
(dotted lines represent metallized OPP film flows).

Founded in 1849 by Charles Pfizer, the company was first located in a modest red-brick building in the Williamsburg section of Brooklyn, New York (USA), that served as office, laboratory, factory and warehouse. The firm's first product was *santonin*, a palatable antiparasitic, which was an immediate success. In 1942 Pfizer responded to an appeal from the US Government to expedite the manufacture of penicillin, the first real defence against bacterial infection, to treat Allied soldiers fighting in World War II. Of the companies pursuing mass production of penicillin, Pfizer alone used the innovative fermentation technology. Building value for its shareholders, Pfizer manufactures and markets some of the most effective and innovative medicines including *atorvastatin calcium*, the most prescribed cholesterol-lowering medicine in the USA, *amlodipine besylate*, the world-leading medicine for hypertension and angina, *azithromycin*, the most-prescribed brand-name oral antibiotic in the USA, and *sildenafil citrate*, a breakthrough treatment for erectile dysfunction.

With a portfolio that includes five of the world's 20 top-selling medicines, Pfizer sets the standard for the pharmaceutical industry. Ten of its medicines are ranked first in their therapeutic class in the US market, and eight earn a revenue of more than one billion dollars annually. Research and development is the lifeblood of Pfizer business. To pursue its heritage of innovation, Pfizer supports the world's largest privately funded biomedical research organization, employing 12 000 scientists worldwide. The research investment was 5.3 billion dollars in 2002.

8.3.1 The Logistics System

The Pfizer logistics system comprises 58 manufacturing sites around the world (see Table 8.2), producing medicines for more than 150 countries.

Table 8.2   Manufacturing sites of Pfizer.

Location	Number of sites
Africa	7
Asia	13
Australia	2
Europe	16
North America	16
South America	4

Table 8.3   Features of some Pfizer plants in Europe.

Country	Number of plants	Number of Articles	Items (millions per year)
Belgium	1	29	6.5
France	1	14	2.4
Germany	1	3	11.4
Italy	3	182	87.1
United Kingdom	1	8	5.0

Because manufacturing pharmaceutical products requires highly specialized and costly machines, each Pfizer plant produces a large amount of a limited number of pharmaceutical ingredients or medicines for an international market (see Table 8.3).

In order to illustrate the main characteristics of a typical Pfizer supply chain, we will examine the supply chain of a cardiovascular product, named ALFA10. The product form is a 5 or 10 mg tablet, packaged in blisters. ALFA10 is based on a patent owned by Pfizer, and every plant involved in its manufacturing is Pfizer's property. ALFA10 is produced in a unique European plant (EUPF plant) for an international market including 90 countries (see Figure 8.4). Every year the plant produces over 117 million blisters. The product expires 60 months after its production and must be stored at a temperature varying between 8 and 25 °C.

The main component of ALFA10 is an *active pharmaceutical ingredient* (API), based on a Pfizer property patent, manufactured in a North American plant. APIs are transferred by air to the European Logistics Center (ELC) located in Belgium which in turn replenishes the European plants on a monthly basis (see Figure 8.5). Freight transportation between the ELC and the manufacturing sites is performed by overland transport providers such as Danzas. The EUPF plant manufactures ALFA10 tablets that are subsequently packaged into 120 blister boxes and sent weekly to a third-party CDC.

The CDC has five docks, a 1250 m^2 receiving zone, a 7700 m^2 storage zone (where 10 000 pallets can be kept in stock), and a 2000 m^2 shipping zone. Products are pal-

Figure 8.4 Pfizer ALFA10 supplied markets (grey area).

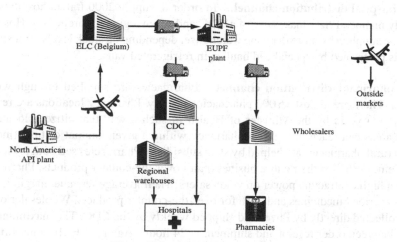

Figure 8.5 ALFA10 supply chain.

letized, and pallets are moved by forklifts and stored onto shelves. The transportation of finished goods is performed in refrigerated trucks by accredited haulers.

8.3.2 The Italian ALFA10 distribution system

ALFA10 sales are fairly stable in Italy, as shown in Figure 8.6. The distribution system is made up of two channels. Hospitals are supplied directly by Pfizer while pharmacies are replenished through wholesalers (see Figure 8.6). Pfizer plants, third-party DCs and wholesalers communicate through a dedicated information system named Manugistics.

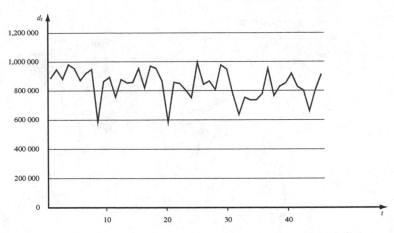

Figure 8.6 Pfizer ALFA10 monthly demand pattern in Italy.

The hospital distribution channel.   In order to supply 2000 Italian hospitals in a timely manner, Pfizer makes use of a CDC and seven regional warehouses. Hospitals may be supplied by more than one warehouse, depending on stock levels. Transportation is performed by specialized haulers in refrigerated vans.

The pharmacy distribution channel.   Pharmacies are supplied through wholesalers. There are almost 16 000 pharmacies in Italy. Pharmacy locations are revised every two years by the Minister of Health in such a way that citizens located in rural areas can reach the nearest pharmacy within a given amount of time (indeed, 5000 rural pharmacies are helped by state subsidies). Pharmacies sell both prescribed medicines (86% of their entire business) and over-the-counter products. Pharmacies have a high contractual power on wholesalers. Their average revenue margin is 27% for prescribed medicines and 33% for over-the-counter products. Wholesaler orders are collected directly by Pfizer and shipped weekly by the CDC. The maximum lead time between order receipt and shipment is 24 hours. Again ALFA10 transportation is performed by specialized haulers. The CDC is able to deliver the product in any Italian location within at most 60 hours. Wholesalers receive orders from pharmacies very frequently (up to four times a day). Pharmacies expect the wholesalers deliver medicines within 4–12 hours.

Compared with other EU countries, the number of Italian pharmaceutical wholesalers is very high (see Table 8.4). In addition, the pharmaceutical distribution business is very fragmented, as shown in Table 8.5. In particular, four pan-European companies have a 42% share.

Each wholesaler has an extensive network of RDCs in order to provide a high level of service to pharmacies. As a result, their average revenue margin is low. As illustrated in the above description, Pfizer makes use of 3PL. The relationship between Pfizer and its partners is regulated by contracts. Audits on a regular basis are performed by Pfizer on all its logistics partners.

Table 8.4 Features of pharmaceutical wholesalers in four major EU countries.

Country	Wholesalers	Wholesaler warehouses
France	14	203
Germany	17	102
Italy	198	302
Spain	101	189

Table 8.5 Classification of Italian pharmaceutical wholesalers.

	Companies	Warehouses	Share
Pan-European groups	4	86	42%
Local wholesalers	164	184	36%
Others (cooperatives, etc.)	30	32	22%

Further distribution channels. Unlike prescribed medicines (such as ALFA10), over-the-counter product distribution is not very critical and is performed directly by Pfizer. Due to the increasing popularity of the Internet among patients and physicians, prescribed medicines are expected to be delivered directly by Pfizer to the pharmacies in years to come, resulting in large savings.

8.4 Freight Rail Transportation at Railion

Railion is an international carrier, based in Mainz (Germany), whose core business is rail transport. Railion is the result of a merger involving DB Cargo AG and NS Cargo NV, and is Europe's first truly international rail company. In The Netherlands, Belgium and Luxembourg, Railion operates under the name Railion Benelux, while in Germany it will continue to operate as DB Cargo for the time being.

Railion transports a vast range of products, such as steel, coal, iron ore, paper, timber, cars, washing machines, computers as well as chemical products. In 2001 the company moved about 500 000 containers. Besides offering high-quality rail transport, Railion is also engaged in the development of integrated logistics chains. This involves close cooperation with third parties, such as trucking firms, maritime transporters, as well as forwarding and transshipment companies. This approach allows Railion to meet the increasingly complex demands of a market which is no longer prepared to settle for the mere carriage of goods, but requires a complete logistics package including all the service aspects this entails.

Railion Benelux offers a variety of transport services, tailored to the type of product to be carried, the destination and the customer's logistics requirements.

- *Scheduled services.* For the transport of smaller volumes of cargo (a few carloads at a time), the scheduled service concept is generally the best and least expensive option. Individual wagons are delivered to and collected from customers, whereupon they are coupled together at the marshalling yards to form complete trains. From there, they are transported directly to destinations (generally outside the Benelux), where individual wagonloads are sorted for delivery to the recipients. If required, Railion can also arrange terminal road services by truck. European scheduled services generally take between 24 and 48 hours, depending on the distance involved and the facilities available at the destination station. The range of different wagon types available includes bulk goods wagons, flat wagons for machinery and plant, special car transporters, tank wagons for chemical products, and covered wagons for palletized goods, industrial and consumer products.

- *Charter trains.* A charter train is often the perfect solution for the transport of large quantities of goods, especially if such shipments take place on a regular basis. A complete train offers tailor-made solutions enabling the delivery of up to several thousand tonnes of cargo. A charter train requires a so-called branch line to reach its final destination. Many industrial plants throughout Europe already have their own branch lines, often as a result of government subsidies.

- *Intermodal trains* (container shuttles). Railion Benelux operates as a sort of wholesaler in the intermodal market, supplying complete trains consisting of various types of container-carrying wagon to the intermodal operators. Intermodal trains almost invariably operate on the basis of a shuttle service with the same composition, and running back and forth between the same destinations. Intermodal operators can either utilize train capacity for their own containers (e.g. shipping lines) or sell 'slots' to other shippers, thereby operating as a 'forwarding agent'. Railion Benelux currently offers shuttle services to 24 European destinations. Table 8.6 lists the main features of the intermodal shuttles.

8.5   Yard Management at the Gioia Tauro Marine Terminal

The Gioia Tauro marine terminal is the largest container transshipment hub on the Mediterranean Sea and one of the largest in the world (see Table 8.7). Medcenter, the company set up to manage container handling at Gioia Tauro, is owned by Contship SpA, which controls 90% of the equity, and by Maersk, a leading international sea carrier.

The terminal is situated on the western coast of the Calabria region (Southern Italy), along major deep-sea vessel routes. Deep sea vessels are large highly automated container ships capable of transporting up to 6000 containers. They include both

Table 8.6 Features of the intermodal shuttles provided by Railion ('r' and 's' stand for roundtrip and single trip, respectively).

Origin	Destination	Frequency per week	TEUs
Rotterdam	Antwerp (Belgium)	11r	81/86
Rotterdam	Athus (Belgium)	5r	30
Rotterdam	Mouscron (Belgium)	5r	43
Rotterdam	Muizen (Belgium)	5r	10
Born	Antwerp (Belgium)	5r	60
Rotterdam	Germersheim (Germany)	6r	81
Rotterdam	Mainz (Germany)	3r	68
Rotterdam	Mannheim/Munich (Germany)	5r	81
Rotterdam	Neuss (Germany)	5r	81
Rotterdam	Milan–Melzo (Italy)	9r	74.5
Rotterdam	Novara (Italy)	12r	78
Rotterdam	Brescia (Italy)	5r	78
Rotterdam	Padova (Italy)	6r	74.5
Rotterdam	Bettembourg (Luxembourg)	4r	75
Rotterdam	Wels (Austria)	2r	77
Rotterdam	Malaszewicze (Poland)	3s	80
Rotterdam	Poznan/Warsaw (Poland)	3r	78
Rotterdam	Prague (Czech Republic)	6r	70
Rotterdam	Basel SBB (Switzerland)	5r	75
Rotterdam	Zurich (Switzerland)	5r	75
Rotterdam	Basel Bad (Switzerland)	2r	75

Table 8.7 The first 20 largest containerized ports in the world.

Port	Traffic in 1999 (TEUs)	Port	Traffic in 1999 (TEUs)
Hong Kong	16 100 000	New York	2 863 000
Singapore	15 900 000	Dubai	2 844 000
Kaohsiung	6 985 000	Felixstowe	2 700 000
Pusan	6 439 000	Tokyo	2 700 000
Rotterdam	6 400 000	Port Klang	2 550 000
Long Beach	4 400 000	Tanjung Priok	2 273 000
Shanghai	4 200 000	Gioia Tauro	2 253 000
Los Angeles	3 828 000	Kobe	2 200 000
Hamburg	3 750 000	Yokohama	2 200 000
Antwerp	3 614 000	Brema	2 180 000

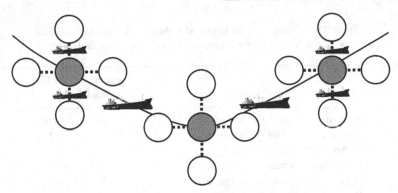

Figure 8.7 Hub and spoke sea transportation system (bold and dotted lines represent deep-sea vessel and feeder routes, respectively; grey and white vertices are hubs and spokes, respectively).

container ships performing around-the-world trips (through the Panama Canal) and Post-Panamax vessels performing North America–Europe, Europe–Asia Pacific and Asia Pacific–North America trips (the Post-Panamax vessel name is due to the fact that they are so large that they cannot traverse the Panama Canal). Because the operating costs of deep-sea vessels are very high, these ships stop at very few transshipment terminals (*hubs*), where they pick up and deliver traffic originating from or arriving at end-of-line ports (*spokes*). Then, smaller vessels (*feeder* or short sea vessels) transport goods between hubs and end-of-line ports (*hub and spoke* system, Figure 8.7).

The Gioia Tauro hub is linked to nearly 50 end-of-line ports on the Mediterranean Sea. When Gioia Tauro began trading in 1996, its traffic amounted to a modest 570 736 TEUs, followed by a dazzling 1.44 million TEUs in 1997, 2.12 million TEUs in 1998, and 2.25 million TEUs in 1999.

Like other hubs, the Gioia Tauro sea terminal (see Figure 8.8) is made up of

- a harbour, where vessels can wait for an available berth;

- a set of quays, where ships can be tied up and loaded or unloaded;

- a yard, where containers and bulk goods can be stored after being unloaded from incoming vehicles and before being loaded onto outgoing vehicles;

- a railway station, where wagons can be loaded or unloaded and convoys can be formed;

- some docks where trucks can be loaded or unloaded;

- a material handling system.

At the Gioia Tauro port, the yard can store nearly 50 000 TEUs (1100 of them can be refrigerated). The storage area is divided into bays. Each bay is made up of 32 rows, each having 16 slots. In each slot, up to three containers are stacked. Empty containers (which occupy approximately 40% of the storage area) have an 8–10 day

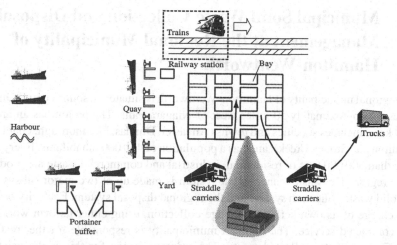

Figure 8.8 A sea container terminal layout.

average dwell time (much more than a full container) and are located in the more remote positions.

The railway station has six tracks where 20 convoys are formed every day (400 000 TEUs are handled annually). The Gioia Tauro port is close to the Salerno–Reggio Calabria highway traversing southern Italy from north to south. The material handling system is made up of 14 portainers, three Gottwald cranes, 51 straddle carriers, five multitrailers, six reach stackers, as well as 11 tractors and 60 trailers. Portainers and Gottwald cranes are used for unloading containers from the vessels. Portainers are cranes moving along tracks parallel to the quayside. Each portainer has a buffer where up to six containers can be stored. When the buffer is full, the portainer has to stop. Gottwal cranes are wheeled vehicles also used for moving containers in the yard. Straddle carriers are usually utilized for moving full containers over relatively short distances (less than 500 m) between railroad, yard and berth. These are wheeled vehicles capable of transporting one or two containers at a time. As a rule, for longer container transfers, different vehicles are used, namely multitrailers. Empty containers are stacked and moved five at a time by reach stackers. In addition, reach stackers are used for moving containers from the yard to the railway station and vice versa. Portainers and straddle carriers are the most important pieces of equipment and can handle seven and 24 containers per hour on average, respectively. Terminal operating cycles consist of a series of container movements, each carried out by several different movers. Once a ship is tied up, sea-side cranes load straddle carriers (or similar movers) that transfer outgoing containers to the terminal yard beside the assigned slot. Then dedicated yard movers insert containers in the right slot. Alternatively, prime movers can perform board–board, board–train or board–truck movements depending on the stowage plan. Similarly, incoming containers can be picked up from the yard or transferred directly from a train or a truck.

8.6 Municipal Solid Waste Collection and Disposal Management at the Regional Municipality of Hamilton-Wentworth

The regional municipality of Hamilton-Wentworth is situated in south-central Ontario (Canada), approximately 50 miles west of Niagara Falls. The region has an area of 1100 km^2, includes six cities and towns (Ancaster, Dundas, Flamborough, Glanbrook Hamilton and Stoney Creek), and has a population of 450 000 inhabitants. Every year, more than 300 000 tons of residential, industrial and commercial waste are produced in the region. The waste management system is made up of two major subsystems: the solid waste collection system and the regional disposal system. Each city or town is in charge of its own kerbside garbage collection, using either its own workforce or a contracted service. The regional municipality is responsible for the treatment and disposal of the collected waste. The primary reason for this is the existence of economies of scale (i.e. the decline of average cost as scale increases) in refuse transportation and disposal.

For the purpose of solid waste management, the region is divided into 17 districts. In 1992, the total cost was approximately 21.7 million dollars. The regional management is made up of a waste-to-energy facility, a recycling facility, a 550 acre landfill, a hazardous waste depot, and three transfer stations located in Dundas, Kenora and Hamilton Mountain. Transfer stations receive waste from municipal collection (or individual deliveries) and move it either to the waste-to-energy facility, to the recycling facility, or to the landfill. The waste picked up through kerbside collection from Flamborough, Dundas and northwest Ancaster goes to the transfer station in Dundas, garbage from Glanbrook, Hamilton Mountain and southeast Ancaster is delivered to the transfer station in Hamilton Mountain, while waste from lower Hamilton and Stoney Creek is delivered directly to the waste-to-energy facility. The transfer stations in Dundas and Hamilton Mountain also receive individual deliveries from local industries and institutions, while the transfer station in Kenora accepts only truckloads of industrial, commercial and institutional waste. The 1992 waste flow allocation pattern is shown in Figure 8.9.

8.7 Demand Forecasting at Adriatica Accumulatori

Adriatica Accumulatori is an electromechanical firm, headquartered in Termoli (Italy), manufacturing car spare parts for the Italian market. In 1993 the results of a survey showed that, although Adriatica Accumulatori car battery sales constantly increased during the previous decade, the company progressively lost market share (see Table 8.8). Until 1993, the company had traditionally based its production and marketing plans on sales forecasts provided by a time series extrapolation technique (see Section 2.4). If applied to the data in Table 8.8, this technique would result in the

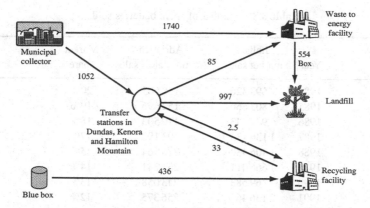

Figure 8.9 The waste flow allocation pattern in the regional municipality of Hamilton-Wentworth (all numbers in the figure are average waste flows in tons per week).

following regression equation (the trend is linear),

$$y = 126\,364.184 + 15\,951.091t, \quad t = 1, 2, \ldots,$$

which would provide the following demand forecasts: 301 826 units in 1994 ($t = 11$) (with a 9.41% increase with respect to 1993), and 317 777 units in 1995 ($t = 12$) (with a 15.19% increase with respect to 1993). However, the results of the survey convinced the company's management that during the previous decade Adriatica Accumulatori had lost an opportunity to sell more, mainly because its forecasts were not related to market demand. Based on this reasoning, it was decided to predict sales by first estimating the Italian market demand and then evaluating different scenarios corresponding to the current market share and increased shares achievable through appropriate marketing initiatives. In order to forecast the Italian market sales, a causal method was used (see Section 2.3). The historical series of national sales of batteries was correlated to the number of cars sold two years before (see Table 8.9). Then the following linear regression model was used,

$$y = a_0 + a_1 x,$$

where y is the Italian demand of spare batteries in a given year, x represents car sales two years before, a_0 and a_1 are two parameters. These parameters were estimated through the least-squares error method, yielding the regression equation $y = 52\,429.797 + 1.924x$, with a correlation index ρ equal to 0.95. Using this equation, the demand of spare batteries in the Italian market in 1994 and 1995 was estimated to be 2 396 003 and 2 676 295 units, respectively. Then, the company's management generated several scenarios based on different market shares. In the case where the firm maintained a market share equal to 11%, the demand would be equal to 263 560 units in 1994 (with a 4.46% decrease with respect to 1993), and 294 392 units in 1995 (with a 6.71% increase with respect to 1993).

Table 8.8 Number of spare batteries sold.

Year	Italian market sales	Adriatica Accumulatori sales	Market share
1984	693 326	138 665	20%
1985	803 666	152 696	19%
1986	947 243	170 503	18%
1987	1 136 433	193 192	17%
1988	1 406 432	210 964	15%
1989	1 666 011	233 241	14%
1990	1 869 683	243 058	13%
1991	2 136 463	256 375	12%
1992	2 316 402	266 386	11%
1993	2 507 929	275 872	11%

Table 8.9 Car sales in Italy.

Year	Number	Year	Number
1982	253 321	1988	886 297
1983	381 385	1989	1 014 975
1984	491 755	1990	1 162 246
1985	634 706	1991	1 167 614
1986	951 704	1992	1 217 929
1987	830 175	1993	1 363 594

The time series technique and the casual method resulted in quite different forecasts. The company therefore decided to analyse in greater detail the logic underlying the two approaches. Because the Italian economy was undergoing a period of quick and dramatic change, the latter method was deemed to provide more accurate predictions than the former technique, which is more suitable when the past demand pattern is likely to be replicated in the future.

8.8 Distribution Logistics Network Design at DowBrands

In 1985 Dow Consumer Products, Inc. acquired a division of Morton Thikol, Inc. giving rise to DowBrands, which produces and markets more than 80 convenience goods all over North America. On that occasion, the management of the new-born company decided to redesign the distribution network. After a preliminary analysis, it was decided that the new distribution system should be made up of CDCs and RDCs.

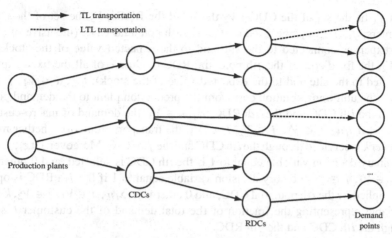

Figure 8.10   Distribution system of DowBrands.

In the proposed system, CDCs receive TL shipments from the production plants and supply the RDCs as well as a restricted number of major supermarkets. RDCs are suburban warehouses from which customers are replenished (see Figure 8.10). Shipments originating from a CDC are TL, while shipments from an RDC may be TL or LTL. In all cases freight transportation is performed by common carriers. Each RDC can be served by a single CDC and each customer can be assigned to a single CDC or RDC. Thirteen potential CDCs and 23 potential RDCs were selected. The demand points were aggregated into 93 sales districts while the products were combined in two macro-products (home products, HP, and food products, FP).

Because customers issue their orders within short notice (a single day or even a few hours) the management of DowBrands decided to impose an upper bound L on the maximum distance of an LTL shipment, but no limit was imposed for TL transportation which is much faster and reliable (see Section 1.2.3).

The distribution system redesign was designed in two stages: first, the curve of the total logistics cost as a function of the service level (represented by L) was defined; then, an efficient configuration was selected on the basis of a qualitative analysis (see Section 1.3). The cost versus level of service curve was drawn as follows. For a pre-established set of values of L, the least-cost configuration was determined by solving an IP model. The outcome was the number and the locations of the CDCs and of the RDCs, the allocation of the RDCs to the CDCs, the assignment of the demand points to the RDCs and to the CDCs, as well as freight routes through the distribution network.

In order to simplify the formulation, for each sales district and for each macro-product, a dummy macro-customer TL and a dummy macro-customer LTL were defined. Therefore, each demand point was represented by four macro-customers: TL-HP, LTL-HP, TL-FP, LTL-FP. Finally, a virtual RDC for each CDC was introduced so that, in the next modelling representation, all macro-customers would be served by an RDC.

Let V_1 be the set of the CDCs; V_2 the set of the RDCs; V_3 the set of the macro-customers; f_i, $i \in V_1$, the fixed cost of the ith potential CDC (inclusive of all the fixed expenses connected to the site and to the expected value of the stock); g_j, $j \in V_2$, the fixed cost of the jth potential RDC (inclusive of all the fixed expenses connected to the site and to the expected value of the stock); t_{ijk}, $i \in V_1$, $j \in V_2$, $k \in V_3$, the unit transportation cost from the production plant to the demand point k through the ith CDC and the jth RDC; d_k, $k \in V_3$, the demand of macro-customer k; $c_{ijk} = d_k t_{ijk}$, $i \in V_1$, $j \in V_2$, $k \in V_3$, the transportation cost whether macro-customer k is serviced through the ith CDC and the jth RDC. Moreover, let z_i, $i \in V_1$, be a binary decision variable equal to 1 if the ith CDC is selected, and 0 otherwise; y_{ij}, $i \in V_1$, $j \in V_2$, a binary decision variable equal to 1 if the jth RDC is opened and supplied by the ith potential CDC, and 0 otherwise; x_{ijk}, $i \in V_1$, $j \in V_2$, $k \in V_3$, a variable representing the fraction of the total demand of the customer k served through the ith CDC and the jth RDC.

The problem was formulated as follows.

Minimize

$$\sum_{i \in V_1} f_i z_i + \sum_{j \in V_2} g_j \sum_{i \in V_1} y_{ij} + \sum_{i \in V_1} \sum_{j \in V_2} \sum_{k \in V_3} c_{ijk} x_{ijk} \tag{8.1}$$

subject to

$$\sum_{i \in V_1} \sum_{j \in V_2} x_{ijk} = 1, \quad k \in V_3, \tag{8.2}$$

$$y_{ij} \leqslant z_i, \quad i \in V_1, \ j \in V_2, \tag{8.3}$$

$$\sum_{i \in V_1} y_{ij} \leqslant 1, \quad j \in V_2, \tag{8.4}$$

$$x_{ijk} \leqslant y_{ij}, \quad i \in V_1, \ j \in V_2, \ k \in V_3, \tag{8.5}$$

$$z_i \in \{0, 1\}, \quad i \in V_1, \tag{8.6}$$
$$y_{ij} \in \{0, 1\}, \quad i \in V_1, \ j \in V_2, \tag{8.7}$$
$$x_{ijk} \in \{0, 1\}, \quad i \in V_1, \ j \in V_2, \ k \in V_3, \tag{8.8}$$

where constraints (8.2) establish that each customer $k \in V_3$ must be served by one and only one CDC–RDC pair, constraints (8.3) impose that a CDC must be opened if an RDC is assigned to it; constraints (8.4) require that each RDC is assigned to a single CDC; constraints (8.5) impose that the transportation service between a CDC–RDC pair is activated if it is used by at least one macro-customer.

Because no capacity constraint is imposed, problem (8.1)–(8.8) satisfies the *single assignment* property (see Section 3.3.1). In order to satisfy the service level constraint, the LTL services j–k between RDC–customer pairs distant by more than a pre-established threshold L are discarded by setting the associated x_{ijk} variables equal to 0.

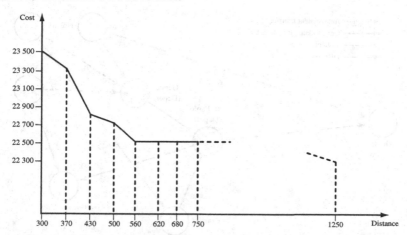

Figure 8.11 Total cost (in thousands of euros)-service level curve at DowBrands.

The solution of problem (8.1)–(8.8) was evaluated through a general-purpose MIP solver for various values of L between 300 and 1200 km (see Figure 8.11). It is worth noting that, as L decreases, at first the cost increases slowly, then it increases sharply. Also, when L becomes very large, there is no need for RDCs. On the basis of these evaluations, the company's management set L equal to 430 km. By implementing this solution the company achieved a saving of about 1.5 million dollars per year compared to the previous configuration.

8.9 Container Warehouse Location at Hardcastle

Hardcastle is a North European leader in intermodal transportation. In 2001 the company operated nearly 240 000 containers, with an annual transportation cost of about 50 million euros.

Like other intermodal transportation companies, Hardcastle manages both full and empty containers. When a customer places an order for freight transportation, Hardcastle sends one or several empty containers of the appropriate type in terms of size, refrigeration, etc., to the pick-up point (see Figure 8.12). The containers are then loaded and sent to destination using a combination of modes (e.g. railway and sea transportation). At the destination, the containers are emptied and sent back to the company unless there is an outgoing load requiring the same kind of container (corresponding to *compensation* between the demand and the supply of empty containers). Unless compensation is possible, empty containers are then moved to a new pick-up point. Relocating empty containers is a resource-consuming activity whose cost should be kept at minimum. Unfortunately, *compensation* between the demand and the supply of empty containers is seldom possible for three main reasons:

- the origin–destination demand matrix is strongly asymmetrical (some loca-

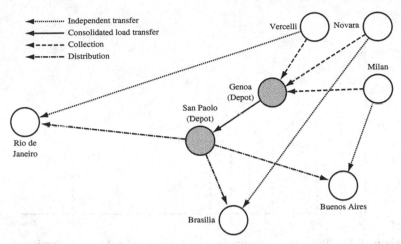

Figure 8.12   Freight transportation at Hardcastle.

tions are mainly sources of materials while some others are mainly points of consumption);

- at a given location, the demand and supply for empty containers do not usually occur at the same time;

- containers may have a large number of sizes and features; as a result, it is unlikely that the containers incoming at a customer facility are suitable for outgoing goods.

For these reasons, the compensation between demand and supply is neglected in the following.

Because of the economies of scale in transportation, it is not convenient to move containers directly from supply to demand points. Instead, containers are sent to a nearby warehouse. Then, on a weekly basis, convoys of empty and full containers are moved between warehouses (see Figure 8.13). Warehouses are often public so that their location can easily be changed if necessary. Prior to its redesign, the logistics system contained 87 depots (64 close to a railway station and 23 close to a sea terminal). Moreover, empty container movements accounted for nearly 40% of the total freight traffic.

The management of the empty containers is a complex decision process made up of two stages (see Figure 8.14):

- at a tactical level, one has to determine, on the basis of forecasted origin–destination transportation demands, the number and locations of warehouses, as well as the expected container flows among warehouses;

- at an operational level, shipments are scheduled and vehicles are dispatched on the basis of the orders collected and of short-term forecasts.

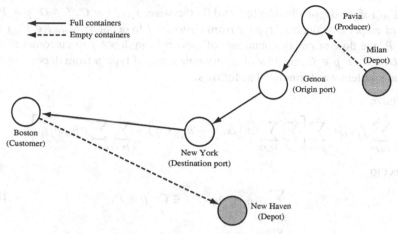

Figure 8.13 Empty container transportation at Hardcastle.

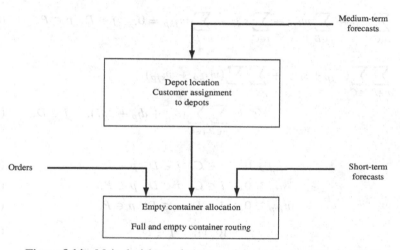

Figure 8.14 Main decisions when managing containers at Hardcastle.

In order to redesign its logistics system, Hardcastle aggregated its customers into 300 demand points. Further, containers types were grouped into 12 types. Let C be the set of customers, D the set of potential depots, P the set of different types of containers, f_j, $j \in D$, the fixed cost of depot j, a_{ijp}, $i \in C$, $j \in D$, $p \in P$, the transportation cost of a container of type p from customer i to depot j; b_{ijp}, $i \in C$, $j \in D$, $p \in P$, the transportation cost of a container of type p from depot j to customer i; c_{jkp}, $j \in D, k \in D, p \in P$, the transportation cost of an empty container of type p from depot j to depot k; d_{ip}, $i \in C$, $p \in P$, the number of containers of type p requested by the customer i; o_{ip}, $i \in C$, $p \in P$, the supply of containers of type p from customer i. Furthermore, let y_j, $j \in D$, be a binary decision variable equal to 1 if the depot j is selected, and 0 otherwise; x_{ijp}, $i \in C$, $j \in D$, $p \in P$, the flow of empty containers of type p from customer i to depot j; s_{ijp}, $i \in C$, $j \in D$, $p \in P$, the flow of empty containers of type p from depot j to customer i; w_{jkp}, $j \in D$, $k \in D$, $p \in P$, the flow of empty containers of type p from depot j to depot k. The problem was formulated as follows.

Minimize

$$\sum_{j \in D} f_j y_j + \sum_{p \in P} \left[\sum_{i \in C} \sum_{j \in D} (a_{ijp} x_{ijp} + b_{ijp} s_{ijp}) + \sum_{j \in D} \sum_{k \in D} c_{jkp} w_{jkp} \right] \qquad (8.9)$$

subject to

$$\sum_{j \in D} x_{ijp} = o_{ip}, \quad i \in C, \ p \in P, \qquad (8.10)$$

$$\sum_{j \in D} s_{ijp} = d_{ip}, \quad i \in C, \ p \in P, \qquad (8.11)$$

$$\sum_{i \in C} x_{ijp} + \sum_{k \in D} w_{kjp} - \sum_{i \in C} s_{ijp} - \sum_{k \in D} w_{jkp} = 0, \quad j \in D, \ p \in P, \qquad (8.12)$$

$$\sum_{p \in P} \sum_{i \in C} (x_{ijp} + s_{ijp}) + \sum_{p \in P} \sum_{k \in D} (w_{jkp} + w_{kjp})$$
$$\leqslant y_j \sum_{p \in P} \sum_{i \in C} (o_{ip} + d_{ip} + 2M), \quad j \in D, \qquad (8.13)$$

$$x_{ijp} \geqslant 0, \quad i \in C, \ j \in D, \ p \in P, \qquad (8.14)$$
$$s_{ijp} \geqslant 0, \quad i \in C, \ j \in D, \ p \in P, \qquad (8.15)$$
$$w_{jkp} \geqslant 0, \quad j \in D, \ k \in D, \ p \in P, \qquad (8.16)$$
$$y_j \in \{0, 1\}, \quad j \in D, \qquad (8.17)$$

where M is an upper bound on the w_{jkp} flows, $j \in D$, $k \in D$, $p \in P$. The objective function (8.9) is the sum of warehouse fixed costs and empty container variable transportation costs (between customers and warehouses, and between pairs of warehouses). Constraints (8.10)–(8.12) impose empty container flow conservation. Constraints (8.13) state that if $y_j = 0$, $j \in D$, then the incoming and outgoing flows from site j are equal to 0. Otherwise, constraints (8.13) are not binding since

$$x_{ijp} \leqslant o_{ip}, \quad i \in C, \ j \in D, \ p \in P,$$
$$s_{ijp} \leqslant d_{ip}, \quad i \in C, \ j \in D, \ p \in P,$$
$$w_{jkp} \leqslant M, \quad j \in D, \ k \in D, \ p \in P.$$

The implementation of the optimal solution of model (8.9)–(8.17) yielded a reduction in the number of warehouses to 48 and a 47% reduction in transportation cost.

8.10   Inventory Management at Wolferine

Wolferine is a division of the industrial group UOP Limited, which manufactures copper and brass tubes. The company's production processes take place in a factory located in London (Ontario, Canada) with highly automated systems operating with a very low work-in-process. The raw materials originate from mines located close to the factory. Consequently, the firm does not need to stock a large amount of raw materials (as a rule, no more than a two-week demand). As far as the finished product inventories are concerned, Wolferine makes use of EOQ models (see Section 4.4.2). In the autumn of 1980, the firm operated with a production level close to the plant's capacity. At that time the interest rate was around 10%. Using this value in the EOQ model, the company set the finished goods inventory level equal to 833 tons. During the subsequent two years, an economic recession hit the industrialized countries. The interest rate underwent continuous and quick variations (up to 20% in August 1981), the demand of finished products went down by 20%, and the price level increased sharply. According to the EOQ model, the finished goods inventory level should have been lower under those conditions. In order to illustrate this result, let n be the number of products; k an order fixed cost (assumed independent from the product); d_i, $i = 1, \ldots, n$, the annual demand of the product i; p the interest rate (increased to take into account warehousing costs); c_i, $i = 1, \ldots, n$, the price of product i; $\bar{I}$ the average stock level at time period t_0 (January 1981). On the basis of Equation (4.28),

$$\bar{I}(t_0) = \frac{1}{2} \sum_{i=1}^{n} \sqrt{\frac{2kd_i}{pc_i}}.$$

We can express parameter p as the sum of a bank interest rate p_1 and of a rate p_2 associated with warehousing costs:

$$p = p_1 + p_2.$$

Moreover, let δ_1, δ_2 and δ_3 be the variation rates (assumed equal for all products) of price, demand and interest rate at time period t, respectively. The average stock level is equal to

$$\bar{I}(t) = \frac{1}{2} \sum_{i=1}^{n} \sqrt{\frac{2[k(1 + \delta_1)][d_i(1 + \delta_2)]}{[p_1(1 + \delta_3) + p_2][c_i(1 + \delta_1)]}}$$

$$= \bar{I}(t_0) \sqrt{\frac{(1 + \delta_2)p}{p_1(1 + \delta_3) + p_2}}. \tag{8.18}$$

According to Equation (8.18), if demand decreases ($\delta_2 < 0$) and the interest rate increases ($\delta_3 > 0$), the stock level should be lower. However, the managers of Wolferine continued to operate as in 1981. As a result, the ITR (see Section 4.1) suddenly decreased. Moreover, to protect the manufacturing process against strikes at the mines, the firm also decided to hold an inventory of raw materials. Consequently, when the recession ended in 1983, the firm had an exceedingly large stock of both raw materials and finished products.

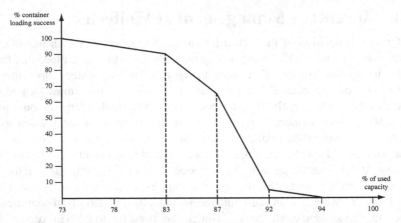

Figure 8.15 Percentage of success in loading an aircraft as a function of
the percentage of load capacity (simulation made by FedEx).

8.11   Airplane Loading at FedEx

FedEx is one of the leading express carriers in the world, with a freight traffic estimated
at about 2 million parcels per day. Its sales offices are located in 187 countries and
the company uses a fleet of 437 airplanes and about 30 000 trucks and vans.

In the USA, parcels whose origin and destination exceed a given distance are
consolidated in containers and sent by air. An airplane may fly between a pair of
destinations or may follow a multi-stop route where containers are loaded or unloaded
at intermediate stops.

In order to use airplane capacity efficiently, a key issue is to devise good loading
plans, taking into account a number of aspects: the load must be balanced around the
centre of gravity of the aircraft, the total weight in the various areas of the aircraft
must not exceed given thresholds in order to limit the cutting forces on the plane,
etc. These aspects are critical, especially for some planes, such as the Airbus A300,
a low fuel consumption aircraft. In addition, when loading an airplane assigned to a
multi-stop route, containers to be unloaded at intermediate stops must be positioned
close to the exit.

In order to allow airplanes to take off on time, allocating containers on-board must
be done in real time, i.e. containers must be loaded on the aircraft as soon as they arrive
at the airport. As a matter of fact, 30% to 50% of the containers are already on board
when the ground staff has a complete knowledge of the features of the containers
to be loaded. Of course, once some containers have been loaded, it may become
impossible to load the subsequent containers. As a result, when new containers arrive
at the airport, it is sometimes necessary to define a new loading plan in which some
containers previously loaded are unloaded. This situation arises frequently when the
total load is close to the capacity of the aircraft, as shown in Figure 8.15, where the
percentage of success in loading an airplane is reported as a function of the percentage
of the load capacity used.

The objective pursued by FedEx consists of loading the largest number of containers as possible. If no container is yet loaded, the following solution procedure is implemented.

Step 1. Let m be the number of containers to be loaded, n the number of positions in the loading area, q the number of areas into which the plane is divided, p_i, $i = 1, \ldots, m$, the weight of container i; P_j, $j = 1, \ldots, n$, the maximum weight that can be loaded in position j; d_j, $j = 1, \ldots, n$, the distance from position j to the centre of gravity O, $M^{\min}$ and $M^{\max}$ the minimum and maximum moments of the loads with respect to O, L_k, $k = 1, \ldots, q$, the total maximum weight that can be placed in area k; f_{jk}, $j = 1, \ldots, n$, $k = 1, \ldots, q$, the fraction of position j contained in area k. Also let x_{ij}, $i = 1, \ldots, m$, $j = 1, \ldots, n$, be a binary decision variable equal to 1 if container i is placed in position j, and 0 otherwise, u_j, $j = 1, \ldots, n$, a binary decision variable equal to 1 if position j is used, and 0 otherwise. A feasible solution is defined by the following set of constraints:

$$\sum_{j=1}^{n} x_{ij} = 1, \qquad i = 1, \ldots, m, \tag{8.19}$$

$$\sum_{i=1}^{m} x_{ij} \leqslant m u_j, \quad j = 1, \ldots, n, \tag{8.20}$$

$$\sum_{i=1}^{m} p_i x_{ij} \leqslant P_j u_j, \quad j = 1, \ldots, n, \tag{8.21}$$

$$\sum_{i=1}^{m} \sum_{j=1}^{n} d_j p_i x_{ij} \leqslant M^{\max}, \tag{8.22}$$

$$\sum_{i=1}^{m} \sum_{j=1}^{n} d_j p_i x_{ij} \geqslant M^{\min}, \tag{8.23}$$

$$\sum_{i=1}^{m} \sum_{j=1}^{n} p_i f_{jk} x_{ij} \leqslant L_k, \quad k = 1, \ldots, q, \tag{8.24}$$

$$x_{ij} \in \{0, 1\}, \quad i = 1, \ldots, m, \ j = 1, \ldots, n, \tag{8.25}$$

$$u_j \in \{0, 1\}, \quad j = 1, \ldots, n. \tag{8.26}$$

Constraints (8.19) guarantee that each container is allocated to a position. Constraints (8.20) state that if a position j, $j = 1, \ldots, n$, accommodates a container, then its u_j variable must be equal to 1. Constraints (8.21) ensure that the total weight loaded in any position does not exceed a pre-established upper bound. Constraints (8.22) and (8.23) impose that the total moment, with respect to point O, is within the pre-established interval. Constraints (8.24) ensure the respect of the weight bounds in each section.

Step 2. If problem (8.19)–(8.26) is infeasible, a container $\bar{i}$ is eliminated from the loading list and the problem is solved again (where, of course, $x_{\bar{i}j} = 0$, $j = 1, \ldots, n$). Step 2 is repeated until a feasible solution is found.

If some containers have already been loaded, the previous procedure is modified as follows. Let $\bar{I}$ be the set of the containers already loaded and let j_i, $i \in \bar{I}$, be the position assigned to container i. Then, additional constraints,

$$x_{ij_i} = 1, \quad i \in \bar{I}, \tag{8.27}$$

are added to (8.19)–(8.26). If this problem is feasible, the procedure stops since the partly executed loading plan can be completed. Otherwise, the constraint (8.27) associated with container $i \in \bar{I}$ allocated to the position closest to the entrance/exit is removed (this corresponds to unloading the container from the aircraft). This step is repeated until a feasible solution is obtained.

8.12   Container Loading at Waterworld

Like other commodities derived from wood, paper is typically produced a long way from the main points of consumption and then often transported by sea in containers. As a rule, the transportation cost accounts for a significant part of the selling price. Moreover, competition among manufacturers is fierce so that profits are very low. In this context, it is crucial to reduce transportation costs as much as possible. To this purpose, a key issue is to consolidate loads efficiently.

Paper is usually transported in rolls having a diameter ranging between 0.5 m and 1.5 m, or in large size sheets. In the first case, rolls are loaded directly into containers, while in the second case, the sheets are loaded onto pallets which are put in containers. Generally, an order is composed of a set of different products of varying size, thickness and density. A typical order can fill several tens of containers.

Waterworld determines its own loading plans through an ad hoc heuristic procedure outlined below.

8.12.1   Packing rolls into containers

Rolls vary in diameter and height. Roll packing amounts to loading rolls into containers with the circular base parallel to the bottom surface. The packing procedure is divided into two steps.

Step 1. In the first step, each roll is characterized by its circular section and the aim is to determine the least number of rectangles (representing bottom surfaces) that can accommodate such sections (referred to as *objects* in the following). Let (X, Y) be a Cartesian reference system with origin in the bottom left corner of a container.

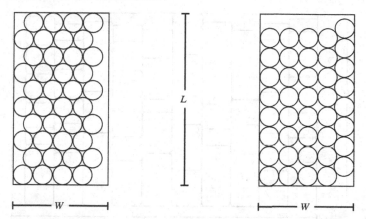

Figure 8.16 Two ways circular objects can be loaded in the Waterworld problem.

The objects are introduced in the container one at a time, the position of each object being defined in such a way that $\lambda_1 x + \lambda_2 y$ is minimized ((x, y) are the coordinates of the centre of the object, λ_1 and λ_2 are two positive constants). The choice of (x, y) must result in a feasible solution, i.e. the object must be entirely kept in the container and must not be overlapping the objects already inserted (see Figure 8.16). The following combinations of parameters λ_1 and λ_2 were used: (a) $\lambda_1 = 1$ and $\lambda_2 = 0$; (b) $\lambda_1 = 0$ and $\lambda_2 = 1$; (c) $\lambda_1 = \lambda_2 = 0.5$.

Step 2. With each cluster of rolls formed in the first step is associated a weight equal to the height of the tallest roll of the cluster. Then, a classical 1-BP (see Section 5.4.3) is solved heuristically in order to determine the least number of containers (each characterized by a capacity equal to container height) that can accommodate all clusters (see Figure 8.17).

8.12.2 Packing pallets into containers

Paper sheets are mounted on pallets in such a way that the total height of two stacked pallets is equal to the container height. These pallet pairs are then packed through a 2-BP heuristic (see Section 5.4.3). The procedure followed by Waterworld is a slight variant of the BL method (see Section 5.4.3).

8.13 Air Network Design at Intexpress

Intexpress is a firm whose core business is express freight delivery all over North America. The services provided to customers are (a) delivery within 24 hours (*next day service*); (b) delivery within 48 hours (*second day service*); (c) delivery within 3–5 days (*deferred service*). In order to provide quick deliveries, long-haul transportation is made by plane.

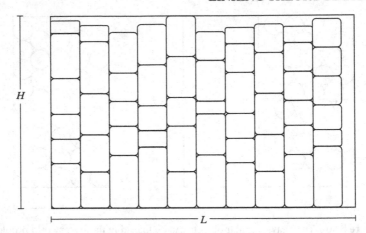

Figure 8.17 Roll packing in the Waterworld example.

The Intexpress logistics system comprises a set of *shipment centres* (SCs), of a single hub, of a fleet of airplanes and of a fleet of trucks. Loads are consolidated both in the SCs and in the hub. In particular, goods originating from the same SC are transported as a single load to the hub while all the goods assigned to the same SC are sent jointly from the hub. Freight is first transported to an originating SC, where it is consolidated; it is then transported to a final SC by air, by truck (*ground service*), or by using a combination of the two modes. Finally, freight is moved from the final SC to the destination by truck. Of course, an SC to SC transfer by truck is feasible only if distance does not exceed a given threshold. Air transportation is performed by a company-owned fleet (*dedicated air service*) or by commercial airlines (*commercial air service*). The outgoing freight is collected in the evening and delivered the morning after. Every day a company-owned aircraft leaves the hub, makes a set of deliveries, then travels empty from the last delivery point to the first pick-up point, where it makes a set of pick-ups and finally goes back to the hub. Each SC is characterized by an 'earliest pick-up time' and by a 'latest delivery time'. Moreover, all arrivals in the hub must take place before a pre-established arrival 'latest delivery time' (*cut-off time*, COT) in such a way that incoming airplanes can be unloaded, sorted by destination, and quickly reloaded on to outgoing aircraft.

Since it is not economically desirable nor technically feasible that airplanes visit all SCs, a subset of SCs must be selected as aircraft loading and unloading points (*air-stops*, ASs). An SC that is not an AS is connected to an AS by truck (*ground feeder route*). Figure 8.18 depicts a possible freight route between an origin–destination pair. Commercial air services are less reliable than dedicated air services and their costs are charged depending on freight weight. These are used when either the origin and its closest SC are so far apart that no quick truck service is possible, or when the overall transportation demand exceeds the capacity of company-owned aircraft.

Planning the Intexpress service network consists of determining

- the set of ASs served by each aircraft of the firm;

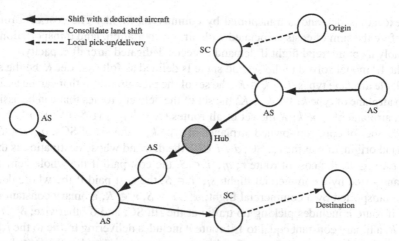

Figure 8.18 A possible freight route between an origin–destination pair at Intexpress.

- truck routes linking SCs (which are not ASs) to ASs;
- the transportation tasks performed by commercial airlines.

The objective pursued is the minimization of the operational cost subject to 'earliest pick-up time' and 'latest delivery time' constraints at SCs, to the COT restriction, etc.

The solution methodology used by Intexpress is made up of two stages: in the first stage, the size of the problem is reduced (*preprocessing phase*) on the basis of a qualitative analysis; in the second stage, the reduced problem is modelled and solved as an IP program. In the first stage,

- origin–destination pairs that can be serviced by truck (in such a way that all operational constraints are satisfied) are allocated to this mode and are not considered afterwards;

- origin–destination pairs that cannot be served feasibly by dedicated aircraft or by truck are assigned to the commercial flights;

- low-priority services (deliveries within 48 hours or within 3–5 days) are made by truck or by using the residual capacity of a company-owned aircraft;

- the demands of origin/destination sites are concentrated in the associated SC.

A *route* is a partial solution characterized by

- a sequence of stops of a company-owned aircraft ending or beginning in the hub (depending on whether it is a collection or a delivery route, respectively).

- a set of SCs (not ASs) allocated to each aircraft AS.

Therefore, two routes visiting the same AS in the same order can differ because of the set of SCs (which are not ASs), or because of the allocation of these SCs to the ASs. If the demand of a route exceeds the capacity of the allocated aircraft, the exceeding demand is transported by commercial flights. The cost of a route is therefore the sum of costs associated with air transportation, land transportation, and possibly a commercial flight if demand exceeds dedicated aircraft capacity.

The IP model solved in the second stage is defined as follows. Let K be the set of available airplane types; U^k, $k \in K$, the set of the pick-up routes that can be assigned to an airplane of type k; V^k, $k \in K$, the set of the delivery routes that can be assigned to an airplane of type k; R the set of all routes ($R = \bigcup_{k \in K}(U^k \cup V^k)$); n^k, $k \in K$, the number of company-owned airplanes of type k; S the set of SCs, o_i, $i \in S$, the demand originating at the ith SC; d_i, $i \in S$, the demand whose destination is the ith SC; c_r, $r \in R$, the cost of route r; q_i, $i \in S$, the cost paid if the whole demand o_i is transported by a commercial flight; s_i, $i \in S$, the cost paid if the whole demand d_i is transported by a commercial flight; α_i^r, $i \in S$, $r \in R$, a binary constant equal to 1 if route r includes picking up traffic at the ith SC, and 0 otherwise; δ_i^r, $i \in S$, $r \in R$, a binary constant equal to 1 if route r includes delivering traffic to the ith SC, and 0 otherwise; γ_i^r, $i \in S$, $r \in R$, a binary constant equal to 1 if the first (last) AS of pick-up (delivery) route r is the ith SC, and 0 otherwise. The decision variables of binary type are v_i, $i \in S$, equal to 1 if demand o_i is transported by commercial flight, and 0 otherwise; w_i, $i \in S$, equal to 1 if demand d_i is transported by commercial flight, and 0 otherwise; x_r, $r \in R$, equal to 1 if (pick-up or delivery) route r is selected, and 0 otherwise.

The integer program is as follows.

Minimize

$$\sum_{k \in K} \sum_{r \in U^k \cup V^k} c_r x_r + \sum_{i \in S}(q_i v_i + s_i w_i) \tag{8.28}$$

subject to

$$\sum_{k \in K} \sum_{r \in U^k} x_r \alpha_i^r + v_i = 1, \quad i \in S, \tag{8.29}$$

$$\sum_{k \in K} \sum_{r \in V^k} x_r \delta_i^r + w_i = 1, \quad i \in S, \tag{8.30}$$

$$\sum_{r \in U^k} x_r \gamma_i^r - \sum_{r \in V^k} x_r \gamma_i^r = 0, \quad i \in S, \; k \in K, \tag{8.31}$$

$$\sum_{r \in U^k} x_r \leqslant n^k, \quad k \in K, \tag{8.32}$$

$$x_r \in F, \quad r \in R, \tag{8.33}$$
$$x_r \in \{0, 1\}, \quad r \in R, \tag{8.34}$$
$$v_i \in \{0, 1\}, \quad i \in S, \tag{8.35}$$
$$w_i \in \{0, 1\}, \quad i \in S. \tag{8.36}$$

The objective function (8.28) is the total transportation and handling cost. Constraints (8.29) and (8.30) state that each SC is served by a dedicated route or by a commercial route; constraints (8.31) guarantee that if a delivery route of type $k \in K$ ends in SC $i \in S$, then there is a pick-up route of the same kind beginning in i. Constraints (8.32) set upper bounds on the number of routes which can be selected for each dedicated aircraft type. Finally, constraints (8.33) express the following further restrictions. The arrivals of the airplanes at the hub must be staggered in the period before the COT because of the available personnel and of the runway capacity. Similarly, departures from the hub must be scheduled in order to avoid congestion on the runways. Let n_a be the number of time intervals in which the arrivals should be allocated; n_p the number of time intervals in which the departures should be allocated; $f_r, r \in R$, the demand along route r; a_t the maximum demand which can arrive to the hub in interval $t, \ldots, n_a$; A_t the set of routes with arrival time from t on; P_t the set of routes with departure time before t; p_t the maximum number of airplanes able to leave before t. Hence, constraints (8.33) are

$$\sum_{k \in K} \sum_{r \in U^k \cap A_t} f_r x_r \leqslant a_t, \quad t = 1, \ldots, n_a, \tag{8.37}$$

$$\sum_{k \in K} \sum_{r \in V^k \cap P_t} x_r \leqslant p_t, \quad t = 1, \ldots, n_p. \tag{8.38}$$

Constraints (8.37) ensure that the total demand arriving at the hub is less than or equal to the capacity of the hub in each time interval $t = 1, \ldots, n_a$, while constraints (8.38) impose that the total number of airplanes leaving the hub is less than or equal to the maximum number allowed by runaway capacity in each time interval $t = 1, \ldots, n_p$.

Other constraints may be imposed. For instance, if goods are stored in containers one must ensure that once a container becomes empty it is brought back to the originating SC. To this end, it is necessary that the aircraft arriving at and leaving from each SC be compatible. In the Intexpress problem, there are four types of airplanes, indicated by 1, 2, 3 and 4. Aircraft of type 1 are compatible with type 1 or 2 planes, while aircraft of type 2 are compatible with those of type 1, 2 and 3. Therefore, the following constraints hold:

$$-\sum_{r \in U^1} x_r \alpha_i^r + \sum_{r \in V^1 \cup V^2} x_r \delta_i^r \geqslant 0, \quad i \in S, \tag{8.39}$$

$$\sum_{r \in U^1 \cup U^2} x_r \alpha_i^r - \sum_{r \in V^1} x_r \delta_i^r \geqslant 0, \quad i \in S, \tag{8.40}$$

$$-\sum_{r \in U^2} x_r \alpha_i^r + \sum_{r \in V^1 \cup V^2 \cup V^3} x_r \delta_i^r \geqslant 0, \quad i \in S, \tag{8.41}$$

$$\sum_{r \in U^1 \cup U^2 \cup U^3} x_r \alpha_i^r - \sum_{r \in V^2} x_r \delta_i^r \geqslant 0, \quad i \in S. \tag{8.42}$$

Moreover, some airplanes cannot land in certain SCs because of noise restrictions or insufficient runaway length. In such cases, the previous model can easily be adapted by removing the routes $r \in R$ including a stop at an incompatible SC.

The variables in the model (8.28)–(8.32), (8.34)–(8.42) are numerous even if the problem is of small size. For example, in the case of four ASs (a, b, c and d), there are 24 pick-up routes ($abcd$, $acbd$, $adbc$, etc.) each of which has a different cost and arrival time at the hub. If, in addition, two SCs (e and f) are connected by truck to one of the ASs a, b, c and d, then the number of possible routes becomes $16 \times 24 = 384$ (as a matter of fact, for each AS sequence, each of the two SCs e and f can be connected independently by land to a, b, c or d). Finally, for each delivery route making its last stop at an AS d, one must consider the route making its last stop in a different SC $g \in S \setminus \{d\}$. Of course, some of the routes can be infeasible and are not considered in the model (in the case under consideration the number of feasible routes is about 800 000).

The solution methodology is a classical branch-and-bound algorithm in which at each branching node a continuous relaxation of (8.28)–(8.32), (8.34)–(8.42) is solved. The main disadvantage of this approach is the large number of variables. Since the number of constraints is much less than the number of variables, only a few variables take a nonzero value in the optimal basic solution of the continuous relaxation. For this reason, the following modification of the method is introduced. At each iteration, in place of the continuous relaxation of (8.28)–(8.32), (8.34)–(8.42), a reduced LP problem is solved (in which there are just 45 000 'good' variables, chosen by means of a heuristic criterion); then, using the dual solution of the problem built in this manner, the procedure determines some or all of the variables with negative reduced costs, introducing the corresponding columns in the reduced problem (*pricing out columns*). Various additional devices are also used to quicken the execution of the algorithm. For example, in the preliminary stages only routes with an utilization factor between 30% and 185% are considered. This criterion rests on two observations: (a) because of the reduced number of company aircraft, it is unlikely that an optimal solution will contain a route with a used capacity less than 30%; (b) the cost structure of the air transportation makes it unlikely that along a route more than 85% of the traffic is transported by commercial airlines.

The above method was first used to generate the optimal service network using the current dedicated air fleet. The cost reduction obtained was more than 7%, corresponding to a yearly saving of several million dollars. Afterwards, the procedure was used to define the optimal composition of the company's fleet (*fleet planning*). For this purpose, in formulation (8.28)–(8.32), (8.34)–(8.42), it was assumed that n^k was infinite for each $k \in K$. The associated solution shows that five aircraft of type 1, three of type 2 and five of type 3 should be used. This solution yielded a 35% saving (about 10 million dollars) with respect to the current solution.

8.14 Bulk-Cargo Ship Scheduling Problem at the US Navy

The Tanker Division of the US Navy is in charge of planning the fuel replenishment of the US naval bases in the world. This task is accomplished by a fleet of 17 tankers owned by the US Navy and, if needed, of specialized companies (*spot carriers*). Every three months a team of the Tanker Division works out a replenishment plan including both a scheduling of the military tanks and a possible list of tasks for the spot carriers. The objective is to minimize the operational expenses of the military tanks plus the hiring costs of additional tankers provided by spot carriers. The main operational constraints are related to delivery times, which must be included in pre-established time windows, and the impossibility for some ships to moor in certain ports. The replenishment plans must also consider the current position and initial status of the fleet. Plans are revised daily in order to consider new demands, or delays due to unfavourable weather conditions.

Each fuel request consists of a type of product, a demand, a delivery point and a delivery time. On the basis of fuel availability at the depots, the team establishes how each request must be satisfied. After this preliminary analysis, the demand is expressed by a set of elementary fuel transfers. An elementary transfer is composed of two or three pick-ups in a same port or in adjacent ports, and of two or three deliveries to adjacent demand points. Typically, a tank is able to make two elementary transfers in a month. The formulation of the model takes into account the fact that because of the operational constraints, a tank can only complete a limited number of elementary transfers (generally, no more than 10) in the planning period.

Let n be the number of elementary transfers; m the number of ports, K ($= 17$) the number of military tanks; S_k, $k = 1, \ldots, K$, the set of the feasible working plans for tank k; a_{ijk}, $i = 1, \ldots, n$, $j \in S_k$, $k = 1, \ldots, K$, a binary constant equal to 1 if working plan j for tank k includes elementary transfer i, and 0 otherwise; f_i, $i = 1, \ldots, n$, the cost paid if transfer i is committed to a spot carrier; g_{jk}, $j \in S_k$, $k = 1, \ldots, K$, the variation of cost incurred if tank k executes working plan j, as opposed to the situation in which the tank lies idle for all the planning period. Further, let x_{jk}, $j \in S_k$, $k = 1, \ldots, K$, be a binary decision variable equal to 1 if working plan j is selected for tank k, and 0 otherwise; y_i, $i = 1, \ldots, n$, a binary decision variable equal to 1 if transfer i is made by a spot carrier, and 0 otherwise.

The problem was formulated as follows.

Minimize

$$\sum_{k=1}^{K} \sum_{j \in S_k} g_{jk} x_{jk} + \sum_{i=1}^{n} f_i y_i \tag{8.43}$$

subject to

$$\sum_{k=1}^{K} \sum_{j \in S_k} a_{ijk} x_{jk} + y_i = 1, \quad i = 1, \ldots, n, \tag{8.44}$$

$$\sum_{j \in S_k} x_{jk} \leqslant 1, \quad k = 1, \ldots, K, \tag{8.45}$$

$$x_{jk} \in \{0, 1\}, \quad j \in S_k, \ k = 1, \ldots, K, \tag{8.46}$$

$$y_i \in \{0, 1\}, \quad i = 1, \ldots, n. \tag{8.47}$$

The objective function (8.43) is equal to the total transportation cost. Constraints (8.44) impose that each transfer is realized by a military tank or by a spot carrier. Constraints (8.45) establish that, for each tank, at most one working plan must be selected. Model (8.43)–(8.47) is therefore a set partitioning problem with some additional constraints (relations (8.45)). The US Navy has almost 30 elementary transfers in each planning period, to which correspond various thousands of feasible working plans. Using a general-purpose IP solver it is generally possible to solve these instances exactly.

The optimal solution yielded an average monthly cost of $341 900, inferior by $467 000 to the manual scheduling. The yearly saving was about 1.5 million dollars. Further cost reductions were obtained by simulating various compositions of the military fleet. This analysis suggested a reduction of the current fleet, no longer using five tanks and using mainly the spot carriers. This solution generated a cost reduction of 3.2 million dollars per year.

8.15   Meter Reader Routing and Scheduling at Socal

Socal (Southern California Gas Company) distributes, with its own pipe network, domestic and industrial gas in an area including all of southern California (USA). The task of surveying user consumption is accomplished by motorized operators. Every day a meter reader makes an initial car trip from the offices of the company up to a parking point; afterwards, he walks along several streets (or segments of streets) where he does some surveying; finally, he reaches the parking point from which he returns to the company by car. According to the working contract, if the duration of a shift exceeds a pre-established maximum value T^{max}, a meter reader receives an additional remuneration, proportional to the overtime. For reasons of equity, Socal prefers solutions in which all service routes have a similar duration. The planning of the meter-reading activities then consists of partitioning the set of street segments into subsets yielding service routes with a duration close to T^{max}.

In 1988 Socal mandated a consulting firm, Distinct Management Consultants, Inc. (DMC) to evaluate the costs and the potential benefits resulting from the use of an optimization software for the planning of meter-reading activities. A pilot study was conducted in a sample region including the townships of Culver City, Century City, Westwood, West Hollywood and Beverly Hills. This area corresponds to about 2.5% of the entire service territory and to 242.5 working days per month. The interest of Socal in this project was prompted by the sizeable expense (about 15 million dollars per year) resulting from meter-reading activities.

DMC built a model of the problem integrating the data contained in the Socal archives with those derived from a GIS. In the case of streets having users on both sides, it was assumed that the meter reader would cover both sides separately. Therefore, the problem was represented by a multigraph $G(V, E)$ in which the road intersections are described by vertices in V and street segments are associated with single edges or to pair of parallel edges. The street sides containing some consumers formed a subset

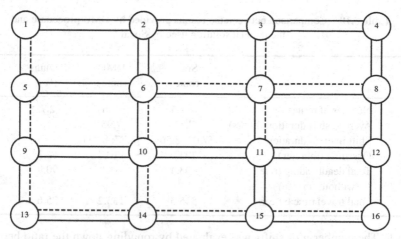

Figure 8.19 An example of a multigraph used for modelling the Socal problem (bold edges are required, while those with a dotted line are not required).

$R \subseteq E$ (see Figure 8.19). With each edge $(i, j) \in E$ was associated a traversing time t_{ij}, and with each edge $(i, j) \in R$, was associated a service time s_{ij} ($> t_{ij}$).

The duration $t(R')$ of a work shift corresponding to a subset of the service edges $R' \subseteq R$ has a cost equal to the sum of three terms: (a) the time spent to deal with administrative issues at the company headquarter (generally 15 min); (b) the travel time by car from the headquarter to a parking site; (c) the duration $T(R')$ of the solution of the RPP (see Section 7.6.2) associated with the service on foot of all edges of R'. Socal prefers work shifts associated with subsets $R' \subseteq R$ whose duration is included between $T^{min} = 7.9$ hours and $T^{max} = 8$ hours (*balanced* work shifts).

A partition $\{R_1, \ldots, R_n\}$ of R is feasible if each pair of parallel service arcs belongs to the same subset R_i, $i = 1, \ldots, n$ (feasible partition). The search for a feasible balanced partition (or, at least, 'almost' balanced) of R can be viewed as the minimization of a penalty associated with a violation of the desired length $[T^{min}, T^{max}]$ of each work shift (*edge partitioning problem*).

Denoting by U and L the penalties associated with the two types of violation, the objective function to minimize is

$$U \sum_{i=1}^{n} \max\{t(R_i) - T^{max}, 0\} + L \sum_{i=1}^{n} \max\{T^{min} - t(R_i), 0\}.$$

DMC solved the problem by means of the following procedure.

Step 0. The workloads associated with $T(R')$, $R' \subseteq R$, are approximated as $W(R') = \sum_{(i,j) \in R'} t_{ij}$. This approach is justified by the presence of consumers on both sides of most of the street segments. An evaluation of the average useful time $\bar{T}$ of a route (the time a meter reader will take on average to travel on foot) was calculated as the difference between T^{max} and the average time required for the administrative tasks and for driving.

Table 8.10   Comparison between the results provided by DMC procedure and the manual solution used by Socal.

	Socal routes	DMC routes	Savings (%)
Number of routes	242,5	236	2.7
Average shift duration (hours)	7.82	7.95	
Shift interval duration	[7.03; 8.56]	[7.77; 8.11]	
Overtime (minutes)	293.4	110.4	62.3
Total deadheading time (without service)	693.4	63.1	90.9
Total travel time by car	545.5	133.2	75.6

Step 1. The number n of shifts was evaluated by rounding down the ratio between the total workload $W(R')$ of the service area and $\bar{T}$.

Step 2. The load of remaining workload $(W(R') - nT^{\max})$ was assigned to an 'incomplete' auxiliary route, built with a heuristic criterion along the border of the service area.

Step 3. The n 'complete' shifts were defined by means of a local search procedure (see Section 7.3.2). The algorithm first selected n seed vertices appropriately spaced out and then progressively generated the routes by adding to each of them couples of parallel edges in order to constantly have balanced work shifts. The solution obtained in this manner was improved, where possible, with some exchanges of couples of edges (or of groups of couples of edges) among the work shifts.

The solution generated by this procedure for the sample district was characterized, compared to the one obtained manually by Socal, by a reduced number of meter readers, by a remarkable reduction of overtime and by an increase in the average useful time of an operator (see Table 8.10). Extrapolating the savings associated with these improvements, DMC estimated at about $870 000 the saving expected after the use of the optimization method for the entire area served by Socal.

The previous analysis convinced the Socal management to order from DMC a decision support system with a user-friendly interface, based on the optimization method just described. The system development required one year and was installed at the beginning of 1991.

8.16   Annotated Bibliography

The ExxonMobil case (see Section 8.2) is derived from:

1. Ferrarese G 2003 Un sistema di supporto alle decisioni per la pianificazione delle spedizioni in un'azienda manifatturiera. Bachelor thesis, University of Lecce, Lecce (Italy).

The Pfizer case (see Section 8.3) is derived from documents courtesly provided by the company.

The Railion case (see Section 8.4) is derived from the company's website: www. railion.nl

The Gioia Tauro case (see Section 8.5) is derived from:

2. Chiodo A 2000 Modelli di ottimizzazione nella gestione dei terminali marittimi. Master thesis, University of Calabria, Rende (CS), Italy.

The Hamilton-Wentworth case (see Section 8.6) is derived from:

3. Huang G H, Baetz B W, Patry G C and Terluk V 1996 Capacity planning for an integrated waste management system under uncertainty: a North American case study. *Waste Management* **15**, 523–546.

The Adriatica Accumulatori case (see Section 8.7) is inspired by:

4. Stampacchia P 1988 Le tecniche di estrapolazione e di correlazione per la previsione delle vendite. In *Il sistema d'impresa* (ed. S Sciarelli), pp. 359–367. Cedam, Milan.

The case of DowBrands (see Section 8.8) is taken from:

5. Powell RE, Gao L and Muggenborg SD 1993 Designing an integrated distribution system at DowBrands. *Interfaces* **23**(3), 107–117.

The case of Hardcastle (see Section 8.9) is taken from:

6. Crainic TG, Dejax P and Delorme L 1989 Models for multimode multicommodity location problems with interdepot balancing requirements. *Annals of Operations Research* **18**, 279–302.

The case of Wolferine (see Section 8.10) is taken from:

7. Bell PC and Noori H 1985 Managing inventories through difficult economic times: a simple model. *Interfaces* **15**(5), 39–45.

The case of FedEx (see Section 8.11) is taken from:

8. Thomas C, Campbell K, Hines G and Racer M 1998 Airbus packing at Federal Express. *Interfaces* **28**(1), 21–30.

The case of Waterworld (see Section 8.12) is taken from:

9. Fraser HJ and George JA 1994 Integrated container loading software for pulp and paper industry. *European Journal of Operational Research* **77**, 466–474.

The case of Intexpress (see Section 8.13) is taken from:

10. Barnhart C and Schneur RR 1996 Air network design for express shipment service. *Operations Research* **44**, 852–863.

The case of the US Navy (see Section 8.14) is taken from:

11. Fisher ML and Rosenwein MB 1989 An interactive optimization system for bulk-cargo ship scheduling. *Naval Research Logistics* **36**, 27–42.

The case of Socal (see Section 8.15) is taken from:

12. Wunderlich J, Collette M, Levy L and Bodin L 1992 Scheduling meter readers for Southern California gas company. *Interfaces* **22**(3), 22–30.

8.17 Further Case Studies

A large number of logistics case studies can be found in operations research and management science journals, such as *Operations Research*, the *European Journal of Operational Research*, the *Journal of the Operational Research Society*, *Omega* and *Interfaces*. In what follows, a few remarkable articles are introduced and summarized.

Various options are examined for shipping tomato paste from Heinz's processing facilities on the west coast of the USA to factories in the Midwest in:

Kekre S, Purushottaman N, Powell TA and Rajagopalan S 1990 A logistics analysis at Heinz. *Interfaces* **20**(5), 1–13.

An MILP model is solved to determine a worldwide manufacturing and distribution strategy at Digital Equipment Corporation in:

Arntzen BC, Brown GG, Harrison TP and Trafton LL 1995 Global supply chain management at Digital Equipment Corporation. *Interfaces* **25**(1), 69–93.

A continuous approximation method is used for designing a service network for the US Postal Service in:

Rosenfield DB, Engelstein I and Feigenbaum D 1992 An application of sizing service territories. *European Journal of Operational Research* **63**, 164–172.

A decision support system is developed to reduce the cost of transporting materials, parts and components from 20 000 supplier plants to over 160 General Motors plants in:

Blumenfeld DE, Burns LD, Daganzo CF and Hall RH 1987 Reducing logistics costs at General Motors. *Interfaces* **17**(1), 26–47.

A system for dispatching and processing customer orders for gasoline and distillates at ExxonMobil Corporation is illustrated in:

Brown GG, Ellis CJ, Graves GW and Ronen D 1987 Real time, wide area dispatch of Mobil Tank Trucks. *Interfaces* **17**(1), 107–120.

A combined location-routing problem is illustrated in:

Rosenfield DB, Engelstein I and Feigenbaum D 1992 Plant location and vehicle routing in the Malaysian smallholder sector: a case study. *European Journal of Operational Research* **38**, 14–26.

A routing problem associated with the collection and delivery of skips is described in:

De Meulemeester L, Laporte G, Louveaux FV and Semet F 1997 Optimal sequencing of skip collection and deliveries. *Journal of the Operational Research Society* **48**, 57–64.

A complex vehicle routing problem occurring in a major Swiss company is analysed in:

Rochat Y and Semet F 1994 A tabu search approach for delivering pet food and flour in Switzerland. *Journal of the Operational Research Society* **45**, 1233–1246.

Index

*Introduction to Logistics Systems Planning
and Control*
G. Ghiani, G. Laporte and R. Musmanno
© 2004 John Wiley & Sons, Ltd
ISBN: 0-470-84916-9 (HB)
0-470-84917-7 (PB)

9 780470 849170